Hands-On Real Time Communications

Hands-On Real Time Communications

A Practical Guide to RTC Protocols in Non-3GPP, 3GPP 4G/5G/6G and IoT Networks

Rolando Herrero
Cyber-Physical Systems and Telecom Networks
Program Director at Northeastern University
Boston

WILEY

Published by John Wiley & Sons, Inc., Hoboken, New Jersey.
Published simultaneously in Canada.

Library of Congress Cataloging-in-Publication Data Applied for:

Hardback ISBN: 9781394239627

Cover Design: Wiley
Cover Image: © naqiewei/Getty Images

Set in 9.5/12.5pt STIXTwoText by Straive, Chennai, India

Contents

Preface

What is the Focus of this book?

Real Time Communication (RTC) has been traditionally associated with telephony and telephony applications. With the emergence of the Internet, however, RTC has been driven to support a wide range of topologies and technologies. Specifically, in a future ruled by the interaction between humans, robots, sensors, actuators, humanoids, and machines in general, RTC plays a crucial role. Although many applications fall under the umbrella of RTC, namely, text messaging, sensor readout transmission, video game controlling, etc., the focus of this book is on media-based RTC.

Because media involves speech, audio, and video, each with different temporal and spectral characteristics, the technologies involved vary greatly. However, a common theme is the use of Internet suite protocols. These protocols enable support for three main topologies that traditionally used different, and sometimes proprietary, mechanisms. The first topology includes mainstream, non-3GPP networks, which do not follow the standards of the *3rd Generation Partnership Project* (3GPP). These networks, such as Wi-Fi and Ethernet-based stacks, support license-free communication, typically without requiring subscriptions. The second topology includes 3GPP networks, which are mainstream cellular networks that follow the standards of the 3GPP. These are wireless, mobile networks that rely on licensed communication channels associated with paid subscriptions. The third and last topology includes *Internet of Things* (IoT) networks. These networks support communication between devices and applications over both licensed and license-free channels. The key difference between IoT and 3GPP/non-3GPP topologies is that IoT incorporates non-human endpoints (devices) into the communication process. Additionally, devices in IoT networks are often power-constrained, meaning protocols optimized for 3GPP and non-3GPP networks may not function as well in IoT architectures.

This book looks at RTC from the perspective of the *Internet Engineering Task Force* (IETF) layered architecture. By grouping functionality into layers, the Internet suite of protocols maximizes the reuse of the technologies involved. Whether it is non-3GPP, 3GPP, or IoT, whether it is Wi-Fi, cellular, or an energy-constrained device, media is sampled, compressed, and transmitted using the same basic principles and mechanisms.

Why Did I Write This Book?

RTC is key to supporting the transmission of media between endpoints by supporting multiple topologies and architectures. However, many of the technologies involved in supporting RTC are legacy mechanisms that have been standardized for many years. On the other hand, some of them are new, particularly in the context of IoT networks. The material introduced in this book is part of the curriculum of the Telecommunication Networks Program at Northeastern University in Boston. Moreover, the content presented in this book is novel as it aims to cover multiple areas:

1) Presents RTC from the perspective of the IETF layered architecture with focus on speech, audio, and video.
2) Introduces media as a type of application that relies on *codecs* (coder/decoders) to compress speech, audio, and video while preserving *Quality of Service* (QoS).
3) Integrates RTC technologies to support 3GPP, non-3GPP, and IoT topologies.
4) Provides a hands-on approach by introducing tools to build, deploy, test, and understand 3GPP, non-3GPP, and IoT networking scenarios.

The first part of this book addresses the first two topics listed above, while the second part focuses on the last topic. The second part supports both hardware virtualization, where network interfaces are emulated, and actual hardware-based network interfaces.

Target Audience

Because this book deals with RTC-based media in 3GPP, non-3GPP, and IoT networks in the context of the IETF layered architecture, it is of importance to those interested in state-of-the-art networking and communication technologies. This includes graduate and undergraduate students pursuing degrees in Electrical Engineering, Computer Engineering, Computer Science, and Information Technology, among others, as well as practitioners, technologists, researchers, and engineers in general. Standardization efforts associated with RTC protocol stacks make this crucial. Moreover, since this book incorporates a second part that presents tools like Netualizer and Wireshark to virtualize these technologies, there are no requirements for dedicated hardware. This enables everyone to analyze and understand the inner workings of RTC protocols. In addition to the networking and communication aspects of media-based RTC, this book also focuses on the application layer, particularly on the details of speech, audio, and video codecs. This makes this book ideal for those in the target audience who are additionally interested in signal processing and information theory.

How is this Book Organized?

This book covers media based RTC focusing on networking and communication protocols as well as the details of the processing and compression of the media itself. To meet these goals, this 7-chapter book is organized in two parts as follows:

1) Part I that includes four chapters provides the theoretical fundamentals of media-based RTC. In this context, Chapter 1 looks at the characteristics of the different types of media including the corresponding codecs and how they pack data. Chapter 2 presents the networking protocol that enables the setup and teardown of media sessions. Chapter 3 focuses on the networking protocols that support the transmission of codec traffic. Finally, Chapter 4 introduces the details of how these networking protocols integrate with non-3GPP, 3GPP, and IoT topologies.
2) Part II that includes three chapters introduces the mechanisms to build the aforementioned topologies. Specifically, Chapters 5, 6, and 7 respectively focus on non-3GPP, 3GPP, and IoT architectures.

March 2024

Rolando Herrero
Boston, MA, USA

Acronyms

3G	3rd Generation
3GPP	3G Partnership Project
4G	4th Generation
5G	5th Generation
6G	6th Generation
6Lo	IPv6 over Networks of Resource-constrained Nodes
6LoBTLE	IPv6 over Low-power Bluetooth Low Energy
6LoWPAN	IPv6 over Low-power Wireless Personal Area Networks
AAC	Advanced Audio Coding
AAL2	ATM Adaptation Layer 2
AC	Alternating Current
ACELP	Algebraic CELP
ACR	Absolute Category Rating
AES	Advanced Encryption Standard
AM	Amplitude Modulation
AMR	Adaptive Multi-Rate
AP	Access Point
APCM	Adaptive PCM
API	Application Program Interfaces
AR	Autoregressive
ARP	Address Resolution Protocol
ARQ	Automatic Request
ASCII	American Standard Code for Information Interchange
ASR	Automatic Speech Recognition
AVC	Advanced Video Coding
AWS	Amazon Web Services
BLE	Bluetooth Low Energy
BOOTP	Bootstrap Protocol
bps	bits per second
BPSK	Binary PSK

BSS	Basic Service Set
BSSID	BSS Identifier
CA	Certification Authority
CBC	Cipher Block Chaining
CBR	Constant Bit Rate
CCR	Comparison Category Rating
CELP	Code-excited LP
CIDR	Classless Interdomain Routing
CIF	Common Intermediate Format
CM	Counter Mode
CSCF	Call Session Control Function
CMR	Codec Mode Request
CNAME	Canonical Name
CNG	Comfort Noise Generation
CoAP	Constrained Application Protocol
codec	Coder/Decoder
CSFB	Circuit-switched Fallback
CSMA/CA	Carrier Sense Multiple Access with Collision Avoidance
CSMA/CD	Carrier Sense Multiple Access with Collision Detection
CSRC	Contributing Source
CSS	Chirp Spread Spectrum
CRC	Cyclical Redundancy Checking
DAC	Digital-to-Analog Converter
DAM	Destination Address Mode
DBPSK	Differential BPSK
DC	Direct Current
DCF	Distributed Coordination Function
DCR	Degradation Category Rating
DCT	Discrete Cosine Transform
DES	Data Encryption Standard
DHCP	Dynamic Host Configuration Protocol
DIFS	Distributed Inter Frame Spacing
DMOS	Degradation MOS
DNS	Domain Name System
DoS	Denial of Service
DPCM	Differential PCM
DPSK	Differential PSK
DRS	Dynamic Rate Shifting
DSA	Digital Signature Algorithm
DSCP	Differentiated Services Code Point
DSSS	Direct Sequence SS
DTMF	Dual-Tone Multi-Frequency

DTLS	Datagram Transport Layer Security
DTX	Discontinuous Transmissions
EBIT	End Bit Position
ECC	Elliptic Curve Cryptography
ECN	Explicit Congestion Notification
EDCF	Enhanced DCF
EID	Extension Identifier
EIRP	Equivalent Isotropic Radiated Power
eMBB	Enhanced Mobile Broadband
EPC	Enhanced Packet Core
ESS	Extended Service Set
ETSI	European Telecommunications Standards Institute
EU-TRAN	Evolved UMTS Terrestrial Radio Access Network
EUI	Extended Unique Identifier
EVRC	Enhanced Variable Rate Codec
EVS	Enhanced Voice Services
FB	Full Band
FCF	Frame Control Field
FCS	Frame Checksum
FDD	Frequency Division Duplex
FDMA	Frequency Division Multiple Access
FEC	Forward Error Correction
FHSS	Frequency Hopping SS
FIPS	Federal Information Processing Standard
FM	Frequency Modulation
fmtp	format specific parameter
FQ	Frame Quality
FSF	Frame Control Field
FSK	Frequency Shift Keying
FT	Field Type
Gbps	Gigabits per second
gNB	Next Generation node B
GOB	Group of Blocks
GOP	Group of Pictures
GSM	Global System for Mobile Communications
HAN	Home Area Network
HBR	High Bit Rate
HC1	Header Compression 1
HC2	Header Compression 2
HD	High Definition
HD TV	HD Television
HEVC	High Efficiency Video Coding

HSS	Home Subscriber Server
HMAC	HASH-based Message Authentication Codec
HTML	Hypertext Markup Language
HTTP	HyperText Transfer Protocol
IBSS	Independent BSS
ICMP	Internet Control Message Protocol
IDE	Integrated Development Environment
IETF	Internet Engineering Task Force
IID	Interface Identifier
IMS	IP Multimedia Subsystem
iLBC	Internet Low Bitrate Codec
IoT	Internet of Things
IP	Internet Protocol
IP TTL	IP Time-to-Live
IPHC	IP Header Compression
iSAC	internet Speech Audio Codec
ISM	Instrumental Scientific and Medical
ITU	International Telecommunications Union
IPv4	Internet Protocol version 4
IPv6	Internet Protocol version 6
JPEG	Joint Photographic Experts Group
Kbps	Kilobits per second
LAN	Local Area Network
LATM	Low-overhead MPEG-4 Audio Transport Multiplex
LBR	Low Bit Rate
LBR	Low-power PAN Border Router
LCEVC	Low Complexity Enhancement Video Coding
LD-CELP	Low Delay CELP
LLC	Link Layer Control
LLN	Low Power and Lossy Network
LoRa	Long Range
LPC	Linear Predictive Coding
LPWAN	Low Power Wide Area Network
LSF	Line Spectral Frequency
LTE	Long Term Evolution
MAC	Media Access Control
MB	Macro Block
Mbps	Megabits per second
MBR	Medium Bit Rate
MCELP	Mixed CELP
MD	Message Digest

MDCT	Modified Discrete Cosine Transform
MF	More Fragments
MIC	Message Integrity Code
MIMO	Multiple Input Multiple Output
MKI	Master Key Identifier
MME	Mobility Management Entity
mMTC	Massive Machine-Type Communication
MOS	Mean Opinion Score
MPEG	Motion Pictures Experts Group
MRF	Media Resource Function
MRFC	MRF Controller
MRFP	MRF Processor
MSS	Maximum Segment Size
MTU	Maximum Transmission Unit
NAL	Network Abstraction Layer
NALU	NAL Unit
NAT	Network Address Translation
NB	Narrowband
NB-Fi	Narrowband Fidelity
NB-IoT	Narrowband IoT
ND	Neighbor Discovery
NFC	Near Field Communication
NFV	Network Function Virtualization
NHC	Next Header Compression
NR	New Radio
NS	Neighbor Solicitation
NT	Netualizer Network Interface
OFDM	Orthogonal Frequency Division Multiplexing
OQPSK	Offset QPSK
OSI	Open Systems Interconnection
PAM	Pulse Amplitude Modulation
PAN	Personal Area Network
PAN ID	PAN Identifier
PCF	Point Coordination Function
PCM	Pulse Code Modulation
PCRF	Policy and Charging Rules Function
PDCP	Packet Data Convergence Protocol
PDN-GW	Packet Data Network Gateway
PG	Prediction Gain
PKI	Public Key Infrastructure
PSK	Phase Shift Keying

PSQM	Perceptual Speech Quality Measure
QAM	Quadrature Amplitude Modulation
QCIF	Quarter CIF
QoS	Quality of Service
QPSK	Quadrature PSK
QUIC	Quick UDP Internet Connection
RAN	Radio Access Network
REST	Representational State Transfer
RGB	Red Green Blue
RLC	Radio Link Control
ROHC	Robust Header Compression
RPE	Regular Pulse Excitation
RPL	Routing for Low Power
RR	Receiver Report
RTC	Real Time Communication
RTCP	Real-Time Control Protocol
RTP	Real-Time Transport Protocol
RTT	Round Trip Time
SAA	Stateless Address Autoconfiguration
SAM	Source Address Mode
SBC	Session Border Controller
SBR	Spectral Band Replication
SBIT	Start Bit Position
SC-FDMA	Single Carrier FDMA
SD-DNS	Service Discovery DNS
SD-WAN	Software Defined WAN
SDAP	Service Data Adaption Protocol
SDES	SDP Security Description for Media Streams
SDN	Software Defined Network
SDP	Session Description Protocol
SDR	Software Defined Radio
SFD	Start Frame Delimiter
SHA	Secure Hash Algorithm
SID	Silence Insertion Descriptor
SIF	Source Intermediate Format
S-GW	Serving Gateway
SIFS	Short Inter Frame Spacing
SIP	Session Initialization Protocol
SIPS	Secure SIP
SNR	Signal to Noise Ratio
sps	samples per second
SSNR	Segmental SNR

SSRC	Synchronization Source
SR	Sender Report
SRC	Source Format
SRTCP	Secure RTCP
SRTP	Secure RTP
SWB	Super Wideband
TCP	Transport Control Protocol
TDD	Time Division Duplex
TDMA	Time Division Multiple Access
TLS	Transport Layer Security
TLV	Type-Length-Value
TTL	Time-to-Live
TTS	Text-to-Speech
UAC	User Agent Client
UART	Universal Asynchronous Receiver Transmitter
UAS	User Agent Server
UDP	User Datagram Protocol
UE	User Equipment
UI	User Interface
UART	Universal Asynchronous Receiver Transmitter
UMTS	Universal Mobile Telecommunications Framework
URI	Uniform Resource Identifier
URL	Uniform Resource Locator
URLLC	Ultra-reliable Low-latency Communication
VAD	Voice Activity Detection
VBR	Variable Bit Rate
VCL	Video Coding Layer
VLBR	Very Low Bit Rate
VoIP	Voice over IP
VoLTE	Voice over LTE
VoNR	Voice over NR
VSELP	Vector Sum Excited LP
VVC	Versatile Video Coding
WAN	Wide Area Network
WB	Wideband
WEP	Wired Equivalent Privacy
Wi-Fi	Wireless Fidelity
WLAN	Wireless Local Area Network
WPA	Wi-Fi Protected Access
WPAN	Wireless Personal Area Network
WSN	Wireless Sensor Network
XML	eXtensible Markup Language

About the Companion Website

This book is accompanied by a companion website:

www.wiley.com/go/herrero/RTCProtocols

This website includes:
- Solutions PDF
- Scripts and Traces
- Slides

Part I

Theoretical Background

1

Media Generation

1.1 Introduction

Real Time Communication (RTC) bridges the gap between devices and people, enabling instant, interactive exchanges. RTC supports a two-way dialog across *User Equipment* (UE) including smartphones, tablets, laptops, and even IoT devices (Herrero 2021). Video conferencing, voice calls, and instant messaging are just some examples of RTC. Other examples include IoT devices generating real time data that drive applications like predictive maintenance, remote monitoring, smart home automation as well as industrial IoT. This means that RTC involves a range of technologies with versatile applications that enable legacy and emerging architectures.

In this context, this chapter is about media generation, and it specifically focuses on the generation of signals that facilitate RTC. Given that signals originate in the analog domain, their conversion into digital form becomes necessary at the UEs (Herrero 2023). The process of digitization involves two essential mechanisms referred to as sampling and quantization (Haykin 2009). These processes lead to the conversion of analog signals with infinite precision into digital data of finite size, enabling storage and transmission. To achieve even greater compression, these streams can be processed through media codecs (Chu 2003).

The conversion of a signal from the analog realm to the digital domain introduces distortion. This distortion is further exacerbated during transmission through a communication channel affected by network impairments like loss and latency. Quantifying the degree of this distortion involves the utilization of quality scores and other relevant metrics.

Hands-On Real Time Communications: A Practical Guide to RTC Protocols in Non-3GPP, 3GPP, 4G/5G/6G and IoT Networks, First Edition. Rolando Herrero.

Companion website: www.wiley.com/go/herrero/RTCProtocols

1.2 Signals

Real time communication signals, capable of carrying any digital data, are our focus in this chapter, specifically how they carry speech, audio, and video. This is because speech, audio, and video signals are the most common types of signals used in RTC applications. Speech signals are used for voice calling, audio signals are used for music and other audio content, and video signals are used for video conferencing and streaming.

1.2.1 Speech

Audio and speech signals are sound signals, characterized by fluctuations in air pressure as they travel. These pressure variations are described as waves, commonly known as sound waves. Furthermore, speech signals play a pivotal role in human communication and constitute a subset of these audio signals. A key fact is that their distinctive nature allows them to be processed and encoded using mechanisms distinct from those employed for standard audio signals (Jurafsky and Martin 2000).

The human auditory system perceives sounds ranging from 20 to 20000 Hertz (Hz) (or 20 kHz), with heightened sensitivity to events occurring between 250 Hz and 5 kHz. This range is where speech signals predominantly reside. Vowel sounds derive their energy mainly from 250 Hz to 2 kHz, while voiced consonants like *b*, *d*, and *m* are strongest between 250 Hz and 4 kHz. Unvoiced consonants like *f*, *s*, and *t* have varying intensities and occupy the 2 to 8 kHz range.

For good speech comprehension, it is especially important to have good hearing in the range of 125 Hz to 4 kHz, where unvoiced consonants reside. Figure 1.1 shows a normalized speech sequence recorded over 25 milliseconds. Interestingly, it displays a strong correlation within about 8 milliseconds, matching the pitch period of the sequence. More details of the characteristics of speech, including the formal definition of the pitch period, are presented in Section 1.4.1.

1.2.2 Audio

Audio signals cover a wider range than speech signals, reaching the full human hearing range and even including the ultrasonic range. In terms of practicality, audio signals typically transmit in one direction, while speech signals spread in all directions. Unlike speech signals, which have a clear organization with specific patterns in frequency and amplitude, audio signals lack such structure and are less complex, with less data, allowing for better compression.

Similar to speech signals, audio signals exhibit various attributes beyond their frequency range. The amplitude of a speech signal corresponds to its volume or

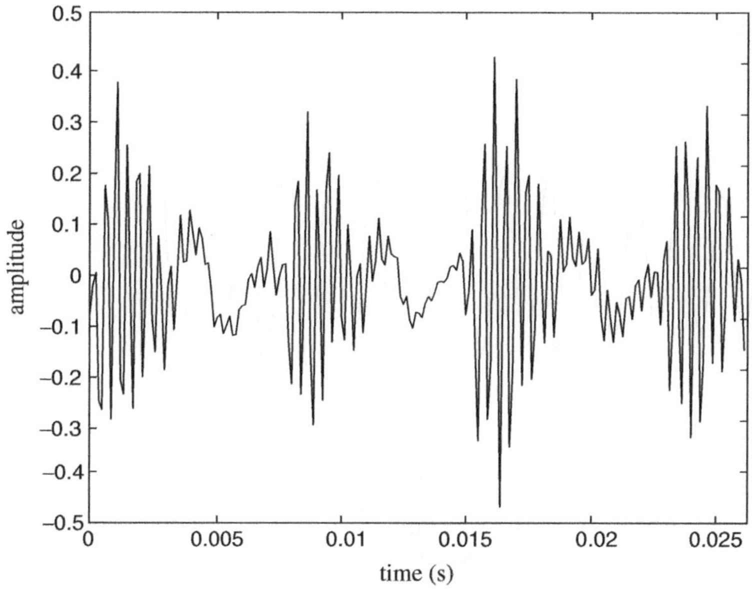

Figure 1.1 Speech Signal.

loudness. This amplitude can vary significantly, influenced by factors like the speaker, surroundings, and the conveyed emotion. Sound pressure amplitude is commonly quantified in units of *pascals* (Pa). In this context, a *micropascal* (μPa) is equivalent to one-millionth of a pascal.

In controlled laboratory settings, the benchmark for the minimum detectable amplitude, known as the threshold of hearing, is approximately 20 μPa for a 1 kHz tone. Some individuals regard 60 μPa as the threshold for experiencing pain, yet this perception is subjective and varies significantly based on individual differences and age. For instance, the sound pressure produced by a jackhammer at a distance of 1 meter reaches a maximum of approximately 2 Pa, while a jet engine at the same distance generates a maximum pressure of 632 Pa. *Sound Pressure Level* (SPL), another parameter of sound field, serves as a metric in acoustic analysis and recording. It is commonly employed to gauge the auditory experience at the listener's location or the positioning of a microphone. The relationship between sound pressure, measured in pascals, and dB SPL is associated with a spectrum spanning from 0.00002 Pa, corresponding to 0 dB SPL, to 200 Pa, equating to 140 dB SPL and even higher (Smith 2010).

Both audio and speech signals exhibit temporal characteristics, such as the zero crossing rate, which refers to the frequency at which the signal crosses the zero amplitude line within a second. Additionally, they manifest time-domain attributes like autocorrelation, illustrating how similar the signal is to itself at

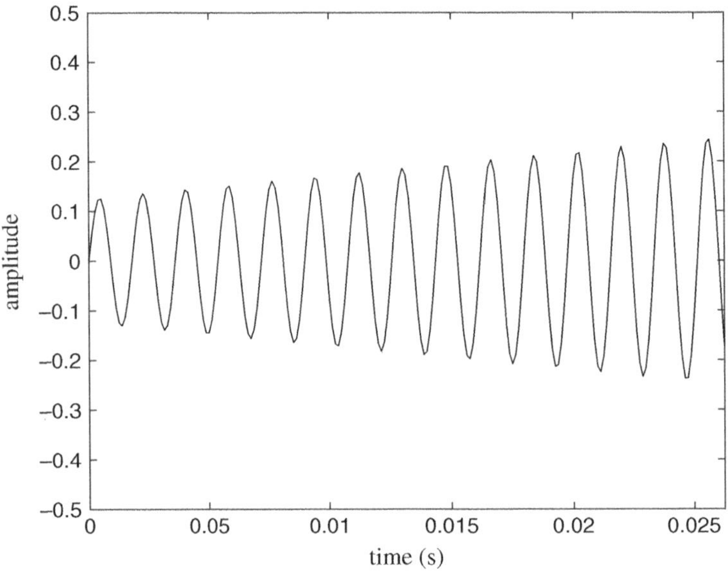

Figure 1.2 Audio Signal.

different time delays. In the frequency domain, these signals possess characteristics including the power spectrum, that show how much energy the signals have at different frequencies. Furthermore, there are *Mel-Frequency Cepstral Coefficients* (MFCCs), which capture the high-frequency content of the signal, and *Linear Predictive Coding* (LPC) coefficients, which model the signal as a linear combination of past samples. Figure 1.1 originally shown in Section 1.2.1 illustrates a speech sequence. In contrast, Figure 1.2 presents a normalized audio sequence of a 2.5 kHz tone with increasing amplitude, showing the differences in temporal and spectral characteristics compared to speech.

1.2.3 Video

The foundation of digital video predominantly originates from technology (now obsolete) introduced during the early stages of black and white television in the 1930s. The receiver component of black and white television featured the *cathode ray tube* (CRT). The CRT utilizes a substantial voltage difference to generate an electric field that propels electrons, giving rise to cathode rays. These rays transport electrons from the cathode to the anode, where, due to their momentum, they collide with a photosensitive screen. The velocity and, therefore, the luminosity on the screen correspond to the voltage applied between the anode and cathode. The CRT also incorporates horizontal and vertical deflection plates, allowing the path of the rays to be altered and ensuring that every point on the photosensitive screen receives illumination.

The image refresh process happens line by line, starting from the upper left corner and progressing to the lower right. Voltage signals applied to each set of deflection plates have a sawtooth waveform. The y-plate signal's period matches the video frame duration, while the x-plate's period aligns with the line duration. This scanning method is known as progressive scanning. The frame frequency, calculated as the reciprocal of the frame duration, is derived from the *Alternating Current* (AC) frequency to minimize interference from transformers and other power fluctuations. In the United States with 60 Hz AC, the standard frame rate has been traditionally 30 *frames per second* (fps). In regions with 50 Hz AC, like most of the world, the standard has been 25 fps. Note that the first and last lines of each frame, as well as the start and end of each line, are transmitted but not displayed. These hidden regions typically carry synchronization data (Benott 2008).

A different approach from progressive scanning is interlaced scanning. Here, each video frame is divided into two fields, scanned in an alternating manner. The first field scans the odd-numbered lines, while the second scans the even-numbered lines. While progressive scanning delivers one complete frame every $\frac{1}{30}$ seconds (assuming a 30 fps frame rate), interlaced scanning transmits half a frame every $\frac{1}{60}$ seconds, resulting in the same field rate (60 Hz in this example). Keep in mind that, interlaced scanning can provide a perceptual advantage by reducing choppiness in certain situations, like scenes with slow or moderate motion, due to its higher apparent frame rate (twice the field rate). However, it can also introduce artifacts under certain conditions.

Figure 1.3 compares progressive and interlaced scanning. Although both methods process the same number of fields per unit time (operating at $\frac{1}{2T}$ field rate), they achieve this in different ways. Progressive scanning displays each complete frame independently, while interlaced scanning displays half frames

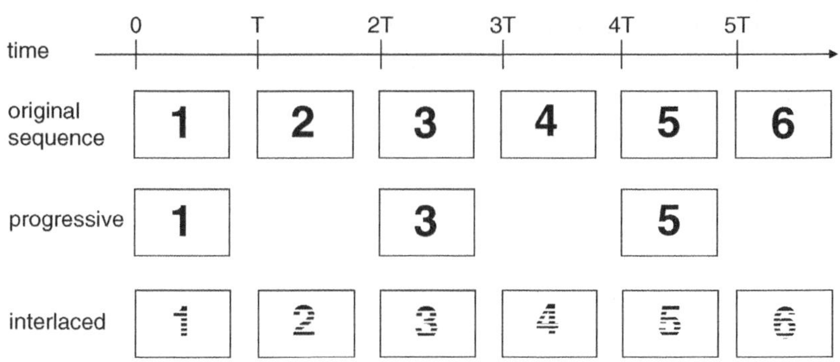

Figure 1.3 Progressive vs Interlaced Scanning.

in an alternating pattern, creating the illusion of a full frame. When viewed in motion under certain conditions, particularly with slow or moderate motion and higher vertical resolutions, interlaced scanning can offer a perception of smoother playback. However, it can also introduce artifacts, especially in fast-moving scenes or at lower resolutions.

Irrespective of the frame rate, all black and white standards utilize a composite analog signal for transmission, incorporating *video, blanking, and synchronization* (VBS) information. This transmission method commonly employs interlaced scanning, where each line's initiation is guided by precisely timed horizontal synchronization pulses embedded between the video signals in the imperceptible portion of each line. The term *blanking* has to do with the pulse responsible for guiding the beam's retrace from the lower right to the upper left corner of the screen.

The black and white television standard in the United States uses a field frequency of 60 Hz, divided into two fields, each containing 262 ½ lines, and a line frequency of 15.75 kHz. Video content uses *Amplitude Modulation* (AM) with a bandwidth of 4.2 MHz, while audio uses *Frequency Modulation* (FM) with a 4.5 MHz carrier frequency.

Unlike the United States standard, the European (and international) black and white television standard operates at a field frequency of 50 Hz. It contains two fields, each composed of 312 ½ lines, resulting in a line frequency of 15.625 kHz. Video content uses AM modulation with a bandwidth of 5 MHz, while audio utilizes FM modulation with a 5.5 MHz carrier frequency.

Since any color can be made by mixing the three primary colors: *red, green, and blue* (RGB), a color TV scheme uses three separate black and white cameras, each with a filter that only lets in one of these primary colors. These cameras capture three separate VBS signals, which can be combined to recreate the original full-color image. However, there is a problem with this approach: old-fashioned black and white TVs cannot display video recorded with this method.

The primary objective is to devise a compatibility scheme that functions bidirectionally–enabling color signals to be displayed on traditional black and white televisions, while also ensuring that monochromatic signals can be exhibited on color televisions.

A linear combination can be applied to the RGB components and transformed into other three components that decompose the information into signals that provide backward compatibility. One such scheme is through a set of linear equations given by

$$
\begin{aligned}
Y &= 0.587G + 0.299R + 0.1145B \\
C_b &= 0.564(B - Y) \\
C_r &= 0.713(R - Y)
\end{aligned}
$$

where R, G, and B are the intensity values of the red, green, and blue components that serve as input to obtain luminance Y, blue chrominance C_b and red chrominance C_r components. An alternative set of equations results from

$$Y = 0.587G + 0.299R + 0.1145B$$
$$U = 0.493(B - Y)$$
$$V = 0.877(R - Y)$$

where the blue and red chrominances are U and V, respectively. Figure 1.4 illustrates how backward compatible signals are generated from the input signals. Each signal involves filtering a specific color of light through optics and then capturing its electrical representation via a regular black and white camera.

The luminance signal conveys the intensity of the color image, allowing black and white TVs to display a grayscale version. For color TVs, the chrominance signals carry the color information needed to render the complete image. Figure 1.5 depicts the composite signal, which combines these elements. Figures 1.6 to 1.8 further break down the Y, U, and V components, respectively, that make up the chrominance signal. Interestingly, human perception is less sensitive to color compared to luminance. This allows less bandwidth to be allocated for transmitting chrominance details, ensuring compatibility with older standards without impacting the critical black and white components.

The VBS signal containing chrominance information is called *Color VBS* (CVBS). Different ways of modulating chrominance signals create various color representation schemes, leading to different video transmission standards. For example, the United States relied on the *National Television Standard Committee* (NTSC) standard, while Europe utilized both *Phase Alternating Line* (PAL) and

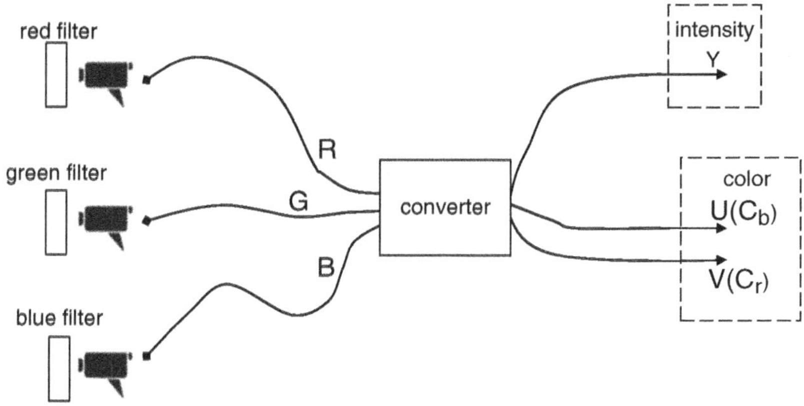

Figure 1.4 RGB to YUV Conversion.

Figure 1.5 Composite Image.

Figure 1.6 Y Component.

Sequential Color with Memory (SECAM), which were once dominant players in analog television broadcasting. Although analog TV is no longer used, these seemingly outdated technologies laid the foundation for many aspects of modern digital video. Details of digital video coding are introduced later in this chapter in Section 1.4.3.

Figure 1.7 U Component.

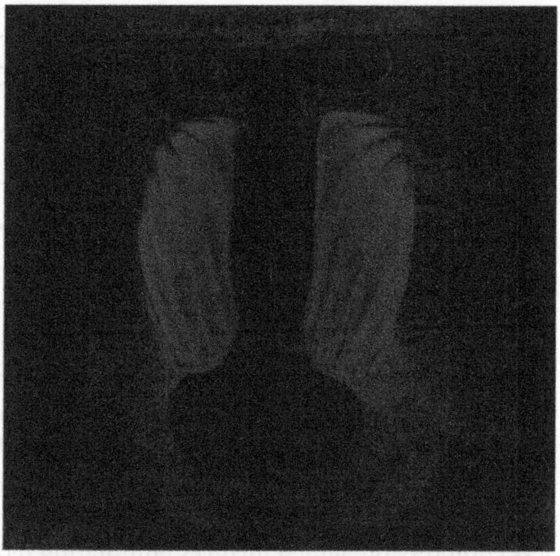

Figure 1.8 V Component.

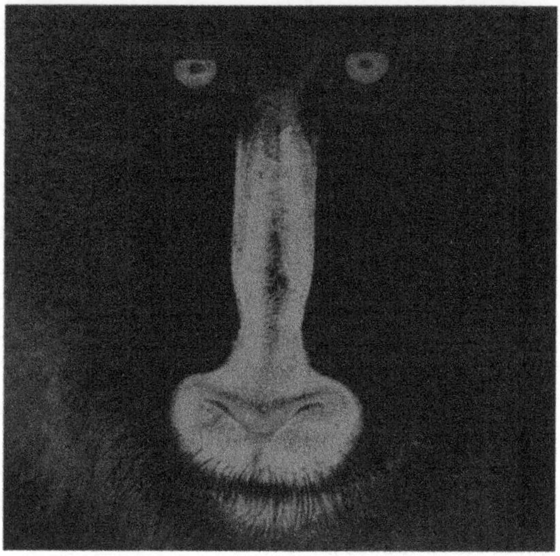

1.3 Sampling and Quantization

As outlined in Section 1.1, information produced by natural sources such as speech, audio, and images exists in the analog domain. In this domain, corresponding signals often include a significant amount of redundant data. This redundant information can be eliminated without appreciably affecting human

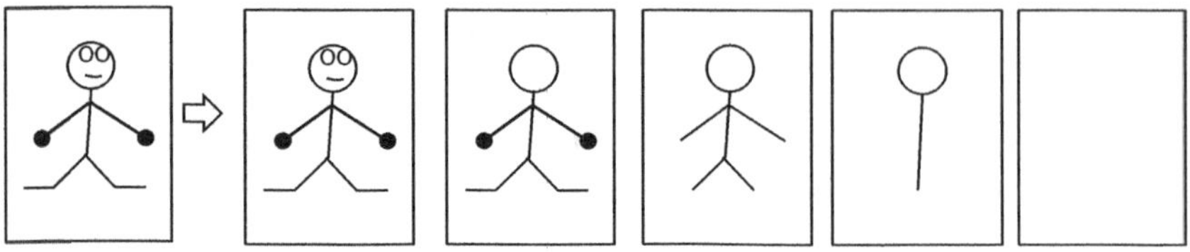

Figure 1.9 Rate–Distortion Example.

perception, a procedure referred to as data compression. It is worth noting that, in contrast to the analog domain, a digital domain exists where analog signals are translated into sequences derived from a countable set of numbers. These sequences can be stored and processed by signal processors and computers. The process of removing information in this context is termed source encoding.

Data compression can be divided into two primary categories. First, there is lossless compression, which involves eliminating redundancy in a reversible manner, ensuring that the reconstructed message remains identical to the original. Second, there is lossy compression, which entails purposefully removing information in a controlled manner. This removal of non-essential data is permanent, rendering the process irreversible. In this scenario, although the reconstructed message differs from the original, the objective is to minimize this divergence to the point of imperceptibility to the recipient.

The compression rate is defined as the ratio of the information present in the compressed message to that in the original message. Generally, lossy compression exhibits a significantly higher compression rate compared to lossless compression.

In the realm of lossy compression, a specific compression rate corresponds to a specific degree of distortion, giving rise to a trade-off known as rate–distortion. An illustration of this concept can be observed in Figure 1.9, where a source image undergoes compression at various rates, leading to distinct levels of distortion. Both rate and distortion are quantified on a unitless scale ranging from 0 to 4. Note that as distortion diminishes, transmission rates tend to increase; conversely, higher distortion levels result in lower transmission rates (Proakis and Manolakis 2006; Haykin 2009).

As indicated in Section 1.1, sampling and quantization are two key mechanisms that must be employed to convert signals from the analog to the digital domain. The subsequent subsections focus on these two key processes.

1.3.1 Sampling

Sampling transforms an analog signal, which is defined continuously across time, into a sequence of discrete samples taken periodically at a fixed interval known as the sampling period, measured in units of time. The reciprocal of the sampling period is the sampling rate, measured in units of *samples per second* (sps). If the sampling rate is too slow, the resulting set of samples might not accurately represent the original analog signal. Conversely, if the rate is excessively high, the set could become unwieldy for efficient processing and modulation for transmission through a channel. Table 1.1 shows how, in audio and speech sampling scenarios, each sampling rate has a specific type that identifies it.

Generally, when an energy signal $g(t)$ (as shown in Figure 1.10) is defined at every instant of time and sampled at a rate of $F_s = \frac{1}{T_s}$ (where T_s denotes the sampling period), the resulting sampled signal resembles Figure 1.11. Mathematically, sampling takes $g(t)$ as input and generates an infinite sequence of samples spaced T_s seconds apart, forming the sequence $\{g(nT_s)\}$. This process, known as ideal or instantaneous sampling, captures the signal's value at specific, infinitesimal

Table 1.1 Sampling Rate Types.

Type	Sampling Rate (sps)
Narrowband (NB)	8000
Wideband (WB)	16000
Super Wideband (SWB)	32000
Fullband (FB)	48000

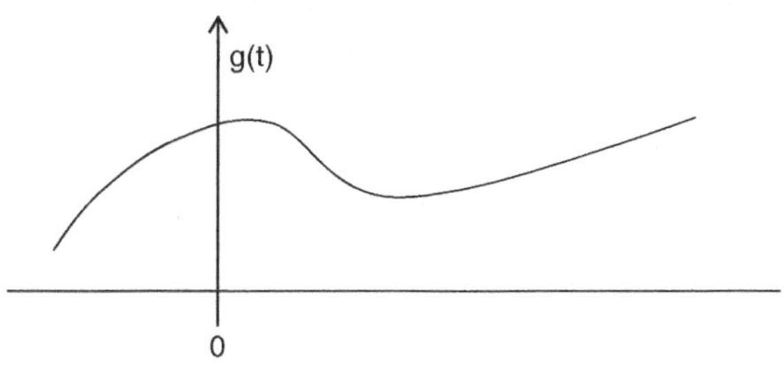

Figure 1.10 Signal $g(t)$.

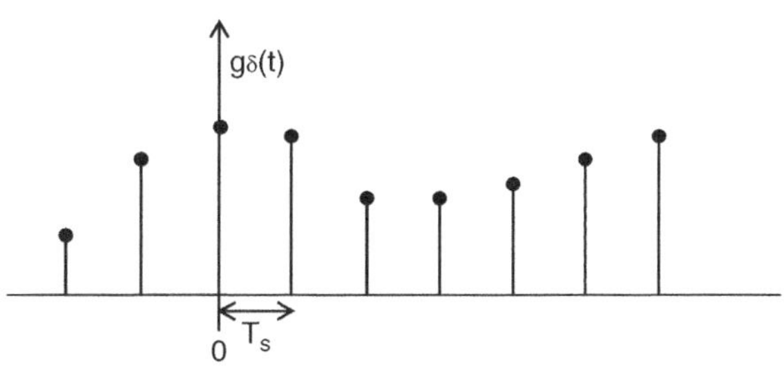

Figure 1.11 Sampled $g(t)$.

instants of time. Note that under ideal sampling, the effect is akin to multiplying the analog signal $g(t)$ by a train of time pulses. Specifically,

$$\begin{aligned} g_\delta(t) &= \sum_{n=-\infty}^{\infty} g(nT_s)\delta(t-nT_s) \\ &= g(0)\delta(t) + \\ &\quad g(T_s)\delta(t-T_s) + \\ &\quad g(2T_s)\delta(t-2T_s) + \cdots \end{aligned} \tag{1.1}$$

where $g_\delta(t)$ is the ideal sampled signal and $\delta(t-nT_s)$ is a delta function positioned at time $t = nT_s$.

In the frequency domain $g_\delta(t)$ becomes $G_\delta(f)$ and it is given by

$$\begin{aligned} G_\delta(f) &\rightleftharpoons F_s \sum_{m=-\infty}^{\infty} G(f-mF_s) \\ &= F_sG(f) + \\ &\quad F_sG(f-F_s) + F_sG(f+F_s) + \\ &\quad F_sG(f-2F_s) + F_sG(f+2F_s) + \cdots \end{aligned} \tag{1.2}$$

where the spectrum is an infinite sequence of shifted versions of the analog signal frequency representation. Whether overlap occurs among these versions hinges upon the sampling rate F_s. To illustrate, consider a scenario where $g(t)$ is band-limited (represented in Figure 1.12), containing no frequency components beyond W Hz. If sampled at a rate of $F_s = 2W$, the resulting spectrum (as shown in Figure 1.13) includes components without any overlap.

Mathematically, from Equation (1.2)

$$G_\delta(f) = F_sG(f) + F_s \sum_{m=-\infty, m\neq 0}^{\infty} G(f-mF_s)$$

where if (1) $G(f) = 0$ for $|f| \geq W$ and (2) $F_s = 2W$, then

$$G(f) = \frac{1}{2W} G_\delta(f) \tag{1.3}$$

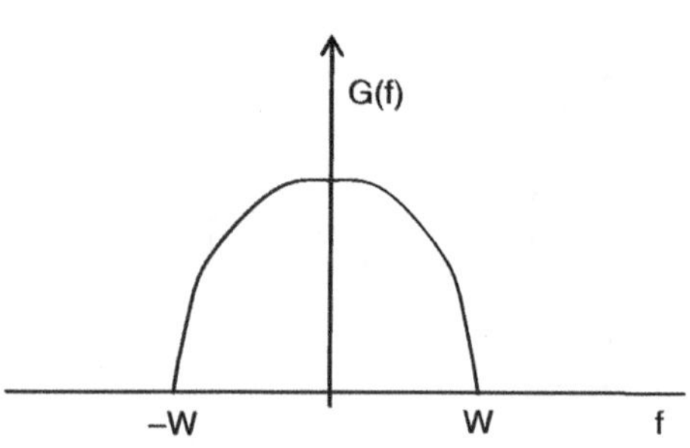

Figure 1.12 Band-limited Signal.

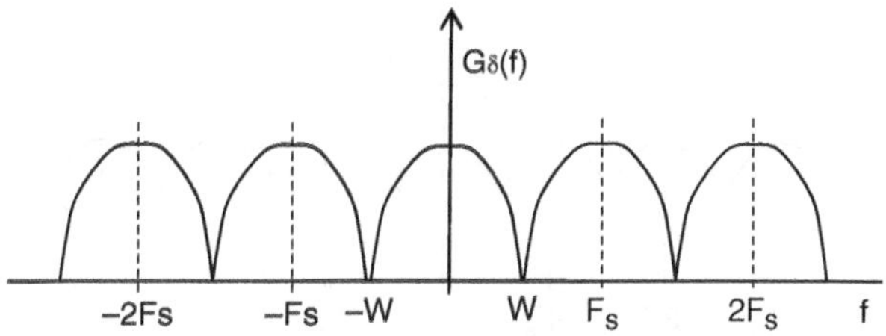

Figure 1.13 Spectrum of Sampled Signal.

if $|f| < W$. In this scenario, the samples $g(\frac{n}{2W})$ of the analog signal taken every $\frac{1}{2W}$ seconds, with n being an integer, contain all the information in $g(t)$.

The inverse operation, that is, recovering the analog signal $g(t)$ out of the samples $g(nT_s)$ is given by

$$
\begin{aligned}
g(t) &= \sum_{n=-\infty}^{\infty} g\left(nT_s\right) \operatorname{sinc}(2Wt - n) \\
&= \sum_{n=-\infty}^{\infty} g\left(nT_s\right) \operatorname{sinc}\left(\frac{t}{T_s} - n\right) \\
&= g(0) \operatorname{sinc}\left(\frac{t}{T_s}\right) + \\
&\quad g(T_s) \operatorname{sinc}\left(\frac{t}{T_s} - 1\right) + \\
&\quad g(2T_s) \operatorname{sinc}\left(\frac{t}{T_s} - 2\right) + \cdots \\
&= g_\delta(t) * \operatorname{sinc}(2Wt)
\end{aligned}
\tag{1.4}
$$

where delayed versions of the sinc function are added together to interpolate $g(t)$ for any value of t. Note that the sinc function is defined as $\operatorname{sinc}(x) = \frac{\sin(x)}{x}$. This convolution in the time domain is analogous to multiplication by a low-pass (LP) filter in the frequency domain, referred to as a reconstruction filter. As a result, to recover the analog signal from its sampled version, $g_\delta(t)$, it is only necessary to process it through this reconstruction filter.

Generally, an energy signal lacking frequency components above W Hz can be accurately described using samples taken periodically every $T_s = \frac{1}{2W}$ seconds. The sampling rate $F_s = \frac{1}{T_s} = 2W$ is known as the Nyquist Rate, while the sampling period $T_s = \frac{1}{F_s} = \frac{1}{T_s}$ is referred to as the Nyquist Interval.

Undersampling occurs when a signal is sampled at a rate below the Nyquist Rate, indicated by $F_s < 2W$. This condition leads to aliasing, a phenomenon where successive shifted replicas of the analog signal's frequency representation in Equation 1.2 overlap. To illustrate, consider the frequency representation of the signal shown in Figure 1.12. When undersampled, as shown in Figure 1.14, the signal exhibits the characteristic distortions of aliasing.

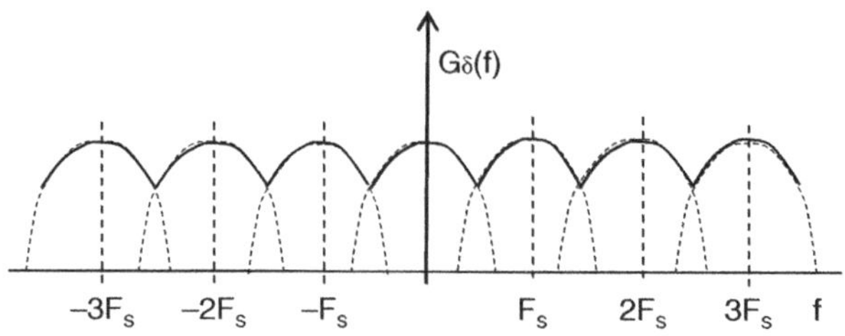

Figure 1.14 Aliasing.

To avoid aliasing, an LP filter, known as an anti-aliasing filter, can be used on the analog signal $g(t)$. This filter suppresses high-frequency components that could otherwise trigger undersampling issues. However, since this filter is not inherently perfect and possesses a transition band, it is important to make slight adjustments to the sampling rate. By ensuring a rate marginally higher than the Nyquist Rate, the absence of spectral overlap can be guaranteed.

When the signal shown in Figure 1.12 is sampled at a rate surpassing the Nyquist Rate, the resulting spectrum of the instantaneous sampled version, shown in Figure 1.15, is achieved.

Reconstruction and anti-aliasing filters typically share analogous characteristics, both possessing a transition band that extends from W to $F_s - W$. Moreover, as the sampling frequency increases, the spectral gap between repetitions of the analog signal's spectrum within the sampled signal's spectrum widens. This phenomenon alleviates constraints on the transition band's width in these filters, illustrated in Figure 1.16. The ultimate extent of this band is dictated by $F_s - 2W$.

Illustrated in Figure 1.17 is an illustrative sampling scenario, examined from the viewpoints of both time and frequency domains. To be precise, the signal designated for sampling at the transmitter undergoes multiplication and convolution by a series of pulses in both the time and frequency realms, respectively. This resultant signal is then transmitted through the communication channel and subsequently reconstructed at the receiver by means of frequency multiplication utilizing an LP filter. Correspondingly, in the temporal domain, this equates to convolution through the time depiction of the LP filter, which manifests as a sinc function.

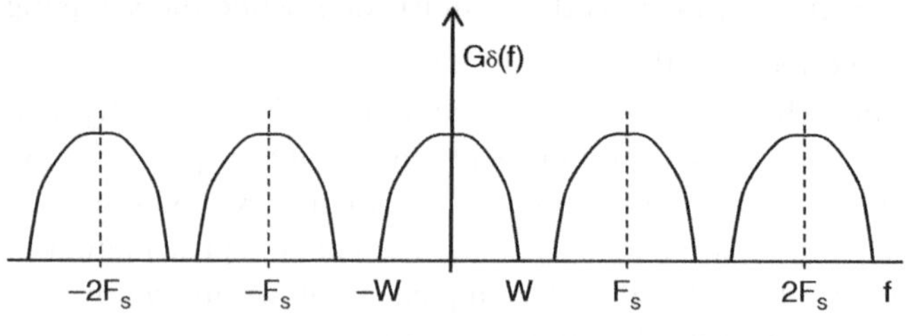

Figure 1.15 Spectrum of Instantaneous Sampled Signal.

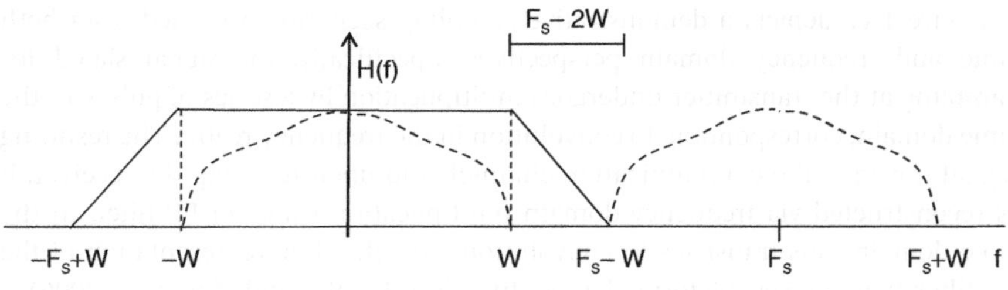

Figure 1.16 Reconstruction Filter.

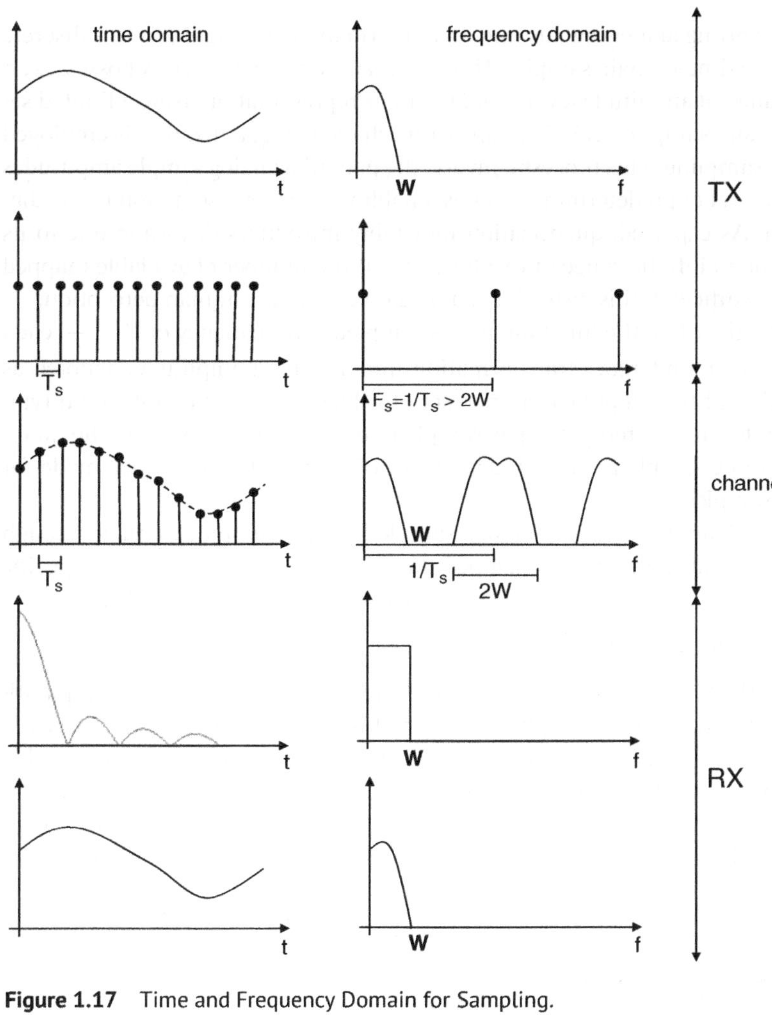

Figure 1.17 Time and Frequency Domain for Sampling.

Figure 1.17 depicts a demonstrative sampling scenario, examined from both time and frequency domain perspectives. Specifically, the signal slated for sampling at the transmitter undergoes multiplication by a series of pulses in the time domain, corresponding to convolution in the frequency realm. The resulting signal traverses the communication channel, and upon reaching the receiver, it is reconstructed via frequency-domain multiplication using an LP filter. In the time domain, this translates to convolution with the time representation of the LP filter, which takes the form of a sinc function (Proakis and Manolakis 2006).

1.3.2 Quantization

While converting analog signals into digital formats, sampling captures a discrete collection of time-domain samples. However, each sample inherently possesses an infinite range of amplitude levels, making their representation using a limited set of digital values impractical. To address this challenge, quantization is employed alongside sampling. This process converts the potential analog sample amplitudes into a finite set of predetermined values, enabling their representation in the digital realm. As expected, quantization inevitably introduces distortion due to its mapping of an infinite range onto a finite set. If the number of available mapped values is insufficient, this distortion can negatively impact human perception.

Accordingly, when the function $m(t)$ is sampled at a frequency of $F_s = \frac{1}{T_s}$, each sample $m(nT_s)$, undergoes transformation into a distinct amplitude, denoted as $v(nT_s)$, selected from a predetermined set of values. Note that quantization typically operates in a memoryless, per-sample manner, guaranteeing that the quantization of one sample remains independent of quantization decisions made for previous samples.

In terms of amplitudes, if an analog sample has an amplitude m, then through quantization, this amplitude is mapped into a number k if m is within a specific range or partition cell, shown in Figure 1.18 and given mathematically by

$$P_k : \{m_k < m \leq m_{k+1}\}, \qquad k = 1, 2, \cdots, L$$

where L is the total number of levels or possible discrete amplitudes in the quantization set and the discrete amplitudes m_k with $k = 1, 2, \cdots, L$ are called decision thresholds. Upon the need to reconstruct the quantized sample as a specific value v, the numerical index k undergoes a conversion process to yield a reconstruction level denoted as v_k. This reconstruction level involves the entire spectrum of conceivable analog amplitudes present within a designated partition referred to as P_k.

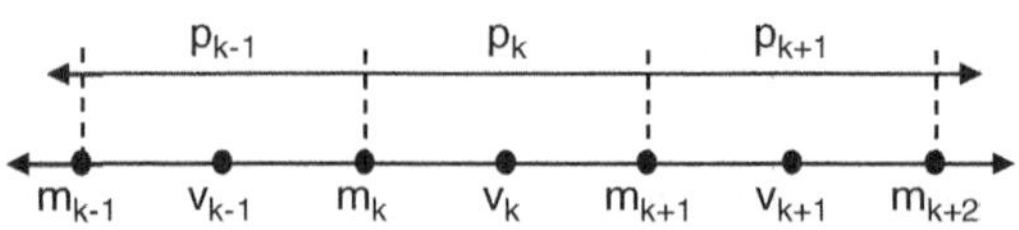

Figure 1.18 Partition Cell.

The interval between two successive levels, represented as $\Delta = v_k - v_{k-1}$, is known as the step size.

In a broad sense, the conversion process from an analog sample to the reconstructed value v is expressed as $v = g(m)$, where the function g(.) is referred to as the quantizer characteristic. Depending on whether it possesses an even or an odd count of levels, the quantizer can take the form of either a midtread or a midrise configuration.

In the process of quantization at the transmitter stage, the analog sample's magnitude is associated with an index denoted as k, which can be transformed into a binary representation, comprising a sequence of bits. This binary sequence is subsequently modulated and transmitted through the communication channel. At the receiver's end, the transmitted signal undergoes demodulation, resulting in a sequence of bits that correspond to the index k. This index is then utilized to reconstruct the original sample's magnitude using the quantizer's characteristics.

Quantization noise, also shown in Figure 1.19 for the case of a ramp signal, is characterized as the disparity between the input signal $m(t)$ and the reconstructed quantized signal $v(t)$. Treating both $m(t)$ and $v(t)$ as random processes, their respective sampling at any given time point yields two *Random Variables* (RVs) denoted as M and V. The quantization noise is then represented as a third random variable, termed as Q, calculated as the difference between M and V. Assuming a uniform midrise quantizer where both M and Q have zero mean and where the distribution of the amplitude of M is in the range $(-m_{\max}, m_{\max})$ then the step size of the

Figure 1.19 Midtrise Noise.

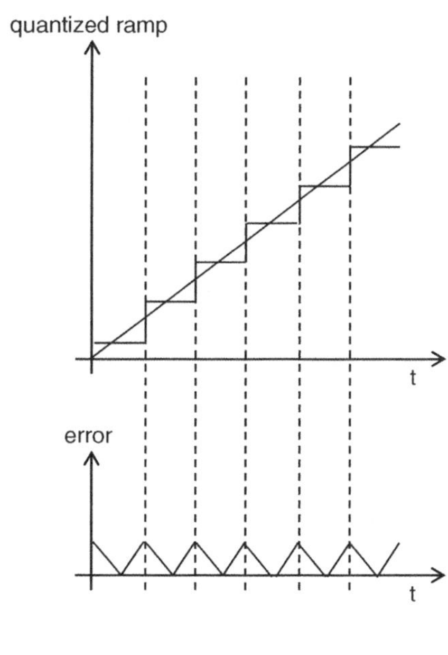

quantizer is given by

$$\Delta = \frac{2m_{\max}}{L} = \frac{2m_{\max}}{2^n} \tag{1.5}$$

where n is the number of bits per sample such that $L = 2^n$ is the even total number of reconstruction levels. The quantization error Q can be thought of as a uniformly distributed RV where an instance q is in the interval $-\frac{\Delta}{2} \leq q \leq \frac{\Delta}{2}$. The *Probability Density Function* (PDF) of Q is therefore given by

$$f_Q(q) = \begin{cases} \frac{1}{\Delta} & -\frac{\Delta}{2} < q \leq \frac{\Delta}{2} \\ 0 & \text{otherwise} \end{cases}.$$

The variance of Q obtained from σ_Q^2 and given by

$$\sigma_Q^2 = E\left[Q^2\right] = \int_{-\frac{\Delta}{2}}^{\frac{\Delta}{2}} q^2 f_Q(q)dq = \frac{\Delta^2}{12} \tag{1.6}$$

is a good indication of the quantization error power, which should be as small as possible to minimize distortion. Replacing Equation 1.6 in Equation 1.5, the power of the quantization error results

$$\sigma_Q^2 = \frac{1}{3} m_{\max}^2 2^{-2n}$$

and assuming that the power of the input analog signal $m(t)$ is P, then the output *Signal-To-Noise Ratio* (SNR) of the quantizer is given by

$$\text{SNR}_O = 10\log_{10}\left(\frac{P}{\sigma_Q^2}\right) = 10\log_{10}\left[\left(\frac{3P}{m_{\max}^2}\right)2^{2n}\right] \tag{1.7}$$

such that SNR increases exponentially with the number of bits per sample n. Unfortunately, increasing n has the undesirable consequence of increasing both the transmission rate and the bandwidth requirements, as more bits are transmitted over the channel.

Example 1.1 Consider a speech signal with a bandwidth of 3000 Hz that is subjected to a 256-level quantizer. What is the transmission rate of this signal when it is digitalized ?

Solution:
The sampling frequency is $F_s = 2 \times 3000 = 6000$ sps. With $L = 256$ level, the number of bits per sample is $n = \log_2 L = 8$ bits per sample. The transmission rate then becomes $R = F_s \times n = 6000 \times 8 = 48$ Kbps.

1.4 Codecs

Codecs depend on particular signal encoding methods that pertain to the specific type of media being considered (Proakis and Manolakis 2006).

1.4.1 Speech Coding

Based on the source's characteristics, further compression can be applied within the context of source encoding to reduce the overall transmission rate. Specifically, when speech is the source, a speech coder or codec can be utilized to convert speech samples into a bitstream using various techniques. These techniques range from simple input–output mapping to more advanced speech models that extract distinct parameters.

Figure 1.20 illustrates a digital speech communication system in which the transmitter relies on a source encoder responsible for anti-aliasing, sampling, and quantization/symbol encoding. This process results in a bitstream that is subsequently processed by the codec encoder. The codec encoder undergoes channel encoding and modulation before transmission over the channel.

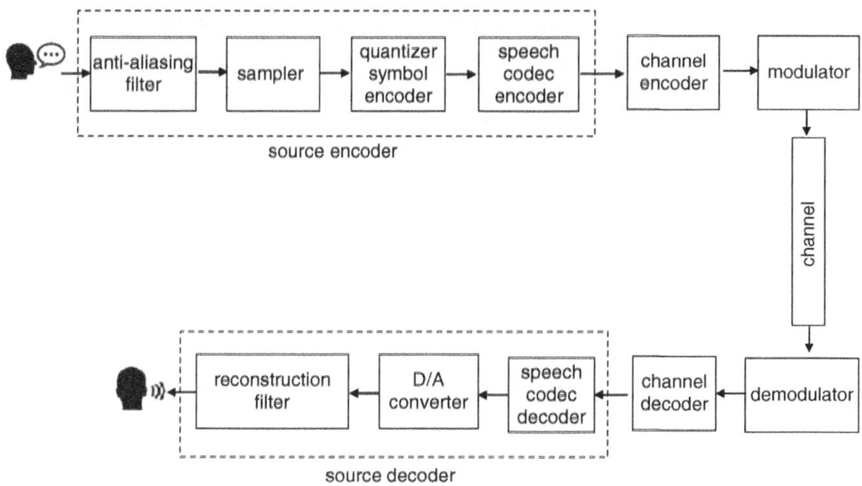

Figure 1.20 Digital Speech Communication System.

Conversely, the receiver performs the inverse operation. After demodulating the modulated wave and decoding the channel, the signal is directed to the codec decoder. The codec decoder restores the signal, which, when subjected to the reconstruction filter, yields the synthetic speech signal. In particular, due to the concentration of speech energy between 300 and 3400 Hz, the speech signal is initially subjected to LP filtering through an anti-aliasing filter with a cutoff frequency of 4000 Hz. To comply with the sampling theorem, a common sampling rate of 8000 sps is often employed. In terms of speech coding, this sampling rate is known as the *narrowband* (NB) sampling rate.

The quantizer used is a midrise uniform scheme with a 16-bit resolution, providing 65536 levels and generally ensuring satisfactory quality for most speech signals. This configuration leads to a bitrate of 8000 sps × 16 bits/sample = 128 Kbps at the input of the codec encoder. Since transmission rates are measured in *bits per second* (bps) units, 1 Kbps is equivalent to 1000 bps. The primary objective of the codec encoder is to compress the raw speech input into an output bitstream lower than 128 Kbps. This scheme is traditionally known as a *Pulse Code Modulation* (PCM) scheme.

Depending on the compression mechanisms employed, a large number of output transmission rates can be achieved. For instance, codecs like ITU-T Recommendation G.711 and AMR (described later in this chapter) exhibit output transmission rates of 64 Kbps and 4.75 Kbps, respectively. This variability is shown in Figure 1.21, where the placement of a codec encoder and decoder in succession maintains an output bitstream of 128 Kbps at the decoder stage.

Within the realm of modern RTC reliant on packet-switched communication networks, the codec encoder plays a key role in buffering raw speech and subsequently compressing it into smaller output frames. To elaborate, the speech codec encoder takes in an input bitstream and generates an output frame. This resultant output frame is then subjected to channel encoding, involving the addition of headers to create packets which are subsequently modulated and sent over the communication channel.

To mitigate the impact of channel encoding overhead on transmission rate, the codec encoder must maintain a relatively larger frame size. However, it is important to have a balance, as an excessively large frame size could introduce latency, negatively affecting the overall user experience. Consequently, there exists a trade-off between frame size and communication latency, ultimately determining

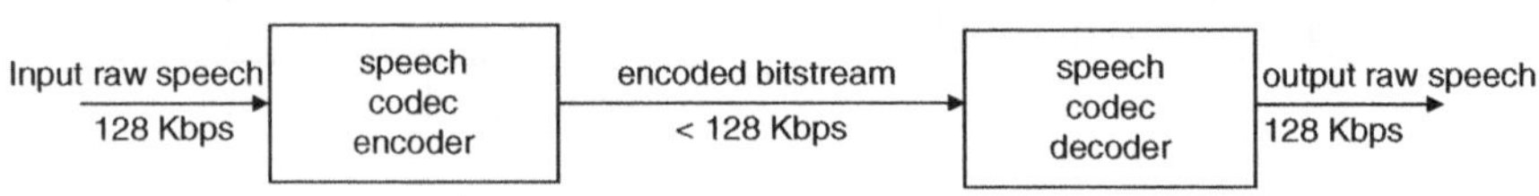

Figure 1.21 Encoder / Decoder.

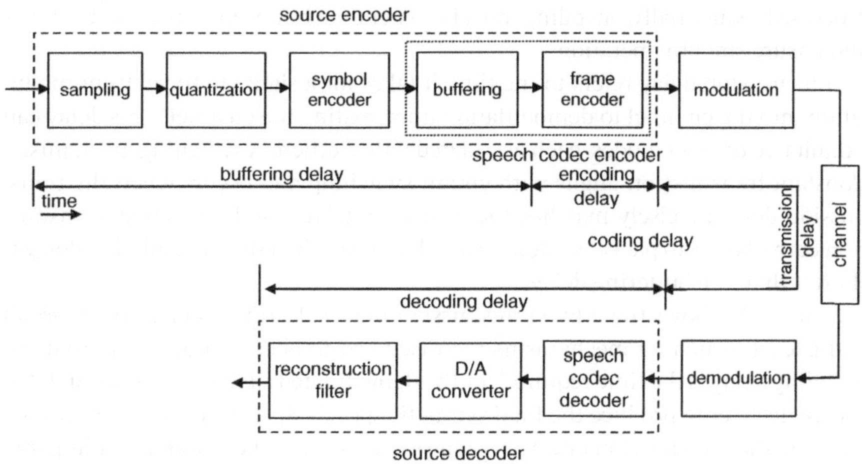

Figure 1.22 Coding Delay.

the optimal frame length. This ideal frame length aims to simultaneously achieve the lowest possible latency and the lowest transmission rate.

The primary objective of a codec is to achieve a substantial reduction in transmission rate through speech compression, while also ensuring quality, resilience against channel errors, commendable performance across diverse languages and ideally compatibility with audio signals like music and tones. From a technical point of view, codecs aim to maintain low computational and memory demands as well as relatively minimal encoding and decoding delays.

The overall delay, as illustrated in Figure 1.22, consists of four distinct components: buffering delay, encoding delay, transmission delay, and decoding delay. Within a speech codec encoder, a buffering element accumulates a sequence of samples into an input frame, which is then compressed by a frame encoding element to generate an output frame. These two elements operate sequentially to achieve speech compression.

The buffering delay, calculated as the time difference between speech entering the source encoder and the input frame buffer reaching the frame encoder, is primarily influenced by the sampling rate and input frame length. For example, with speech sampled at 8000 sps and a frame length of 160 samples, the buffering delay is $\frac{160}{8000} = 20$ ms. Common buffering delays typically range within 10, 20, and 30 milliseconds.

The encoding delay, which is the time required to compress an input frame into an output frame, depends on both the complexity of the compression algorithm and the processing speed of the system. In a pipelined setup, where frames are processed sequentially, encoding must be faster than buffering to prevent backlogs and ensure smooth operation.

Transmission delay refers to the time it takes for a signal to travel from modulation into the channel to demodulation upon exiting the channel. This delay can exhibit two distinct modes, each associated with a different switching mechanism. Constant transmission aligns with circuit switching and occurs when the transmission delay precisely matches the buffering delay. Similarly, bursty transmission corresponds to packet switching and happens when the transmission delay is shorter than the buffering delay.

Figure 1.23 shows these two transmission modes. Further details about circuit and packet switching mechanisms are discussed in Section 1.5.1. Note that the decoding delay is the time required for the demodulated signal to pass through the source decoder to produce the final synthetic speech. Speech codecs are classified based on the transmission rate of the compressed speech they produce, falling into the following categories:

- *High Bit Rate* (HBR) codecs: Exceed 15 Kbps.
- *Medium Bit Rate* (MBR) codecs: Operate between 5 and 15 Kbps.
- *Low Bit Rate* (LBR) codecs: Range from 2 to 5 Kbps.
- *Very Low Bit Rate* (VLBR) codecs: Fall below 2 Kbps.

Waveform codecs, as opposed to other codec types based on their coding mechanism, preserve the shape of the original speech waveform in the synthesized output. This waveform fidelity grants them adaptability to diverse input signal types, with SNR serving as a reliable quality metric. Due to their inherent characteristics, waveform codecs typically align with the HBR classification (Chu 2003). Parametric codecs, in contrast to waveform codecs, operate by extracting key representative

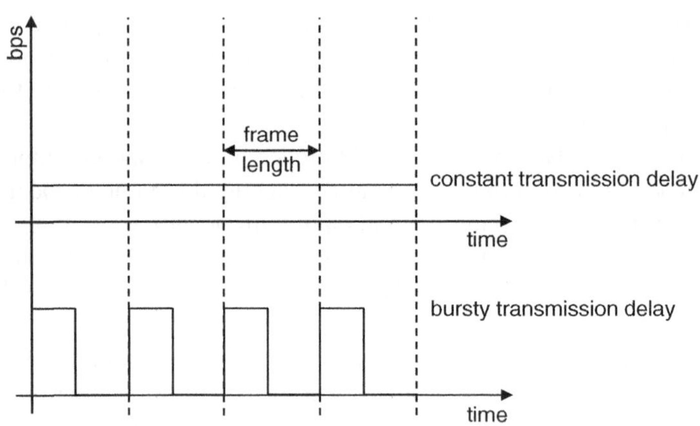

Figure 1.23 Transmission Delay; Constant vs Burst.

parameters from speech sequences. These parameters, rather than the complete waveform, are encoded and transmitted, leading to lower bit rates. However, this approach typically results in a loss of waveform fidelity, making SNR an unsuitable quality metric for parametric codecs. Due to their focus on parameter extraction, they often fall into the LBR or VLBR categories. Hybrid codecs, bridging the gap between waveform and parametric approaches, offer a balanced solution for speech compression. They extract parameters from speech sequences, similar to parametric codecs, but integrate enhancements to preserve the shape of the original waveform in the synthesized output. This combination of techniques often positions hybrid codecs within the MBR category. Waveform and parametric encoders diverge in their treatment of raw frames. Waveform encoders process each sample within a frame individually, while parametric encoders exhibit a different approach, generating compressed output frames distinct from their original inputs.

Understanding and extracting parameters from speech relies heavily on recognizing different signal types. Voiced signals result from the periodic vibration of vocal cords, creating a succession of sounds marked by a characteristic periodicity. This translates to a periodic waveform with a well-defined pitch period and its inverse, the pitch frequency. This frequency typically ranges from 50 to 500 Hz, depending on the speaker's gender. In contrast, unvoiced signals lack any such periodicity and display a noise-like character.

Figure 1.24 contrasts unvoiced and voiced signals within a speech sequence of the phrase *Thank You*. The syllable *Th* exemplifies unvoiced signals with its random, noise-like waveform, while the syllable *U* exhibits the periodic nature of voiced signals. While speech is generally considered non-stationary due to its inherent variability, it can be approximated as stationary over short intervals for analytical purposes. These intervals, characterized by relatively stable signal properties, align with the frame sizes employed by speech codecs. By processing speech on a per-frame basis, codecs manage its non-stationary nature. Frame sizes are further optimized to minimize latency and overhead associated with transport headers.

In speech processing with an NB sampling rate, frame lengths of 10, 20, and 30 milliseconds are commonly employed. These lengths correspond to frame sizes of 80, 160, and 240 samples, respectively.

The speech model foundational to parametric codecs is illustrated in Figure 1.25. This model conceptualizes speech production as a process initiated by airflow from the lungs, which then excites the vocal tract. Airflow is represented as a white noise source, while the vocal tract is modeled as an adaptive, time-varying filter. The filter parameters mirror the acoustic properties of various vocal tract structures, including the nasal cavity, trachea, nostrils, and others. Techniques like LPC are employed to accurately determine these filter parameters.

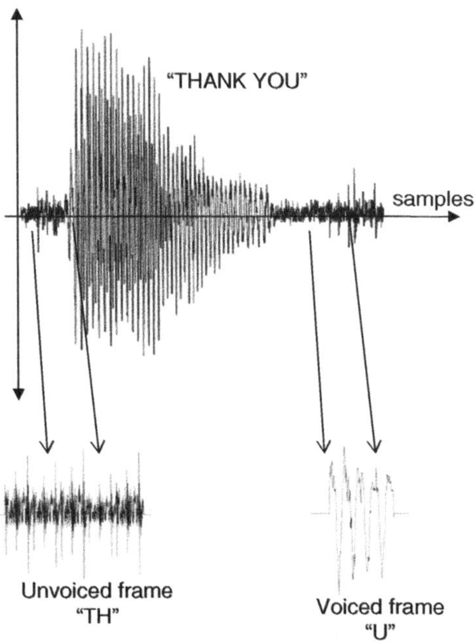

Figure 1.24 The Phrase "Thank You".

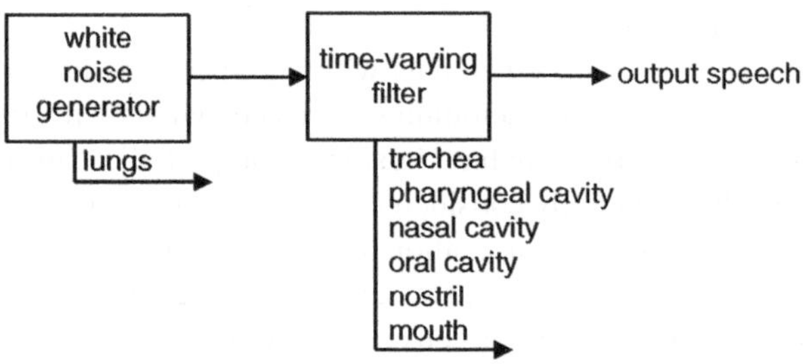

Figure 1.25 Speech Production Model.

1.4.1.1 Waveform Codecs

As indicating in the preceding section, codecs that maintain the original form of the source waveform are referred to as waveform codecs. One of the mechanisms used by these codecs is known as *compression-expansion* (companding). Companding is applied to PCM streams to adjust the speech's dynamic range while preserving the fundamental signal structure.

In traditional PCM systems, analog domain companding takes place prior to numerical conversion, as illustrated in Figure 1.26. This analog companding process can be digitally simulated on a per-frame basis by converting 16-bit samples into 8-bit companded samples using dedicated conversion tables. The underlying concept involves the symbol encoder transforming each of the 65536 sampled and quantized symbols into signed 16-bit numerical representations, accurately

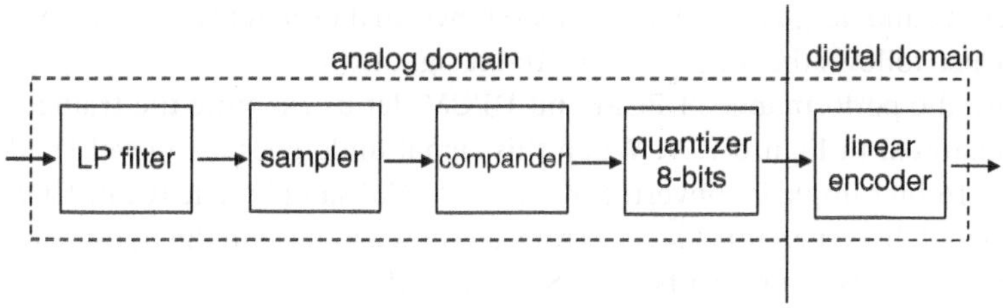

Figure 1.26 μ-Law Analog Companding.

reflecting the analog amplitude of each sample. Note that the maximum positive and negative deviations of an analog sample are mapped to values 32768 and −32767, respectively.

The continuity of analog speech waveforms leads to minimal value variation between consecutive samples, opening a pathway for efficient transmission rate reduction through differential encoding. This technique capitalizes on the narrower range of differences between adjacent samples compared to the range of the individual samples themselves. By operating within a reduced input dynamic range, differential encoding achieves a higher output resolution while maintaining the same number of mapping levels, conserving transmission bandwidth. This technique known as *Differential Pulse Code Modulation* (DPCM) supports this quantization principle and enables efficient speech transmission. A first approximation of DPCM consists of calculating the differential or error signal $e(t)$ as

$$\begin{aligned} e(n) &= x(n) - x_p(n) \\ &= x(n) - x(n-1) \end{aligned} \tag{1.8}$$

where $x(n)$ and $x_p = x(n-1)$ are the current and the predicted 16-bit input samples, respectively. The error signal is then quantized as $\hat{e}(n)$ in accordance with a number of possible levels and transmitted over the channel. At the receiver, the quantized error is used to generate the output sample $\hat{x}(n)$ given by

$$\begin{aligned} \hat{x}(n) &= \hat{e}(n) + \hat{x}_p(n) \\ &= \hat{e}(n) + \hat{x}(n-1) \end{aligned} \tag{1.9}$$

where, for the same sample number, $\hat{x}_p(n)$ is different than $x(n)_p$ at the transmitter. The result of this scheme is that for a 2-level quantizer, the resulting $\hat{x}(n)$ is different than $x(n)$.

A solution to this problem consists on a scheme where the quantized error $\hat{e}(n)$ is related to the quantized prediction error $\hat{x}_p(n)$ as

$$\begin{aligned} \hat{x}(n) &= \hat{e}(n) + \hat{x}_p(n) \\ &= \hat{e}(n) + \hat{x}(n-1) \end{aligned} \tag{1.10}$$

on both, encoder and decoder. The figure also shows, that even if the same 2-level is used, $\hat{x}(n)$ also follows the original input 16-bit signal $x(n)$.

To compare the performance of PCM and DPCM, let us examine the triangular signal illustrated in Figure 1.27. When this signal undergoes a 16-level PCM scheme, each 16-bit sample is converted into a 4-bit PCM sample. The reconstruction of this signal, however, introduces errors, as shown alongside the output signal in Figure 1.28. This error results in an SNR of 24 dB.

Likewise, when subjecting the identical signal to a 16-level DPCM scheme, where an error is computed for each 16-bit sample and subsequently quantized into a 4-bit DPCM sample, the resulting reconstruction error is evident. This error is displayed alongside the output signal in Figure 1.29, leading to a corresponding SNR of approximately 49.42 dB. In particular, the SNR disparity between DPCM and PCM amounts to roughly 25 dB, representing an improvement of approximately 340 times in favor of DPCM. This enhancement in performance is achieved by diminishing the dynamic range of the quantized signal while retaining the same number of quantization levels, thereby upholding the transmission rate.

Applying a 16-level DPCM scheme to the same triangular signal yields distinct results. In this case, an error is calculated for each 16-bit sample and quantized into a 4-bit DPCM sample. As shown alongside the output signal in Figure 1.29, the reconstruction error is reduced, leading to a significantly higher SNR of approximately 49.42 dB. This disparity of 25 dB between DPCM and PCM, translating to a 340-fold improvement in SNR, is about the effectiveness of differential encoding.

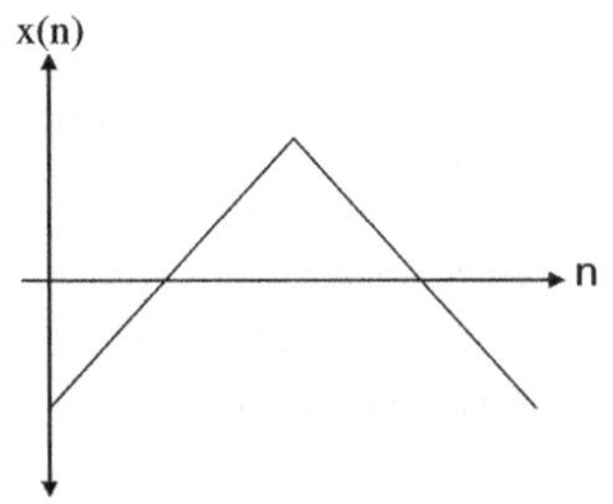

Figure 1.27 Input Signal.

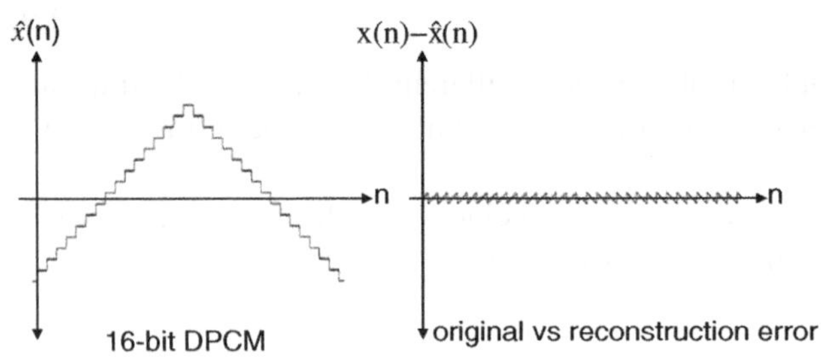

Figure 1.28 PCM Quantized Signal and Reconstruction Error.

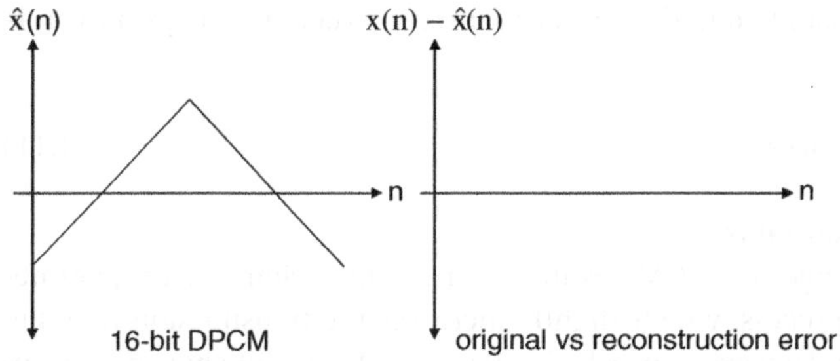

Figure 1.29 DPCM Quantized Signal and Reconstruction Error.

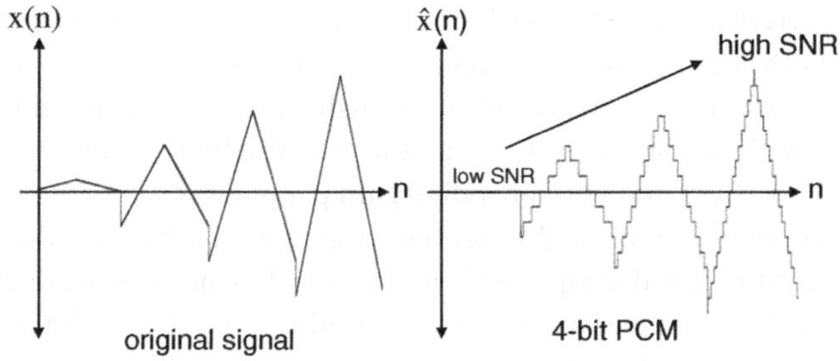

Figure 1.30 4-bit PCM.

DPCM achieves this performance gain by reducing the dynamic range of the quantized signal while preserving the number of quantization levels, thereby maintaining transmission rate efficiency.

Figure 1.30 illustrates an issue inherent to 4-bit PCM mapping. Due to the limited resolution, samples with low amplitudes undergo significant distortion during encoding, resulting in a low SNR. As the signal's amplitude increases, the relative error decreases, leading to a consequent augmentation in SNR.

Adaptive PCM (APCM) offers a method to address quantization issues that result from narrow dynamic ranges. This approach introduces adaptability by first amplifying the signal, then applying uniform quantization. This amplification occurs on a per-frame basis, and the resulting gain is encoded alongside the quantized signal samples. During decoding, the codec reverses this process by attenuating the amplified signal segments.

To elaborate further, consider a speech frame $x(n)$ comprising N 16-bit samples. The gain is determined by calculating the ratio between the dynamic range of the signal in relation to the quantizer's maximum range. This adaptive process enables to counteract quantization challenges and enhance performance. This gain, $i_g(m) \leq 1$, that is calculated for each frame m, is used to amplify the signal in accordance with $i_a(n) = \frac{x(n)}{i_g(m)}$. $N+1$ numbers, the N signal samples $i_a(n)$, and the gain $i_g(m)$ are transmitted to the decoder for reconstruction. The gain $i_g(m)$ is calculated as

$$i_g(m) = k_1 \sum_{n=1}^{N} x^2(n) + k_2 \tag{1.11}$$

where k_1 and k_2 are constants.

The primary challenge of APCM results from the inclusion of gain parameters in the encoding process, which slightly increases the transmission rate due to the additional data. However, speech signals typically exhibit short periods of stationarity, meaning the gain affecting one frame often closely resembles the gain affecting the previous frame. This observation leads to *Backward APCM* (BAPCM), where the gain for the current encoding frame is estimated based on the gain of the previous frame. Since both the encoder and decoder have access to the previous frame, the gain value for any given frame is simultaneously available to both end-points. This modification allows for efficient gain adjustment while minimizing the transmission overhead associated with encoding gain parameters.

Differential pulse code modulation is a differential scheme with a fixed prediction mechanism where the predicted sample is always identical to the last encoded or decoded sample. This can be further extended by assuming a prediction that is adaptive and relies on a linear combination of past samples

$$\hat{x}_p = \sum_{i=1}^{M} a_i \hat{x}(n-i) \tag{1.12}$$

where M is the prediction order and a_i denotes prediction parameters. If adaptive quantization, like that of APCM, is combined with an adaptive prediction, a new scheme known as *Adaptive Differential Pulse Code Modulation* (ADPCM) is obtained. Both the ADPCM and DPCM encoder and decoder are very similar in that the error amplitude signal $i_a(n)$ is transmitted alongside gain parameters $i_g(m)$ computed on a frame-by-frame basis. Note that these $i_g(m)$ values extend the dynamic range of the error signal samples. Furthermore, adaptive prediction coefficients, denoted as a_i and also determined on a per-frame basis, are conveyed as $i_p(l)$. The decoder processes $i_a(n)$, $i_p(l)$, and $i_g(m)$ to yield $\hat{x}(n)$ samples. Just like any encoding and decoding mechanism within a codec, a trade-off arises between frame size–dictating how frequently these parameters are updated–and latency. Larger frames reduce overhead but concurrently increase the delay experienced between the transmitter and receiver.

Another potential enhancement to ADPCM involves integrating a backward-based estimation technique for both the gain $i_a(m)$ and the prediction coefficients $i_p(l)$. Specifically, estimation of error gain and prediction parameters is conducted using information from the preceding transmitted frame, which is accessible to both the encoder and decoder. These estimates are then applied to the frame currently undergoing encoding or decoding. This approach gives rise to a scheme known as *Backward ADPCM* (BADPCM). Like any backward-oriented encoding mechanism, BADPCM raises a concern. Gain and prediction parameters are calculated based on previous signal samples, and these calculated values then influence the determination of current samples. This creates significant potential for error propagation. Consequently, BADPCM possesses a lower level of reliability compared to ADPCM.

1.4.1.2 LPC

Parametric codecs rely on the extraction of essential parameters from a speech sequence. These parameters are then utilized in the source decoder to synthesize the speech signal. They are also closely related to the correlation coefficients that are inherent to speech signals. Specifically, speech can be modeled as an *autoregressive* (AR) process where samples $x(n)$ are linked together as follows

$$x(n) + a_1 x(n-1) + \cdots + a_M x(n-M) = v(n) \tag{1.13}$$

and $a_1, a_2, \cdots, a_M$ are the AR parameters and $v(n)$ represents a white noise signal. Note that, therefore, the current sample value $x(n)$ is given by

$$x(n) = -a_1 x(n-1) - a_2 x(n-2) - \cdots - a_M x(n-M) + v(n) \tag{1.14}$$

as a linear combination of its own regressed values of the signal, $x(n-1), \cdots x(n-M)$, and the error term $v(n)$ that is a white noise signal plays the role of the source of air flow in the vocal tract.

Now, the following question becomes key. How can the parameters a_i that will enable the synthesis of the speech signal at the receiver be estimated? LPC analysis offers a mechanism for estimating these parameters from past speech samples. Specifically, given a frame of speech, this analysis predicts a given sample $\hat{x}(n)$ as

$$\hat{x}(n) = -\sum_{i=1}^{M} \hat{a}_i x(n-i) \tag{1.15}$$

where M is the prediction order and $\hat{a}_i$ are LP coefficients that estimate the a_i parameters. Now, the prediction error is $e(n) = x(n) - \hat{x}(n)$ so the idea is to select those $\hat{a}_i$ that minimize the error. One approach is by means of the mean-squared prediction error J as

$$J = E\{e^2(n)\} = E\left[\left(x(n) + \sum_{i=1}^{M} \hat{a}_i x(n-i)\right)^2\right] \tag{1.16}$$

where minimization results from performing partial derivatives $\frac{\partial J}{\partial a_i} = 0$ that lead to the normal equation given by

$$\sum_{i=1}^{M} \hat{a}_i R_x(i-k) = -R_x(k) \tag{1.17}$$

where $R_x(k) = \sum_{n=0}^{N} x(n)x(n-k)$ is the autocorrelation of $x(n)$ obtained over the whole N-sample frame. The normal equation can be represented as a set of M equations with M unknowns

$$\begin{aligned}
\hat{a}_1 R_s(1) + \hat{a}_2 R_s(2) \cdots \hat{a}_M R_s(M) &= -R_s(0) \\
\hat{a}_1 R_s(0) + \hat{a}_2 R_s(1) \cdots \hat{a}_M R_s(M-1) &= -R_s(1) \\
&\vdots \; \vdots \\
\hat{a}_1 R_s(-1) + \hat{a}_2 R_s(0) \cdots \hat{a}_M R_s(M-2) &= -R_s(2) \\
\hat{a}_1 R_s(1-M) + \hat{a}_2 R_s(2-M) \cdots \hat{a}_M R_s(0) &= -R_s(M)
\end{aligned} \tag{1.18}$$

that can be solved to obtain $\hat{a}_i$. One metric of interest is called *Prediction Gain* (PG) that is defined as

$$\begin{aligned}
\text{PG} &= 10\log_{10}\left(\frac{\sigma_x^2}{\sigma_e^2}\right) \\
&= 10\log_{10}\left(\frac{E\{x^2[n]\}}{E\{e^2[n]\}}\right)
\end{aligned} \tag{1.19}$$

and provides a ratio in dB between the energy of the input signal and the energy of the error measured by means of their corresponding variances. In general, higher PG are responsible of better signal estimation, and thus, less distortion.

To illustrate the significance of the PG, Figure 1.31 shows two speech frames, one characterized as voiced and the other as unvoiced, in terms of their temporal attributes. As expected, the voiced frame exhibits periodicity and higher energy compared to the noise-like, unvoiced frame. Furthermore, Figure 1.32 presents

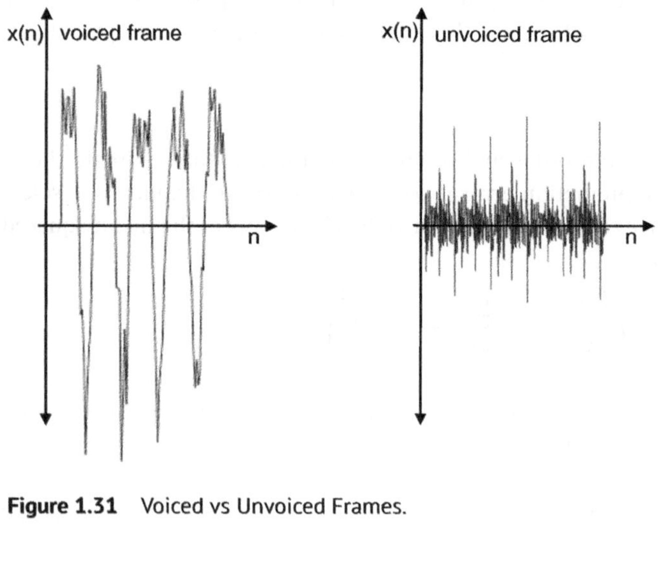

Figure 1.31 Voiced vs Unvoiced Frames.

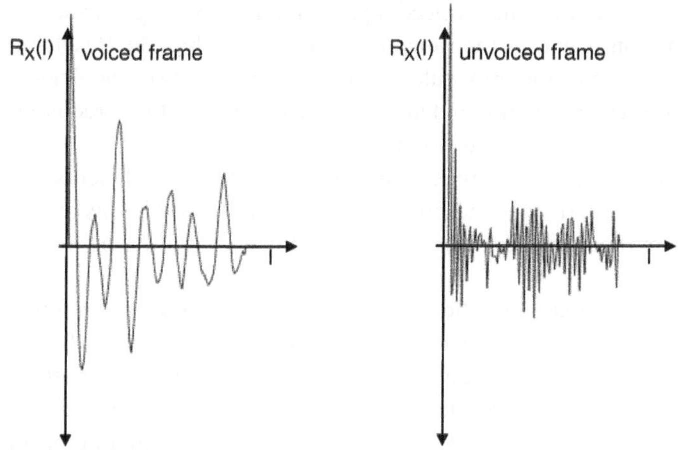

Figure 1.32 Autocorrelation.

the corresponding autocorrelation functions, denoted as $R_x(l)$ and dependent on lag, for these two speech frames. As expected from the nature of the signals themselves, voiced frames demonstrate pronounced correlation among samples, while unvoiced frames manifest randomness and minimal correlation when samples are spaced more than 5 time units apart.

When LPC is applied to the voiced and unvoiced speech frames shown in Figure 1.31 and PG is assessed in relation to the prediction order M, the results shown in Figure 1.33 are obtained. For unvoiced frames, PG increases up to $M \leq 5$ and then levels off due to the limited correlation among samples that are more distantly spaced. In opposition, for voiced frames, PG grows until $M \leq 10$, followed by a plateau until around $M \approx 50$, where it resumes growth due to the correlation among samples separated by a pitch period.

This plot confirms the observations evident in the autocorrelation plot of Figure 1.32. Unvoiced signals display minimal correlation when samples are

Figure 1.33 PG for Voiced and Unvoiced Frames.

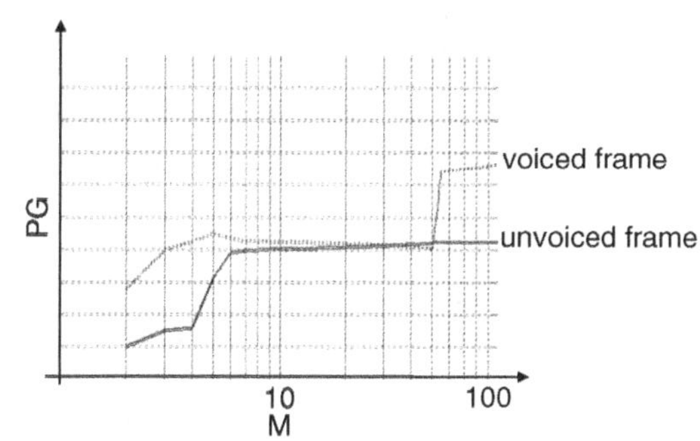

more than 5 time units apart, while voiced signals manifest strong correlation when sample separation is within the range of 10 or greater than 50. Resolving the normal equation requires the prior calculation of autocorrelation coefficients $R_x(l)$. The timing of these calculations determines which of the two prediction mechanisms is employed (i.e., internal vs external).

In internal prediction, both the computation of autocorrelation coefficients and the LPC process itself are performed within the frame being encoded. This results in an encoded stream that includes both the LPC coefficients and the associated error signal.

Conversely, in external prediction, the encoder performs LPC on a given frame while relying on autocorrelation coefficients calculated from the previous frame. This approach assumes sufficient speech stationarity to ensure minimal variations in autocorrelation coefficients across frames. External prediction is associated with backward encoding, where past frames contribute to the computation of the current frame. Consequently, the encoded stream consists solely of the error signal, as the pre-computable LPC coefficients need not be transmitted to the decoder.

As with any backward scheme, error propagation becomes a critical concern. This issue can be mitigated by maintaining manageable frame sizes to guarantee speech stationarity while simultaneously reducing end-to-end latency.

Because PG, as shown in Figure 1.33, grows in two regions, when $M \leq 10$ and when $M \approx 50$, and remains flat for all other values of M, rather than performing an order 50 linear prediction given by

$$\hat{x}(n) = -\sum_{i=1}^{M\approx 50} \hat{a}_i x(n-i) \tag{1.20}$$

prediction can be divided into two components, short-term prediction to address the correlation within 10 samples and long-term prediction to address the correlation due to the speech pitch period. Specifically, with this modification the estimation becomes

$$\hat{x}(n) = -\sum_{i=1}^{M\approx 10} \hat{a}_i x(n-i) - \hat{a}_T x(n-T) \tag{1.21}$$

where now the prediction is order $M \approx 10$ and therefore only about 10 LP coefficients are needed besides the long-term LP coefficient $\hat{a}_T$ where $T \approx 50$. With this new scheme, a considerable improvement is done by reducing the number of encoded LP coefficients from around 50 to about 11. Depending on the speech codec under consideration, $8 \leq M \leq 12$ and $T \approx 50$ will be dynamically estimated before linear prediction is performed. In general, since the pitch period varies much faster than short-term LP coefficients, T and the long-term LP coefficient $\hat{a}_T$ must be estimated on a much faster basis.

Incorporating both short-term and long-term LPC results in a structure where frames are decomposed into approximately four or more subframes. Since short-term prediction is performed less frequently than long-term prediction, the LP coefficients a_i, with $i \leq M$, are computed on a per-frame basis, while T and a_T are computed on a per-subframe basis.

The LPC coefficients are real numbers that require quantization using a floating point scheme for encoding. Unfortunately, quantization can introduce small errors that potentially destabilize the predictor, negatively impacting the overall SNR. To address this issue, an alternative approach involves transforming LPC coefficients into a less quantization-sensitive format. Specifically, the *Line Spectral Frequency* (LSF) representation enables the conversion of LPC coefficients into LSF coefficients, which are then quantized for encoding.

1.4.1.3 LBR Codecs

While several speech codecs exist, some distinguish themselves by incorporating supplementary mechanisms that bolster overall compression efficiency. One such example is *Linear Prediction Codec 10* (LPC-10), an early and now obsolete speech codec based on LPC principles. It achieves exceptionally low transmission rates, at the expense of producing artificial-sounding speech. Operating with a frame size of 180 samples, yielding a transmission rate of 2.4 Kbps, this codec employs a rigid speech production model that categorizes frames as either voiced or unvoiced.

Like most codecs, both the encoder and decoder in LPC-10 employ preemphasis and deemphasis filters to manage high-frequency components. The preemphasis filter amplifies these components, recognizing their greater importance in speech signal intelligibility compared to lower frequencies. This results in a net improvement of the SNR. After quantization, which introduces noise into the signal, the deemphasis filter selectively attenuates the noise while preserving the amplified speech signal.

The speech production model, shown in Figure 1.34, employs three parameters, energy, zero crossing, and PG–to ascertain the voiced or unvoiced nature of a given speech segment. In the case of voiced speech, energy tends to be concentrated at lower frequencies, the signal exhibits periodic zero crossings attributed to the speech pitch, and the PG value is considerably higher compared to that of unvoiced speech.

It is important to note that this codec's binary choice between voiced and unvoiced speech segments results in either white noise or a series of pulses as input, a significant departure from the AR production model. This simplification contributes to the characteristic distortion and artificial quality often perceived in speech synthesized using LPC-10.

As previously noted, ADPCM codecs leverage LPC analysis to calculate coefficients for each frame. These coefficients, along with gain values and the resulting

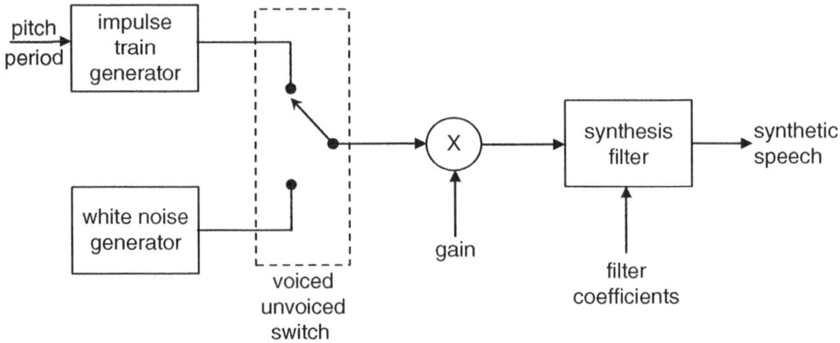

Figure 1.34 LPC-10 Speech Production Model.

error signal, are quantized to form the encoded data stream. Significantly, the error signal's samples demonstrate a relatively confined dynamic range, enabling the controlled removal of certain samples without causing perceptible distortion that would compromise audio quality.

Achieving enhanced compression becomes possible by employing a more efficient error signal encoding approach. *Code-Excited LP* (CELP), as illustrated in Figure 1.35, operates by transmitting an index that references a specific error signal or excitation within a predefined codebook or dictionary. Specifically, the CELP encoder analyzes the error signal and searches its codebook for an excitation that closely resembles it. The index associated with the chosen excitation, along with a scale factor or gain, are then encoded as components of the output stream. Conversely, the CELP decoder utilizes the index to retrieve the corresponding excitation from the codebook and subsequently rescales it using the transmitted gain value. Once the error signal is reconstructed, speech synthesis proceeds in a manner analogous to other ADPCM codecs.

More important, the synthetic speech generation process involves encoding both long-term LPC parameters, indispensable for pitch synthesis, and short-term LPC coefficients, utilized for formant synthesis. A very important requirement is ensuring absolute codebook consistency between the encoder and decoder,

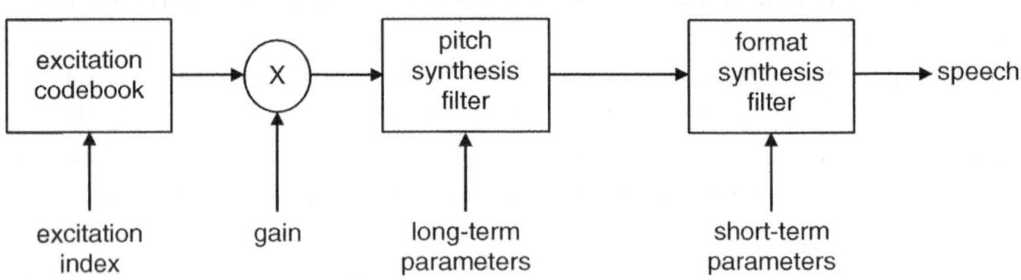

Figure 1.35 CELP.

even if these codebooks undergo dynamic construction and synchronization throughout speech processing.

The question is, how does a CELP encoder determine the index of the excitation that most closely aligns with the error signal? It accomplishes this through a technique called *analysis-by-synthesis*. This process involves systematically comparing the input error signal with a range of scaled excitations, ultimately selecting the one that yields the minimal energy of discrepancy.

For further optimization, CELP has integrated several enhancements, including techniques such as *Vector Sum Excited LP* (VSELP), which constructs codewords through the mixing of a predetermined set of basis vectors. Another CELP variant that has widespread adoption due to its simplified codebook structure is *Algebraic CELP* (ACELP). Rather than relying on extensive codebooks, ACELP employs binary addition and shifting operations to build excitation codewords. As an alternative to reduce codebook complexity, *Mixed CELP* (MCELP) mixes aperiodic, randomly perturbed pulses with noise sequences to generate excitation vectors.

In contrast, *Low delay CELP* (LDCELP) relies on small frames and subframes to facilitate external prediction. This strategy enables LPC coefficients specific to one frame to be calculated based on statistics from the preceding frame. As a result, the necessity for transmitting LPC coefficients, which can be computed concurrently by both the encoder and decoder, is eliminated. Additionally, as these coefficients are not transmitted, there is no requirement to support short- and long-term predictions, allowing for comprehensive LPC of order $M = 50$ to be conducted for improved compression efficiency.

The speech codecs discussed earlier exhibit a *Constant Bit Rate* (CBR), wherein a fixed transmission rate is determined by the ratio of bits within a frame to the frame's duration in seconds. However, in many scenarios, it is possible to encode the speech frame using fewer bits, making a fixed transmission rate less efficient. Addressing this challenge, *Variable Bit Rate* (VBR) codecs offer a solution by dynamically selecting the most efficient encoding scheme for each frame based on an analysis of speech and background noise.

1.4.1.4 ITU-T Recommendation G.711

The *International Telecommunications Union* (ITU) ITU-T Recommendation G.711 presents an HBR waveform codec encoder that transforms the 16-bit samples into 8-bit samples utilizing the established μ-Law and A-Law curves. This conversion process reduces each sample's size by half, resulting in an output transmission rate of 64 Kbps for the codec encoder. Note that unlike traditional PCM, the compression in ITU-T Recommendation G.711 occurs on a per-frame basis. Specifically, an input frame containing 80, 160, or 240 16-bit samples is converted into a corresponding output frame with 80, 160, or 240 8-bit samples, respectively (ITU-T 1988a).

Tables 1.2 and 1.3 show the procedure for converting 16-bit samples to 8-bit samples according to ITU-T Recommendation G.711 for both μ-Law and A-Law encoding. Consider a 16-bit sample *1000 0000 1001 0000*. When subjected to μ-Law encoding, the entry *s0000000wxyz'a* → *s000wxyz* applies due to the presence of the sequence *0000000* following the initial bit. In this instance, $s = 1$, $wxyz = 1001$, resulting in an output 8-bit value of *1000 1001*. Conversely, if A-Law is employed for the same input sample, the entry *s00000001wxyz'a* → *s000wxyz* comes into play. Here, $s = 1$ and $wxyz = 0010$, yielding an output 8-bit value of *1000 0010*.

Figures 1.36 and 1.37 show an ITU-T Recommendation G.711 μ-Law compression scheme and a linear compression scheme, respectively. These schemes involve the conversion of 16-bit samples into 8-bit samples with values spanning from $-2^7 + 1 = -127$ to $2^7 = 128$. When a sinusoidal signal with a varying amplitude A is subjected to each of these schemes, the resulting SNR plots shown

Table 1.2 A-Law.

16-bit Raw Input Sample	μ-law Code
s0000000wxyz'a	s000wxyz
s0000001wxyz'a	s001wxyz
s000001wxyz'ab	s010wxyz
s00001wxyz'abc	s011wxyz
s0001wxyz'abcd	s100wxyz
s001wxyz'abcde	s101wxyz
s01wxyz'abcdef	s110wxyz
s1wxyz'abcdefg	s111wxyz

Table 1.3 A-Law.

16-bit Raw Input Sample	A-law Code
s00000001wxyz'a	s000wxyz
s0000001wxyz'ab	s001wxyz
s000001wxyz'abc	s010wxyz
s00001wxyz'abcd	s011wxyz
s0001wxyz'abcde	s100wxyz
s001wxyz'abcdef	s101wxyz
s01wxyz'abcdefg	s110wxyz
s1wxyz'abcdefgh	s111wxyz

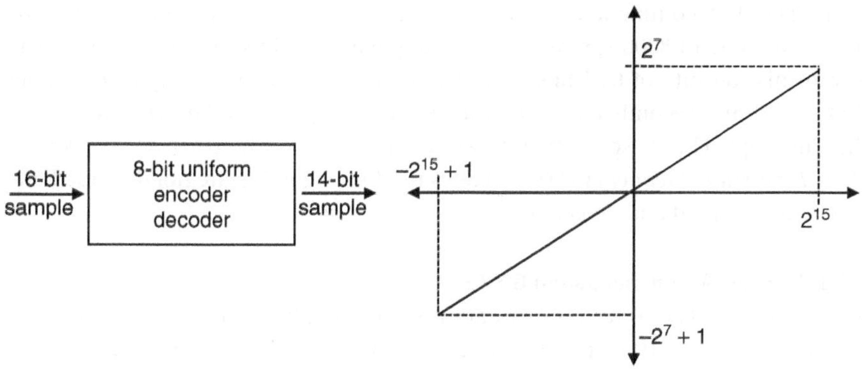

Figure 1.36 8-bit Uniform Quantization.

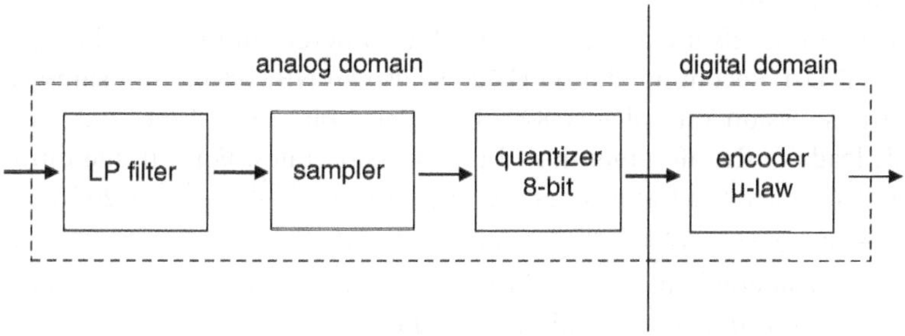

Figure 1.37 8-bit ITU-T Rec. G.711 Quantization.

Figure 1.38 Performance Comparison; Linear vs ITU-T Rec. G.711.

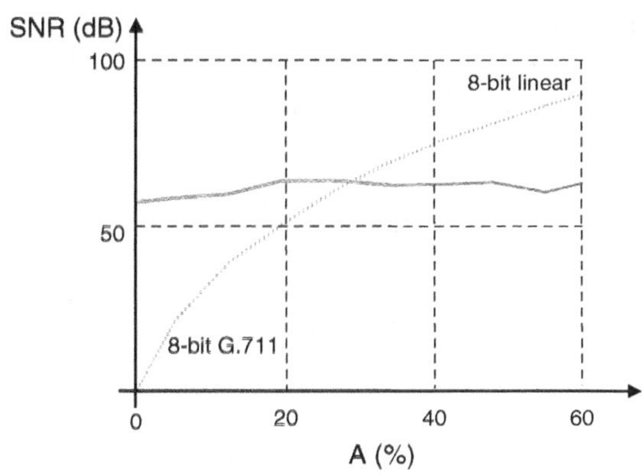

in Figure 1.38 are generated. Note that when the input dynamic range is about 30% of the threshold, a point is reached where lower amplitudes, commonly associated with speech signals, exhibit superior performance under the ITU-T Recommendation G.711 μ-Law compression compared to the linear compression.

The ITU-T Recommendation G.711 constitutes an NB codec, where sampling occurs at a rate of 8000 sps. With each sample being 8 bits in size, this results in a transmission rate of 64 Kbps. Because the codec processes samples in frames that are 10 milliseconds long, the frames are $\frac{64\ \text{Kbps} \times 0.01\ \text{s}}{8} = 80$ bytes long. Note that although ITU-T Recommendation G.711 is a speech codec, tones like the *Dual-Tone Multi-Frequency* (DTMF) ones used for signaling are not degraded and can be transmitted and detected.

1.4.1.5 ITU-T Recommendation G.726

Similar to how ITU-T Recommendation G.711 adopts PCM, the ITU-T Recommendation G.726 HBR speech codec employs ADPCM, offering four distinct transmission rates (16, 24, 32, and 40 Kbps) based on quantization levels. ITU-T Recommendation G.726 utilizes a second-order LPC predictor as a foundational component of its process.

The quantization levels are associated with four different numbers of bits per sample. Since ITU-T Recommendation G.726 is also an NB codec, n-bit samples result in a transmission rate of $n \times 8000$ bps. For 2-bit, 3-bit, 4-bit, and 5-bit samples, this leads to the aforementioned transmission rates. For a frame duration of 10 milliseconds, the frame sizes are respectively $\frac{16\ \text{Kbps} \times 0.01\ \text{s}}{8} = 20$ bytes, $\frac{24\ \text{Kbps} \times 0.01\ \text{s}}{8} = 30$ bytes, $\frac{32\ \text{Kbps} \times 0.01\ \text{s}}{8} = 40$ bytes and $\frac{40\ \text{Kbps} \times 0.01\ \text{s}}{8} = 50$ bytes. Because of the considerably high transmission rates supported by ITU-T Recommendation G.726, tone distortion is minimal, and DTMF tones can be transmitted and detected (ITU-T 1990).

Example 1.2 The ITU-T Recommendation G.726 codec samples speech at an NB rate. Assuming input 16-bit raw samples, what is the compression rate (the ratio between the compressed and non-compressed media rate) for 40 Kbps?

Solution:

If $F_s = 8000$ sps and $n = 16$ bits, then the raw transmission rate is $R = F_s \times n = 8000 \times 16 = 128$ Kbps. The compression rate is then $k = \frac{40}{128} = 0.3125$.

1.4.1.6 ITU-T Recommendation G.723.1

The ITU-T Recommendation G.723.1 introduces an MBR codec that relies on ACELP to support two transmission rates of 5.3 and 6.3 Kbps. Because of the low transmission rates, it is only suitable for speech, and signaling DTMF tones are distorted such that they cannot be detected. As such, this codec requires out-of-band mechanisms to support the transmission of DTMF tones.

The ITU-T Recommendation G.723.1 codec is also an NB codec with a frame duration that is 30 milliseconds long. At 8000 sps, each frame includes 8000 sps $\times$ 0.03 s = 240 samples. The frame size for 5.3 Kbps is $\frac{5300\ \text{bps} \times 0.03\ \text{s}}{8} \approx 20$ bytes long. Similarly, the frame size for 6.3 Kbps is $\frac{6300 \text{ bps} \times 0.03 \text{ s}}{8} \approx 24$ bytes long. The transmission rates are further improved by relying on the silence compression speech scheme specified in Annex A of the ITU-T Recommendation G.723.1. This mechanism enables the support of *Discontinued Transmissions* (DTX) through *Comfort Noise Generation* (CNG) that results from *Silence Insertion Descriptor* (SID) frames (ITU-T 1996a).

1.4.1.7 ITU-T Recommendation G.729 Annexes A, D, and E

The ITU-T Recommendation G.729 Annex A delineates a well-known MBR ACELP codec, which sustains a transmission rate of 8 Kbps. Like other MBR codecs, identifying DTMF tones is unfeasible due to the distortion engendered by compression. Being an NB codec and employing a 10-millisecond frame duration results in a frame size of 10 bytes, calculated as $\frac{8000 \text{ bps} \times 0.01 \text{ s}}{8}$.

Annex A of the ITU-T Recommendation G.729 is commonly coupled with Annex B to facilitate DTX and CNG, executed through the transmission of SID frames. In addition, ITU-T Recommendation G.729 Annexes D and E respectively lower and increment the transmission rate to 6.4 Kbps and 11.8 Kbps for 8-byte and 15-byte payload sizes (ITU-T 1996b).

1.4.1.8 GSM 6.10

An important example of an early MBR codec is the GSM 6.10 codec, which introduces a technique whereby the error signal is subjected to analysis and downsampling to reduce the overall transmission rate. This method, referred to as multipulse excitation, involves encoding a limited number of non-zero samples by indicating both their positions and values.

In Figure 1.39, a modification of multipulse excitation called *Regular Pulse Excitation* (RPE) is shown. This method involves downsampling the error signal into multiple sub-sets, each defined by distinct initial offsets. The set with the greatest energy content is chosen, and its sample values, along with an identifier for the selected set, are quantized and encoded to replace the original error signal.

Table 1.4 shows the bit allocation for GSM 6.10, where short- and long-term predictions are performed over 20-millisecond frames. The GSM 6.10 codec therefore specifies a transmission rate given by $\frac{260 \text{ bits}}{0.02 \text{ s}} = 13$ Kbps. Note that this MBR codec cannot be used with DTMF tones as they cannot be detected due to the compression distortion (ETSI 1992).

1.4.1.9 AMR

The *Adaptive Multi-Rate* (AMR) codec is a multipulse excitation CELP NB codec with the capability to accommodate eight distinct transmission rates, facilitating both LBR and MBR speech encoding. The primary concept underlying this codec is its ability to dynamically select the optimal transmission rate for ensuring the

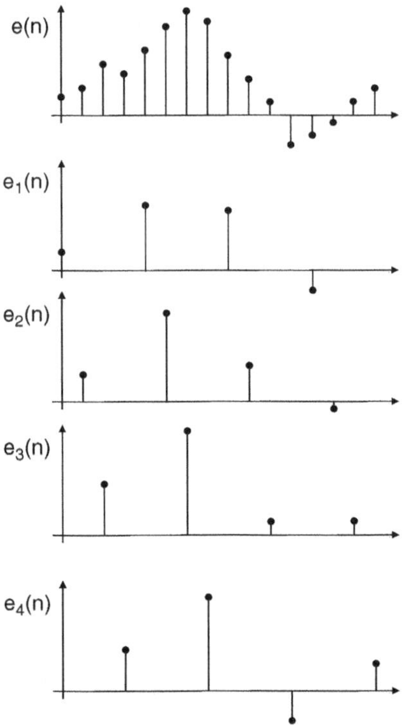

Figure 1.39 RPE.

Table 1.4 GSM 6.10 Bit Allocation

Parameter	Parameters per Frame	Resolution	Total Bits per Frame
LPC	8	6,6,5,5,4,4,3,3	36
Pitch Period	4	7	28
Long-Term Gain	4	2	8
Position	4	2	8
Peak Magnitude	4	6	24
Sample Amplitude	4-13	3	156
Total			260

fulfillment of QoS goals. These eight rates include 4.75 Kbps, 5.15 Kbps, 5.90 Kbps, 6.70 Kbps, 7.40 Kbps, 7.95 Kbps, 10.20 Kbps, and 12.20 Kbps.

For frame durations of 20 milliseconds, the corresponding frame sizes are approximately 12 bytes for 4.75 Kbps, 13 bytes for 5.15 Kbps, 15 bytes for 5.90 Kbps, 17 bytes for 6.70 Kbps, 19 bytes for 7.40 Kbps, 20 bytes for 7.95 Kbps, 26 bytes for

10.20 Kbps, and 31 bytes for 12.20 Kbps (calculated as $\frac{r \text{ bps} \times 0.02 \text{ s}}{8}$, where r is the corresponding AMR rate). The AMR codec, however, introduces signal distortions that hinder its ability to detect DTMF tones. Consequently, external methods are required to transmit such tones using out-of-band mechanisms. AMR supports DTX and also provides an additional transmission rate of 1.8 Kbps, specifically designed for transmitting SID frames. AMR is key to transmission of speech and audio in the context of the 4G technologies described in Section 4.3 (3GPP 2008a).

1.4.1.10 ITU-T Recommendation G.728

The ITU-T Recommendation G.728 describes an HBR CELP codec that supports a transmission rate of 16 Kbps. Although it is an LPC-based codec, its rate is large enough that this NB codec does not distort the media stream and can, therefore, be used to detect DTMF tones. Each ITU-T Recommendation G.728 frame is 20 milliseconds long and carries $\frac{16000 \text{ bps} \times 0.02 \text{ s}}{8} = 40$ bytes (ITU-T 1992).

1.4.1.11 ITU-T Recommendation G.722

The ITU-T Recommendation G.722 introduces an HBR *wideband* (WB) codec associated with a WB sampling rate of 16000 sps. The codec relies on an ADPCM mechanism that operates at 48, 56, and 64 Kbps. As an HBR codec, DTMF tones exhibit little distortion and they can be detected when carried in ITU-T Recommendation G.722 streams. For 20-millisecond frames, each frame carries $\frac{48000 \text{ bps} \times 0.02 \text{ s}}{8} = 120$ bytes, $\frac{56000 \text{ bps} \times 0.02 \text{ s}}{8} = 140$ bytes, and $\frac{64000 \text{ bps} \times 0.02 \text{ s}}{8} = 160$ bytes for 48, 56, and 64 Kbps (ITU-T 1988b).

1.4.1.12 iLBC

The *Internet Low Bitrate Codec* (iLBC) is an MBR codec that relies on LPC techniques to accomplish compression. It is specified in the IEFF RFC 3951. It is an NB codec that supports two frame durations: 20 and 30 milliseconds. As an MBR codec, it introduces distortion that prevents DTMF tones from being detected. For 20-millisecond duration frames, the data transmission rate is 15.2 Kbps, resulting in a frame size of 38 bytes calculated as $\frac{15200 \text{ bps} \times 0.02 \text{ s}}{8}$. Similarly, when considering frames lasting 30 milliseconds, the transmission rate becomes 13.334 Kbps, and the corresponding frame size is approximately 50 bytes, derived from $\frac{13334 \text{ bps} \times 0.03 \text{ s}}{8}$ (Hagen et al. 2004).

1.4.1.13 Speex

Speex represents a CELP-based codec that accommodates a broad spectrum of sampling and transmission rates, supporting LBR, MBR, and HBR configurations. The supported rates span from 2 Kbps to 44 Kbps. In configurations geared toward higher sampling and transmission rates, the codec is generally able to detect DTMF tones. Note that Speex extends its support to NB sampling at 8000 sps, WB sampling at 16000 sps, *super wideband* (SWB) sampling at 32000 sps, and *Fullband* (FB) sampling at 48000 sps.

The frame duration associated with Speex is consistently set at 30 milliseconds. Consequently, the frame size ranges from around 8 bytes, calculated as $\frac{2000 \text{ bps} \times 0.03 \text{ s}}{8}$, to 165 bytes, derived from $\frac{44000 \text{ bps} \times 0.03 \text{ s}}{8}$ (Valin 2016).

1.4.1.14 EVRC and EVRC-B

The *Enhanced Variable Rate Codec* (EVRC) stands as a CELP NB speech codec, accommodating three distinct transmission rates: MBR 8.55 Kbps, LBR 4 Kbps, and 800 bps. This last rate, the lowest, serves the purpose of facilitating DTX by allowing the conveyance of SID frames. When considering a frame duration of 20 milliseconds, these rates translate to frame sizes of approximately 22 bytes (for $\frac{8550 \text{ bps} \times 0.02 \text{ s}}{8}$), 15 bytes (for $\frac{4000 \text{ bps} \times 0.03 \text{ s}}{8}$), and 2 bytes (for $\frac{800 \text{ bps} \times 0.02 \text{ s}}{8}$), respectively. Similar to AMR, the aim is for these rates to be dynamically chosen to fulfill QoS objectives. It is important to note that this codec introduces a level of distortion that hinders its ability to detect DTMF tones. EVRC-B is an enhancement to EVRC that relies on an additional 2 Kbps rate that supports 5-byte-long frames (3GPP2 2004).

1.4.1.15 GSM-EFR

GSM Enhanced Full Rate (GSM-EFR) defines an ACELP MBR NB speech codec, enabling a transmission rate of 12.2 Kbps. In frames lasting 20 milliseconds, each frame's size is approximately 31 bytes, calculated as $\frac{12200 \text{ bps} \times 0.02 \text{ s}}{8}$. Due to the distortion introduced by GSM-EFR at this transmission rate, detecting DTMF signals becomes challenging, requiring the use of out-of-band mechanisms for their successful transmission (ETSI 1999).

1.4.1.16 LPC-10

The LPC-10 codec, designated as *Federal Information Processing Standard* (FIPS) publication 137, represents an earlier LBR codec that has been deprecated but it remains important due to its historical relevance. Operating with a frame duration of 22.5 milliseconds, and a transmission rate of 2.4 Kbps, it results in a frame size of approximately 7 bytes ($\frac{2400 \text{ bps} \times 0.0225 \text{ s}}{8}$). Due to its nature as an NB LBR codec, the distortion introduced by LPC-10 prevents the accurate detection of DTMF tones. Other details of LPC-10 are presented in Section 1.4.1.3 (FIPS 1984).

1.4.1.17 AMR-WB/ITU-T Recommendation G.722.2

The speech codec known as *Adaptive Multi-Rate Wideband* (AMR-WB), which is standardized as ITU-T Recommendation G.722.2, is built upon the ACELP technique. This codec is designed to facilitate high-quality speech communication and offers a range of nine transmission rates: 6.6 Kbps, 8.85 Kbps, 12.65 Kbps, 14.25 Kbps, 15.85 Kbps, 18.25 Kbps, 19.85 Kbps, 23.05 Kbps, and 23.85 Kbps.

When considering a frame duration of 20 milliseconds, the corresponding frame sizes are approximately 17 bytes, 23 bytes, 32 bytes, 36 bytes, 40 bytes, 46 bytes, 50 bytes, 58 bytes, and 60 bytes, respectively. These sizes are calculated using the formula $\frac{r \text{ bps} \times 0.02 \text{ s}}{8}$, where r represents the given AMR-WB rate. These characteristics categorize the codec as both MBR and HBR.

Like AMR, AMR-WB operates adaptively, dynamically selecting the optimal transmission rate based on specified QoS objectives. It should be noted, however, that the codec introduces a level of distortion that precludes the detection of DTMF tones. The codec's functionality includes support for DTX modes, accomplished through the transmission of SID frames. AMR-WB is key to transmission of speech and audio in the context of the 5G technologies described in Section 4.3 (3GPP 2008b).

1.4.1.18 RTAudio

Real Time Audio (RTAudio) is a CELP-based speech codec that supports an NB 8.8 Kbps MBR rate and WB 18 Kbps as well as 24 Kbps HBR rates. With a frame duration set at 20 milliseconds, the corresponding sizes for these frames are calculated as follows: 22 bytes for 8.8 Kbps, 45 bytes for 18 Kbps, and 60 bytes for 24 Kbps (as $\frac{r \text{ bps} \times 0.02 \text{ s}}{8}$ where r is the corresponding transmission rate). Note that, all of these bit rates introduce a level of distortion that renders proper detection of DTMF tones unfeasible within the signal, requiring the out-of-band transmission of such tones.

1.4.1.19 EVS

The *Enhanced Voice Services* (EVS) codec is a speech codec based on MBR and HBR ACELP techniques. It introduces a primary mode that offers support for various sampling rates, including NB, WB, SWB, and FB. When using NB sampling, there are seven potential transmission rates: 5.9 Kbps, 7.2 Kbps, 8 Kbps, 9.6 Kbps, 13.2 Kbps, 16.4 Kbps, and 24.4 Kbps. WB sampling allows for 12 rates: 5.9 Kbps, 7.2 Kbps, 8 Kbps, 9.6 Kbps, 13.2 Kbps, 16.4 Kbps, 24.4 Kbps, 32 Kbps, 48 Kbps, 64 Kbps, 96 Kbps, and 128 Kbps. In the case of SWB sampling, there are nine feasible rates: 9.6 Kbps, 13.2 Kbps, 16.4 Kbps, 24.4 Kbps, 32 Kbps, 48 Kbps, 64 Kbps, 96 Kbps, and 128 Kbps. Finally, for FB sampling, seven rates are available: 16.4 Kbps, 24.4 Kbps, 32 Kbps, 48 Kbps, 64 Kbps, 96 Kbps, and 128 Kbps. In addition, under WB sampling, the *AMR-WB InterOp* (AMR-WB IO) mode can accommodate all AMR-WB rates detailed in Section 1.4.1.17.

For frames lasting 20 milliseconds, the frame sizes are determined by the formula $\frac{r \text{ bps} \times 0.02 \text{ s}}{8}$, where r denotes the given rate. The resulting frame sizes are as follows: 15 bytes for 5.9 Kbps, 19 bytes for 7.2 Kbps, 20 bytes for 8 Kbps,

24 bytes for 9.6 Kbps, 33 bytes for 13.2 Kbps, 41 bytes for 16.4 Kbps, 61 bytes for 24.4 Kbps, 80 bytes for 32 Kbps, 120 bytes for 48 Kbps, 160 bytes for 64 Kbps, 240 bytes for 96 Kbps, and 320 bytes for 128 Kbps. Moreover, EVS supports DTX and CNG through the transmission of SID frames. While HBR transmission rates allow for DTMF tone detection, it is advisable to use external mechanisms for transmitting these tones to ensure reliable detection. As AMR-WB, EVS is key to transmission of speech and audio in the context of the 5G technologies described in Section 4.3 (3GPP 2018).

1.4.1.20 Lyra

Lyra, a speech codec developed by Google, employs *Machine Learning* (ML) techniques for feature extraction from the input signal to generate an encoded stream. Lyra is compatible with NB, WB, SWB, and FB sampling rates, producing LBR and MBR transmission rates of 3.2 Kbps, 6 Kbps, and 9.2 Kbps. The codec handles DTX and CNG through the use of SID frames, while DTMF tone detection requires out-of-band mechanisms. In the case of 20-millisecond frames, the frame sizes are 8 bytes, 15 bytes, and 23 bytes for rates of 3.2 Kbps, 6 Kbps, and 9.2 Kbps, respectively (calculated as $\frac{r\ \text{bps} \times 0.02\ \text{s}}{8}$, where *r* represents the specified rate). The latest version of Lyra, V2, is optimized to improve audio quality and latency by relying on a *Residual Vector Quantizer* (RVQ) (Google 2023).

1.4.2 Audio Coding

Speech is a distinct form of audio signal, known for predominantly containing its spectral components at frequencies below 4 kHz. It can be represented through an AR process. In contrast, more intricate audio signals, such as music and tones, possess a more consistent spectral distribution that stretches up to 20 kHz. Due to their intricacy, these signals cannot be efficiently compressed using LPC techniques. As a result, audio encoders are capable of handling speech signals, while speech encoders are not suitable for processing general audio signals.

In contrast to speech, audio undergoes compression through methods involving spectral transformations like the *Modified Discrete Cosine Transform* (MDCT), which are similar to the techniques employed in video compression (in Section 1.4.3). Once transformed into the frequency domain, redundancy within the spectral samples can be eliminated by capitalizing on the human auditory system's perceptual limitations. Note that the phenomenon of audio masking influences the threshold of ear perception, shown in Figure 1.40, and dictates the extent to which data can be discarded without compromising quality. The perceptual threshold graph illustrates that the highest sensitivity of the human ear is within the sub-4 kHz range. For instance, a tone A remains audible as long as it lies above this threshold.

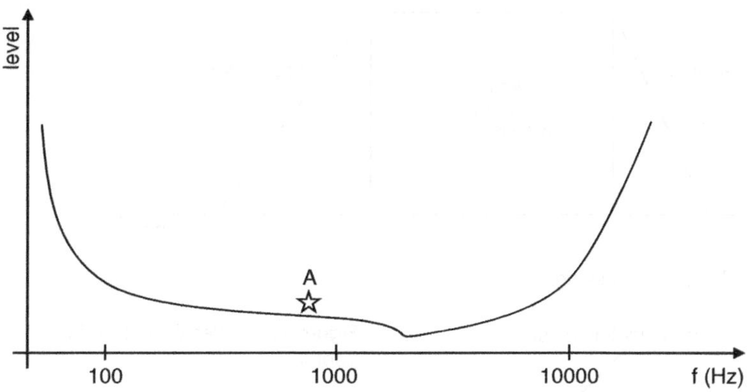

Figure 1.40 Ear Perception Threshold with Audible Tone A.

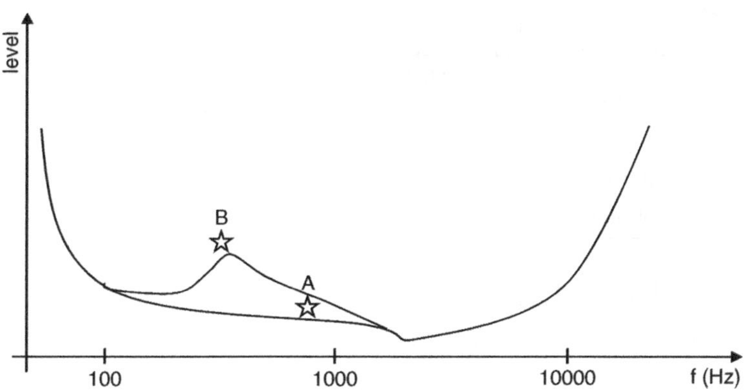

Figure 1.41 Ear Perception Threshold (Tone A is masked by tone B).

When two tones, such as tone A and tone B illustrated in Figure 1.41, have frequencies that are in close proximity, the louder tone conceals the presence of the other through a phenomenon referred to as frequency masking. Similarly, in the time domain, if a faint signal follows a powerful one or a strong signal is succeeded by a feeble one, the dominant signal eventually eclipses the other due to temporal masking (Figure 1.42).

The psychoacoustic model derived from the perception threshold serves as a cornerstone for audio encoders. This model, in turn, plays a pivotal role in shaping a perceptual encoder, which encodes incoming audio based on the sensitivity of the human ear across various spectral regions. The process, as shown in Figure 1.43, involves two sequential operations: (1) the incoming signal is passed through a bank of 32 subband filters, referred to as a bank filter and (2) the resulting 32 waves are individually subjected to quantization using quantizers that follow the

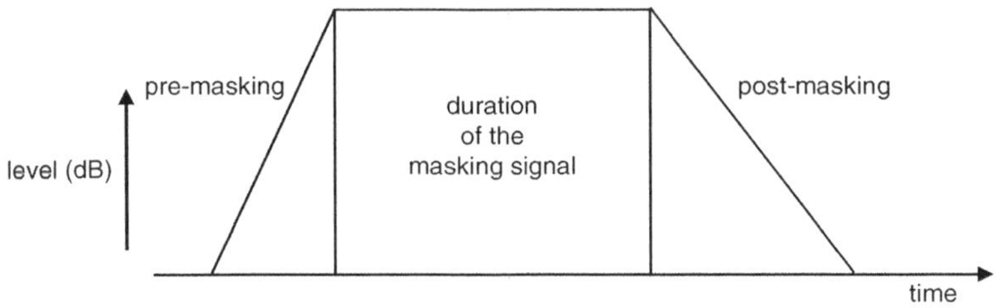

Figure 1.42 Temporal Masking.

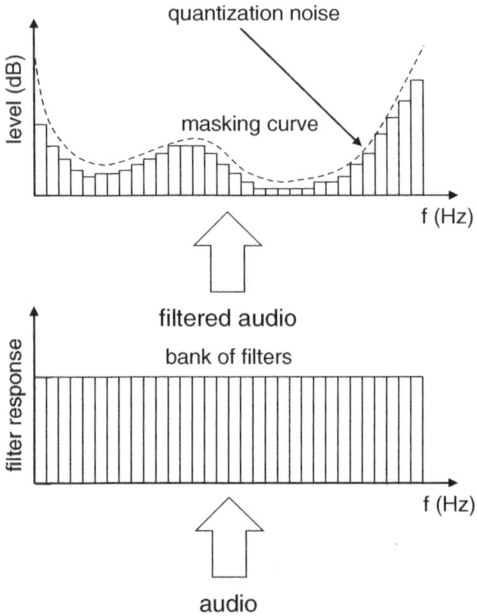

Figure 1.43 Perceptual Audio Coding.

perception threshold specific to each filter's spectral location. The underlying concept hinges on the notion that higher perception thresholds allow for greater permissible quantization errors. This implies that subband samples originating from regions with larger thresholds can be encoded with fewer bits than those from areas with smaller thresholds, thereby reducing the overall transmission rate.

A perceptual encoder can be used for stereo audio where the compression of audio originating from distinct left and right channels is performed. To achieve this, each channel undergoes the previously mentioned filter bank process. By leveraging the psychoacoustic model's assessment of masking effects, quantizers are precisely configured to operate at appropriate quantization levels. The quantizer outputs are then translated into symbol-encoded codewords, which are subsequently multiplexed to form the compressed audio stream.

Similarly, a perceptual decoder designed for stereo audio functions in a reverse manner. It starts by demultiplexing the compressed audio stream into distinct codewords, which are subsequently transformed into analog samples through a *Digital-to-Analog Converter* (DAC). These analog samples are then combined to reconstruct the audio streams for both the left and right channels.

Observing the impact of subband filtering on the overall transmission rate under the context of sampling, it can be seen that subband filtering does not influence the transmission rate. Under sampling without the inclusion of subband filtering, every sample is encoded using B bits, and W signifies the signal's bandwidth, leading to a transmission rate of Rate $= 2WB$. Similarly, when the same signal is subjected to sampling with the addition of 4-band filtering, each subband holds a bandwidth of $\frac{W}{4}$, yet the cumulative transmission rate remains unchanged, maintaining Rate $= 4 \cdot 2 \cdot \frac{W}{4} \cdot B = 2WB$.

As previously indicated, an aspect of audio compression pertains to the incorporation of multiple channels, resulting in distinct operational modes. These modes include stereo compression, where the left and right channels are individually encoded with varying quantization levels; joint stereo, which considers the redundancy present between the left and right channels during encoding; dual channel, involving the independent encoding of channels as uncorrelated waveforms; and mono, which entails encoding only a single channel.

1.4.2.1 AAC

The *Advanced Audio Coding* (AAC) standard establishes an audio codec relying on the MDCT method, allowing for sampling rates of up to 96000 sps and supporting 48 channels in joint stereo mode. This results in the existence of both CBR and VBR scenarios, which are linked to variable frame lengths. Among the three profiles supported by the standard, *AAC-Low Complexity* (AAC-LC) is capable of accommodating up to 16 channels (ISO/IEC 2006).

1.4.2.2 ITU-T Recommendation G.718

ITU-T Recommendation G.718 introduces a codec that adapts its behavior based on the sampling rates employed. When operating at NB sampling rates, it functions as a speech codec, utilizing CELP techniques for compression. In contrast, at WB sampling rates, it transitions to an audio codec, employing MDCT for compression. In speech coding modes, the codec achieves distortion levels that prevent the detection of DTMF tones. Additionally, it supports features such as DTX and CNG through the transmission and detection of SID frames.

This recommendation supports a range of transmission rates, including both MBR and HBR options: 8 Kbps, 12 Kbps, 16 Kbps, 24 Kbps, and 32 Kbps. Specifically, NB sampling rates are compatible with 8 Kbps and 12 Kbps, whereas WB sampling rates are compatible with all five transmission rates. For 20-millisecond frames, these rates result in frame sizes of 20 bytes for 8 Kbps, 30 bytes for 12 Kbps, 40 bytes for 16 Kbps, 60 bytes for 24 Kbps, and 80 bytes for 32 Kbps (calculated as $\frac{r \text{ bps} \times 0.03 \text{ s}}{8}$, where *r* is the transmission rate under consideration) (ITU-T 2008).

1.4.2.3 ITU-T Recommendation G.722.1

The ITU-T Recommendation G.722.1 outlines an HBR WB audio codec that utilizes an MDCT mechanism to achieve data transmission rates of 24 Kbps and 32 Kbps. This transform-based codec exhibits minimal distortion when handling DTMF tones, and these tones are typically conveyed through separate out-of-band mechanisms. The frame duration is 20 milliseconds and, therefore, the frame sizes amount to 60 bytes and 80 bytes for transmission rates of 24 Kbps and 32 Kbps, respectively (calculated as $\frac{r \text{ bps} \times 0.03 \text{ s}}{8}$ where *r* is the transmission rate under consideration).

1.4.2.4 iSAC

The *internet Speech Audio Codec* (iSAC) stands as a creation from the developers behind the iLBC codec, known as *Global IP Solutions* (GIPS). iSAC serves as a complementary counterpart to iLBC by extending its capabilities to cover both speech and audio content. This is achieved through transform coding mechanisms similar to MDCT, where speech and audio signals undergo WB and SWB sampling at rates of 16000 sps and 32000 sps, respectively.

For WB sampling, the transmission rates span from approximately 10 Kbps to 32 Kbps, employing two different frame durations: 30 milliseconds and 60 milliseconds. As a result, frame sizes for a 30-millisecond duration range from about 38 bytes (calculated as $\frac{10000 \text{ bps} \times 0.03 \text{ s}}{8}$) to 120 bytes (calculated as $\frac{32000 \text{ bps} \times 0.03 \text{ s}}{8}$). These values double when considering 60-millisecond frames.

Likewise, for SWB sampling, the transmission rates vary between roughly 10 Kbps and 56 Kbps, utilizing a fixed frame duration of 30 milliseconds. This leads to frame sizes ranging from around 38 bytes to 210 bytes (calculated as $\frac{56000 \text{ bps} \times 0.03 \text{ s}}{8}$).

1.4.2.5 SILK

SILK, a speech-oriented audio codec, utilizes LPC speech compression techniques to accommodate NB, *Medium Band* (MB) at 12000 sps, WB and SWB sampling rates. As a result, it achieves MBR and HBR VBR transmission rates ranging from 6 Kbps to 40 Kbps. When considering a frame duration of 20 milliseconds, the frame sizes vary between 15 bytes (computed as $\frac{6000 \text{ bps} \times 0.02 \text{ s}}{8}$) and 100 bytes (calculated as $\frac{40000 \text{ bps} \times 0.02 \text{ s}}{8}$) (Vos et al. 2010).

1.4.2.6 Opus

The Opus codec extends the functionality of the SILK codec, as introduced in Section 1.4.2.5. It incorporates an MDCT mechanism known as the *Constrained*

Energy Lapped Transform (CELT) to enable audio compression. In addition to supporting NB, MB, WB, and SWB sampling rates, Opus also accommodates FB sampling rates. As a result, Opus provides VBR transmission rates ranging from an MBR of 6 Kbps to an HBR of 512 Kbps. The frame sizes vary between 15 bytes (calculated as $\frac{6000\text{ bps} \times 0.02\text{ s}}{8}$) and 1280 bytes (calculated as $\frac{512000\text{ bps} \times 0.02\text{ s}}{8}$) (Valin et al. 2012).

1.4.2.7 LC3

The *Low Complexity Communication Codec* (LC3) is an audio codec based on CELT, offering support for NB, WB, SWB, and FB, along with a 44100 sps sampling rate option. This codec achieves HBR VBR rates ranging from 16 Kbps to 320 Kbps for 10-millisecond frames. Consequently, the frame sizes fall within the range of 200 bytes (calculated as $\frac{16000\text{ bps} \times 0.01\text{ s}}{8}$) to 4000 bytes (computed as $\frac{320000\text{ bps} \times 0.01\text{ s}}{8}$) (Bluetooth 2018).

1.4.3 Video Coding

In the realm of digital technology, it can be thought that digital video emerges when an analog video input is transformed into a bitstream by a source encoder. This resulting bitstream can be duplicated, encrypted, stored, and adapted for transmission through a channel. Given that analog video's LP bandwidth is approximately 5 MHz, using a sampling rate of 10 Msps is deemed safe. However, this leads to excessively high raw video transmission rates. The initiative to establish standards for video digitization began in 1982 with the now obsolete ITU-R Recommendation BT.601. This recommendation, based on a 6-MHz video signal bandwidth and quantization of samples at 8 or 10 bits, relies on a sampling frequency of $F_s = 13.5$ Msps, generating minimum transmission rates of $R = 8F_s = 108$ Mbps.

The resultant digital video frames consist of either 525 or 625 lines, each containing a corresponding number of 720 luminance or 360 chrominance samples. These samples are arranged in an orthogonal sampling configuration. Nonetheless, directly sampling and quantizing analog to digital video is bounded by the quality of the original analog video. Moreover, digital video obtained through this process is not well-suited for editing, encryption, compression, or other subsequent forms of processing.

The widespread use of *Charged-Coupled Devices* (CCD) has resulted in the development of cameras that, unlike traditional analog video sampling, offer high-quality video at a reasonable cost. Furthermore, digital cameras produce images in digital raw video formats that are exceptionally well-suited for digital processing and transmission. Among these formats, one called 4:2:2 encodes four luminance samples, along with two red chrominance and two blue chrominance samples. In this format, each luminance frame measures 720×480 pixels^2 (United States) or 720×576 pixels^2 (Europe and rest of the world), while each chrominance frame measures 360×480 pixels^2 (United States) or 360×576 pixels^2 (Europe). At 8-bit per pixel or sample, the United States-based transmission rate is given by

$$\begin{aligned} \text{rate} &= 720 \times 480 \times 30 \times 8 + \\ &\quad 360 \times 480 \times 30 \times 8 + \\ &\quad 360 \times 480 \times 30 \times 8 \\ &= 165.888 \text{ Mbps} \end{aligned} \tag{1.22}$$

and the European-based transmission rate is identical and equal to

$$\begin{aligned} \text{rate} &= 720 \times 576 \times 25 \times 8 + \\ &\quad 360 \times 576 \times 25 \times 8 + \\ &\quad 360 \times 576 \times 25 \times 8 \\ &= 165.888 \text{ Mbps} \end{aligned} \tag{1.23}$$

because although the United States format includes fewer lines (480 vs 576), these lines are encoded more often (30 fps vs 25 fps) leading to a ratio $\frac{576}{30} = \frac{480}{25} = 19.2$.

Generally, by reducing both vertical and horizontal resolutions of the 4:2:2 format, various raw video format derivatives can be generated. For instance, the 4:2:0 raw video format involves encoding an average red chrominance sample and an average blue chrominance sample for every four luminance samples. The effective vertical and horizontal resolution of each chrominance frame is halved. More precisely, each luminance frame measures 720×480 pixels^2 (United States) or 720×576 pixels^2 (Europe), while each chrominance frame measures 360×240 pixels^2 (United States) or 360×288 pixels^2 (Europe). At 8-bit per pixel or sample, the United States-based transmission rate is given by

$$\begin{aligned} \text{rate} &= 720 \times 480 \times 30 \times 8 + \\ &\quad 360 \times 240 \times 30 \times 8 + \\ &\quad 360 \times 240 \times 30 \times 8 \\ &= 124.416 \text{ Mbps} \end{aligned} \tag{1.24}$$

and the European-based transmission rate is identical and equal to

$$\begin{aligned} \text{rate} &= 720 \times 576 \times 25 \times 8 + \\ &\quad 360 \times 288 \times 25 \times 8 + \\ &\quad 360 \times 288 \times 25 \times 8 \\ &= 124.416 \text{ Mbps} \end{aligned} \tag{1.25}$$

where again, although the United States format includes fewer lines (480 vs 576), these lines are encoded more often (30 fps vs 25 fps) leading to the same ratio $\frac{576}{30} = \frac{480}{25} = 19.2$. 4:2:0 images are generated out of 4:2:2 ones and they serve as input to most digital video compression schemes. When 4:2:0 is spatially downsampled by half, on both vertical and horizontal dimensions, a format known as *Source Intermediate Format* (SIF) is obtained. In the context of digital video, this format provides a comparable, now obsolete, Video Home System (VHS)-like quality. At 8-bit per pixel or sample, the United States-based transmission rate is given by

$$\begin{aligned} \text{rate} &= 360 \times 240 \times 30 \times 8 + \\ &\quad 180 \times 120 \times 30 \times 8 + \\ &\quad 180 \times 120 \times 30 \times 8 \\ &= 31.104 \text{ Mbps} \end{aligned} \tag{1.26}$$

and the European-based transmission rate is identical and equal to

$$\begin{aligned} \text{rate} &= 360 \times 288 \times 25 \times 8 + \\ &\quad 180 \times 72 \times 25 \times 8 + \\ &\quad 180 \times 72 \times 25 \times 8 \\ &= 31.104 \text{ Mbps} \end{aligned} \tag{1.27}$$

where the overall rate is a quarter of that of 4:2:0. The *Common Intermediate Format* (CIF) was designed as a trade-off between United States and European based SIF formats. The frame rate was set at 30 fps and the luminance and chrominance resolutions at 360×288 pixels^2 and 180×144 pixels^2, respectively. At 8-bit per pixel or sample, the CIF transmission rate is given by

$$\begin{aligned} \text{rate} &= 360 \times 288 \times 30 \times 8 + \\ &\quad 180 \times 144 \times 30 \times 8 + \\ &\quad 180 \times 144 \times 30 \times 8 \\ &= 37.324 \text{ Mbps} \end{aligned} \tag{1.28}$$

slightly higher than the SIF transmission rate. By reducing the spatial resolution of CIF by half, *Quarter Common Intermediate Format* (QCIF) is obtained. The frame rate is also reduced to 15 fps. At 8-bit per pixel or sample, the QCIF transmission rate is given by

$$\begin{aligned} \text{rate} &= 180 \times 144 \times 15 \times 8 + \\ &\quad 90 \times 144 \times 15 \times 8 + \\ &\quad 90 \times 144 \times 15 \times 8 \\ &= 4.665 \text{ Mbps} \end{aligned} \tag{1.29}$$

much lower than the CIF transmission rate. When the frame rate is reduced to 7.5 fps, the corresponding QCIF transmission rate results

$$\begin{aligned}\text{rate} = {} & 180 \times 144 \times 7.5 \times 8 + \\ & 90 \times 144 \times 7.5 \times 8 + \\ & 90 \times 144 \times 7.5 \times 8 \\ = {} & 2.332 \text{ Mbps}\end{aligned} \tag{1.30}$$

which although lower, it is still very high specially when considering that it is associated with very poor quality.

For good quality, *High Definition* (HD) raw formats can be used instead. Although there are multiple HD formats, two of them are fairly common; the 720p scheme is a progressive scanning mechanism that provides a resolution of 1280×720 pixels^2 and the 1080i scheme is an interleaved scanning mechanism that provides a resolution of 1920×1080 pixels^2. All these formats do not make any special downsampling of the chrominance components as they are considered 4:4:4. Under 4:4:4 for every four luminance samples, four blue chrominance and four red chrominance samples are also encoded. Because of the huge amount of information that is transmitted, transmission rates typically range between 1 and 3 Gbps based on the resolution and frame rate under consideration. For a 1080i resolution at 60 fps with 8-bit per pixel or sample, the transmission rate is given by

$$\begin{aligned}\text{rate} = {} & 1920 \times 1080 \times 60 \times 8 + \\ & 1920 \times 1080 \times 60 \times 8 + \\ & 1920 \times 1080 \times 60 \times 8 \\ = {} & 3 \times 1920 \times 1080 \times 60 \times 8 \\ = {} & 2985.984 \text{ Mbps} \approx 3 \text{ Gbps}\end{aligned} \tag{1.31}$$

where it is clear that a compression scheme becomes vital. Other *HD Television* (HDTV) resolutions include the generic 4K resolutions that are typically associated with dimensions involving a horizontal size of 4000 pixels. Although they are standardized by multiple organization, the ITU-R Recommendation BT.2020 specifies the *Ultra HD TV1* (UHDTV1) resolution as 3840×2160. The same recommendation also specifies an 8K UHDTV2 resolution with dimension 7680×4320.

Video compression heavily depends on space-frequency transformations similar to those achieved through Fourier transforms. More specifically, a primary technique called the *Discrete Cosine Transform* (DCT) decomposes an input signal into sets of sine and cosine functions. These functions, in turn, can be entirely described using amplitude and phase coefficients that influence these harmonic components. From a processing perspective, an analog signal is first represented as a sequence of numbers, for example by means of sampling and quantization, then when subjected to the DCT, a new sequence of numbers or coefficients is obtained as

$$X_k = \sum_{n=0}^{N-1} x_n \cos\left(\frac{\pi}{N}\left(n + \frac{1}{2}\right)k\right) \tag{1.32}$$

where x_n is the input sequence and X_k is the output transformed sequence with $k = 0, \cdots N - 1$. Because images have two dimensions, a more appropriate transformation is a 2D DCT one where the DCT is applied independently on the vertical and horizontal dimensions and an $N \times N$ input image is transformed into an $N \times N$ output image as follows

$$X_{k_1,k_2} = \sum_{n_1=0}^{N-1}\sum_{n_2=0}^{N-1} x_{n_1,n_2} \cos\left(\frac{\pi}{N_1}\left(n_1 + \frac{1}{2}\right)k_1\right) \cos\left(\frac{\pi}{N_2}\left(n_2 + \frac{1}{2}\right)k_2\right) \tag{1.33}$$

where x_{n_1,n_2} and X_{k_1,k_2} are input and output pixels, respectively.

As shown in Figure 1.44, due to the significant computational demands of applying a 2D DCT to a large image, images are divided into multiple 8 × 8 blocks. This segmentation enables the execution of individual 8 × 8 DCT operations on each block. Within this process, each of the 64 input samples is transformed into 64 output samples that depict the significance of components within a set comprising of 8 × 8 basis blocks. The basis blocks situated closer to the upper-left corner of the set exhibit lower spatial frequencies compared to those positioned nearer to the lower-right corner.

As shown in Figure 1.45, when considering an 8 × 8 input block comprising quantized 8-bit samples, the resulting 8 × 8 output represents a set of coefficients that characterize the significance of basis blocks. The nature of high-resolution images tends to make these blocks exhibit minimal transitions, causing the weight of low spatial frequency components to generally outweigh that of high spatial

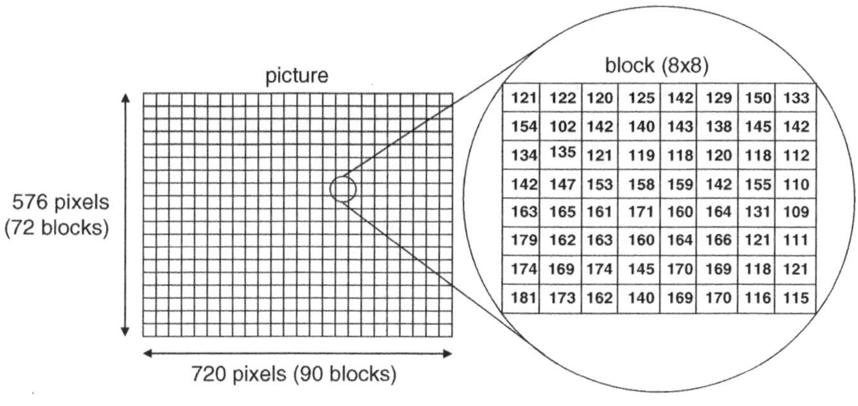

Figure 1.44 8 × 8 Blocks.

space domain block (8x8)

121	122	120	125	142	129	150	133
154	102	142	140	143	138	145	142
134	135	121	119	118	120	118	112
142	147	153	158	159	142	155	110
163	165	161	171	160	164	131	109
179	162	163	160	164	166	121	111
174	169	174	145	170	169	118	121
181	173	162	140	169	170	116	115

DCT →

← IDCT

frequency domain block (8x8)

2011	23	12	−13	8	7	−3	0
61	34	23	14	−11	4	0	1
45	32	−21	12	−5	2	0	4
34	−54	25	11	18	3	0	1
24	−19	26	−21	17	0	−6	4
28	12	16	0	0	0	0	−1
14	9	−7	6	−1	3	2	3
8	−6	1	2	−4	1	7	−1

horizontal frequency

vertical frequency

Figure 1.45 DCT of a Block.

frequency components. Consequently, the output coefficients generated by the 2D DCT experience a reduction in amplitude as frequency increases.

Additionally, it is observable that the quality of reconstruction progressively deteriorates as more components are selectively eliminated from the output sequence during the transition from the input sequence. In general, the magnitude of a coefficient correlates with the distortion introduced upon its removal. Consequently, to reduce transmission rates, it becomes feasible to encode exclusively those coefficients of sufficient size to impact human perception. Specifically, smaller coefficients that signify high spatial frequency components, which are commonly less influential, can be omitted without significantly impacting overall quality.

Hence, every coefficient can be matched against a threshold that progressively diminishes as the frequency rises. In the event that a coefficient's value falls below the threshold, it is substituted with zero. Once this thresholding procedure is enacted upon the coefficients, the resultant values are subjected to quantization using fewer levels as the vertical and horizontal frequencies ascend. This process is referred to as thresholding and quantization.

Due to the two-dimensional nature of each block, including both horizontal and vertical dimensions, and considering that transmission is a unidirectional process, a mechanism for sequential traversal and serialization of blocks becomes necessary. A favored approach is achieved through zig-zag serialization which arranges coefficients based on their spatial frequency characteristics, progressing from the lowest to the highest. It is worth noting that the very first coefficient located at the top-left corner signifies the DC component and is not tied to spatial frequency. Typically, this *Direct Current* (DC) component is encoded using DPCM, wherein the present DC sample is differentially encoded using the previous DC sample as a reference.

Due to the effects of thresholding and quantization, zig-zag serialization typically yields a unidimensional sequence that contains an extended series of zeros. This sequence can be efficiently encoded utilizing *Run Length Coding* (RLC). In the context of RLC, if a source exhibits extended sequences of identical symbols, the encoding process involves denoting the symbol itself along with the sequence's length, rather than encoding the entire sequence of symbols. It is important to note that the transmission rate can be adjusted based on thresholding and quantization parameters, potentially at the expense of image quality. This manipulation is carried out through a mechanism termed rate-control. Note that the acronym RLC is also used to describe 3GPP Radio Link Control in Chapter 4.

The *Joint Photographic Experts Group* (JPEG) is responsible for creating the widely adopted JPEG image compression standard. The input for the JPEG algorithm can take the form of either a raw image in luminance/chrominance format or an RGB format. Lossless JPEG employs LPC mechanisms, akin to those used in speech compression, to achieve compression rates of approximately 3. Conversely, lossy JPEG leverages the DCT mechanism, detailed earlier in this section, to attain compression rates of around 10.

In essence, lossy JPEG encoding entails the following steps: (1) decomposing the raw image into 8×8 luminance and chrominance blocks, (2) applying the DCT to all resulting luminance and chrominance blocks, (3) employing thresholding and quantization on the DCT coefficients, (4) traversing each block using a zig-zag pattern to maximize sample correlation, (5) employing RLC to reduce the size of the encoded sequence of symbols, and (6) employing Huffman encoding to achieve a transmission rate as close as possible to the entropy of the sequence. Conversely, the process of JPEG decoding involves a sequence of steps that reverse these six aforementioned processes.

Up until now, compression has primarily focused on individual images. When addressing video compression, a preliminary approach might involve compressing each individual frame in a manner similar to JPEG images. However, this technique proves to be inefficient as it overlooks the substantial correlation present between consecutive frames. The *Motion Pictures Experts Group* (MPEG) has introduced the MPEG standards, which serve as a comprehensive specification including various standards for audio and video multiplexing and compression. The initial MPEG standard is the MPEG-1 standard, which acts as a foundational basis for most video compression methodologies.

MPEG-1 Part 2 video compression yields a transmission rate of approximately 1 Mbps. It builds upon JPEG-based image compression but incorporates additional mechanisms to exploit the substantial correlation between successive frames, reducing the quantity of information that requires encoding. Specifically, with the utilization of prediction combined with motion compensation, a video frame is encoded in relation to a preceding frame by encoding the discrepancy between the two frames using a motion estimator.

The motion estimation process executed at the encoder involves a trade-off between the processing speed and video quality. Faster processing speed generally results in reduced latency. The latency is also influenced by the number of video frames or pictures grouped together for motion estimation. In practical terms, when correlations between video frames are identified, these frames need to be buffered to determine their interdependencies before the encoding process can start. The cumulative latency stemming from this grouping approach is directly proportional to the number of frames within the group, with the rate of this increase being inversely related to the video frame rate.

A *Group of Pictures* (GOP) includes various frame types categorized based on how they can be predicted from preceding frames. Specifically, there are three main types: (1) Intraframe pictures (I frames) that are encoded without external reference using a mechanism similar to JPEG compression, (2) Interframe pictures (P or predicted frames), predicted from a previous I or P reference using motion estimation techniques, and (3) Bidirectional pictures (B frames), predicted as an interpolation of future and past I and P pictures.

I frames offer the highest quality but have the lowest compression rate, followed by P frames and then B frames. Typically, the encoder dynamically compresses raw video frames and determines the type of compressed frame to generate based on rate-control and quality prerequisites. More precisely, the encoder selects the GOP size and determines which types of frames to include, resulting in a specific transmission rate.

It is important to note that at the initiation of a GOP, the initial frame is always an I frame because there are no preceding pictures for reference. When accessing a video stream randomly, the process typically occurs on a GOP-to-GOP basis, ensuring complete decoding of a picture without introducing artifacts.

Shown in Figure 1.46 is an illustrative GOP, which can be characterized by two parameters, M and N. In this context, N signifies the GOP size, set at $N = 12$ for this instance, while M represents the gap between I and P frames, which is $M = 3$ in this particular scenario. Fundamentally, the compressed GOP originates from a set of 12 raw frames, transformed into a sequence comprising compressed I, B, B, P, B, B, P, B, B, P, B, B, and I pictures, where the last I picture pertains to the subsequent GOP. However, the sequence of these raw frames must follow a certain order that aligns with the prediction hierarchy. In specific terms, the raw frames require encoding or decoding in a sequence that results in I, P, B, B, P, B, B, P, B, B, I, B, and B for transmission. This rearrangement of pictures is required due to buffering and gives rise to additional latency.

The MPEG video compression standard includes hierarchies of both subsets and supersets of GOPs. More precisely, a GOP can be further divided into frames (also

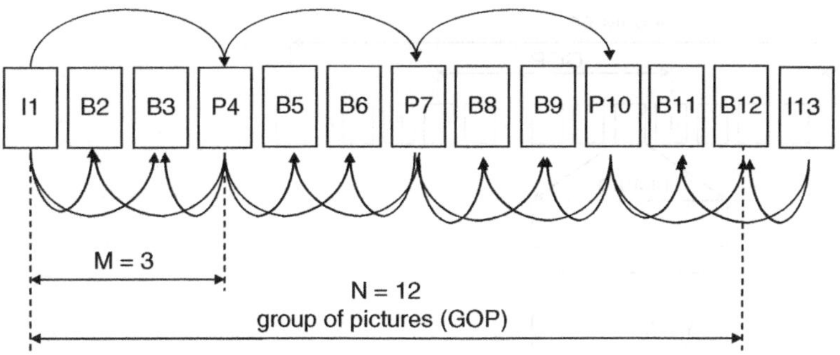

Figure 1.46 GOP Example.

known as pictures), each categorized as intra, inter, or bidirectional as previously outlined. These pictures, in turn, are divided into slices, which consist of consecutive sequences of *macroblocks* (MBs) typically organized in rows. Macroblocks generally comprise 2 × 2 blocks, serving as the foundational units for motion estimation. Much like the JPEG case, each block measures 8 × 8 pixels[2]. Note that MB is also used to denominate Medium Band in Subsection 1.4.2.5.

Conversely, GOPs act as the fundamental building blocks for video sequences. All these interrelated components and their connections are represented in Figure 1.47. As a video frame is composed of numerous MBs, and considering that each picture can adopt a 4:2:0 format, it follows that each MB also conforms to the 4:2:0 format. With this arrangement, each MB includes 16 × 16 square luminance pixels and two sets of 8 × 8 square chrominance pixels.

In most instances of video compression, where MBs play a pivotal role, the picture's resolution is designed to be an integer multiple of the MB's resolution. For instance, the aforementioned CIF format relies on a resolution of 360 × 288, but since 360 is not an exact multiple of 16, the adapted resolution that accommodates video compression becomes 352 × 288.

Motion estimation involves the task of identifying identical objects in both a reference frame and the frame undergoing compression, as illustrated in Figure 1.48. Once the object is recognized, the only pertinent information to encode pertains to the object's position within the reference frame, the motion vector detailing the object's translation, and the encoding of the image portions that become exposed through this motion. As previously mentioned, object identification occurs on a per-MB basis through a search area where the reference frame is compared against the frame being considered, thereby determining the motion vector. Given that this matching process is influenced by color, it requires simultaneous execution across all chrominance and luminance components of an MB.

Since the object in the compressed frame is substantially correlated but not entirely identical to that in the reference frame, in P frames, the disparities

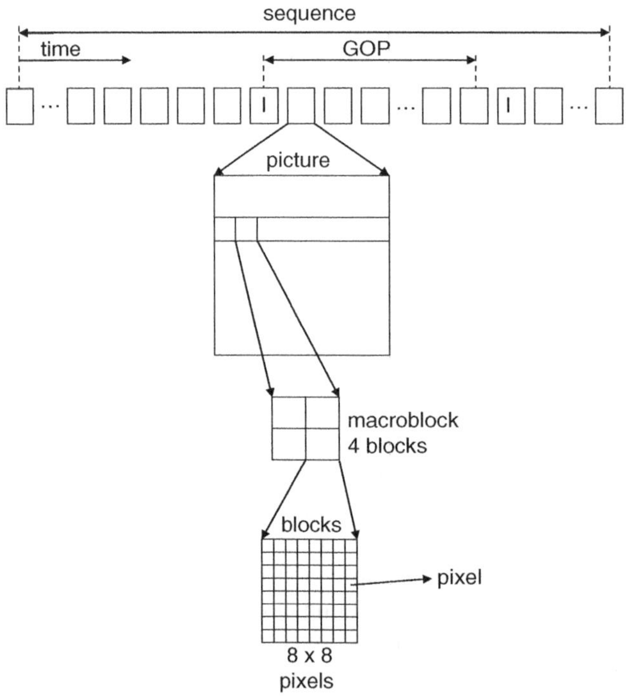

Figure 1.47 Video Frames.

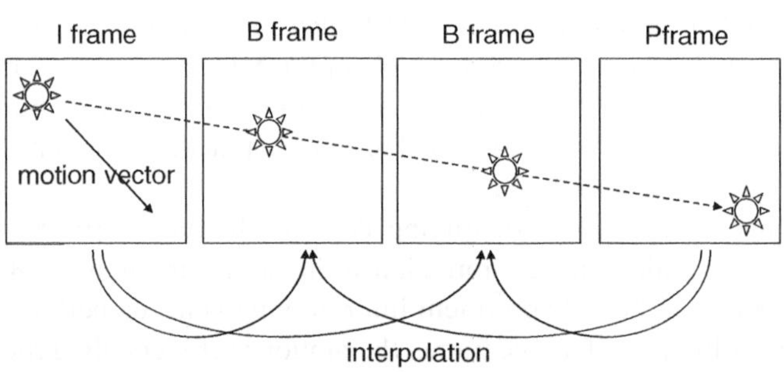

Figure 1.48 Motion Estimation.

between them must also be encoded. This is achieved through a process referred to as motion compensation. On the other hand, B frames emerge from spatial interpolation between the two reference frames, resulting in minimal information being encoded. In scenarios where the distortion resulting from B frame encoding surpasses a certain threshold, encoding is executed with P frames. Likewise, when

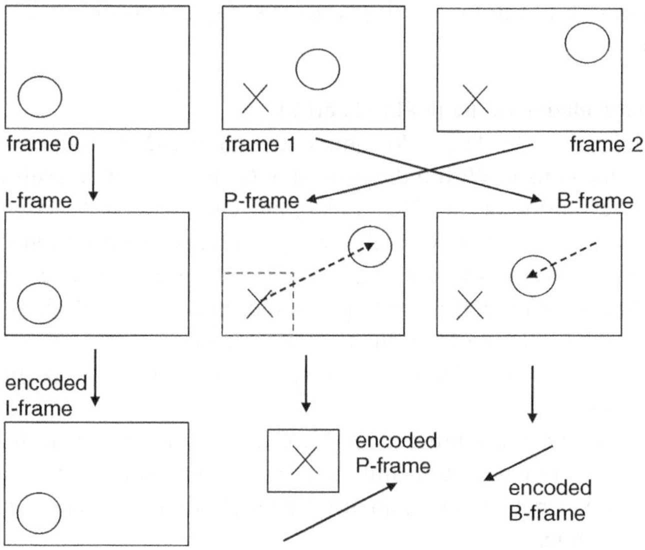

Figure 1.49 Sequence Encoding.

the distortion originating from P frame encoding exceeds a specific threshold, I frame encoding is engaged instead.

For instance, consider Figure 1.49, which illustrates a sequence featuring a circle moving along a straight path, with the intention of encoding three frames. These frames are rearranged in such a way that the initial frame is designated as an I frame, while the second and third frames are interchanged, facilitating the processing of P frames before B frames. In this arrangement, the I frame is subjected to conventional still image compression for the entire frame.

As for the P frame, it undergoes regular still image compression for the portion of the image that differs from the I frame. Additionally, a vector is incorporated to signify the direction and magnitude of the circle's movement between frame 0 and frame 2. Lastly, the B frame focuses exclusively on encoding the vector that represents the circle's movement intensity and direction between frame 2 and frame 1. Quality manipulation plays an important role in achieving rate control. Specifically, quantization parameters (such as the number of levels) and prediction schemes are dynamically chosen using a feedback mechanism to achieve a targeted transmission rate.

1.4.3.1 ITU-T Recommendation H.261

The ITU-T Recommendation H.261 specifies, a now obsolete, early VBR video codec standard that supports transmission rates between 40 Kbps and 2 Mbps to enable both QCIF and CIF resolutions. Frame rates 7.5, 15, and 30 fps are supported (ITU-T 1993).

1.4.3.2 ITU-T Recommendation H.262 (MPEG-2 Part 2)

MPEG-2 Part 2, standardized by ITU-T Recommendation H.262, introduces a generic video coding mechanism. Within the context of MPEG-2, a set of levels is defined, linked to the resolution of the video undergoing compression. Precisely, there are four attainable levels: a low level corresponding to a resolution of CIF (360×288), a main level corresponding to 4:2:0 (720×576) resolution, a high level with 1440 resolution corresponding to standard HDTV resolution of (1440×1152), and a high level corresponding to widescreen HDTV resolution of up to (1920×1152). This leads to VBR transmission rates between 3.5 Mbps and 6 Mbps at 7.5, 15, 25, and 30 fps.

The picture types I, B, and P are present in MPEG-2. Additionally, there is a spatially scalable profile that involves transmitting a base quality picture and supplementary information layers to enhance quality. The high profile is specifically designed for HDTV encoding.

In the MPEG-2 main profile and level, interlaced 4:2:0 formatted frames with dimensions of 720×576 and captured at a frame rate of 25 fps are compressed using I, P, and B encoding (ITU-T 1995).

1.4.3.3 ITU-T Recommendation H.263

The ITU-T Recommendation H.263 enhances the previous ITU-T Recommendation H.261 and ITU-T Recommendation H.262 standards, to specify a VBR video codec that supports 7.5, 15, and 30 fps. Transmission rates between 40 Kbps and 2 Mbps are typical. Resolutions include: 128×96 (Sub Quarter Common Intermediate Format SQCIF). QCIF, CIF, 704×576 (4CIF), and 1408×1152 (16CIF).

In turn, ITU-T Recommendation H.263v2 (or H.263+) improves the original ITU-T Recommendation H.263 by adding additional annexes to the specification. Among many things, H.263+ supports custom size picture encoding (ITU-T 2005).

Example 1.3 Consider a 128 Kbps CIF ITU-T Recommendation H.263 video stream with 4:2:0 frame encoding. If the frame rate is 15, what is the overall compression rate (the ratio between the compressed and non-compressed media rate)?

Solution:

The raw video rate is given by $R = 15 \times 360 \times 288 \times 8 + 2 \times 180 \times 144 \times 8 =$ 12.85 Kbps. The compression rate is then given by $k \approx \frac{12.8}{128} = 0.1$.

1.4.3.4 ITU-T Recommendation H.264 (MPEG-4 Part 10)

MPEG-2 Part 2 has been enhanced as MPEG-4 Part 10, also known as ITU Recommendation H.264 or *Advanced Video Coding* (AVC). This new standard enhances compression efficiency by at least 50%. The primary innovation of H.264/AVC lies in its dual-layer structure: (1) the *Video Coding Layer* (VCL) integrates interframe compression, capitalizing on temporal statistical dependencies between consecutive frames, and intraframe compression, exploiting spatial dependencies within a given frame, and (2) the *Network Abstraction Layer* (NAL) packages VCL data into units, ensuring reliable transportation through data networks.

The efficiency of ITU-T Recommendation H.264 is attributed to three factors: (1) incorporation of more sophisticated prediction modes that can vary within a frame and reference a greater number of successive frames, (2) adoption of an alternative transform known as the integer transform, which is faster and employs block sizes of either 4 × 4 or 8 × 8 pixels, and (3) utilization of adaptive Huffman and adaptive algebraic coding as entropy encoding mechanisms.

The standard defines levels that support the following resolutions: QCIF, CIF, 720 × 576, 1280 × 720, 2048 × 1024, 3672 × 1536, 4096 × 2304, and 8192 × 4320. Frame rates range from 15 fps to 120 fps leading to transmission rates between 64 Kbps and 800 Mbps (ITU-T 2009).

1.4.3.5 ITU-T Recommendation H.265

The ITU-T Recommendation H.265, also known as *High Efficiency Video Coding* (HEVC), is a video compression standard that extends the ITU-T Recommendation H.264. As with this latter standard, the ITU-T Recommendation H.265 supports the following resolutions: SQCIF, QCIF, CIF, 640 × 360, 720 × 576, 960 × 540, 1280 × 720, 1920 × 1080, 2048 × 1080, 3840 × 2160, 4096 × 2160 (4K), 7680 × 4320 (8K), and 8192 × 4320 (also 8K). Transmission rates of 30 Kbps to 800 Mbps for frame rates between 15 fps and 300 fps are possible (ITU-T 2019).

1.4.3.6 ITU-T Recommendation H.266 (MPEG-I Part 3)

MPEG-I Part 3, standardized as ITU-T Recommendation H.266 and also known as *Versatile Video Coding* (VVC), supports the same resolutions and frame rates as ITU-T Recommendation H.265, including 15360 × 8640 (which falls under the umbrella of 16K resolutions), but improves performance and provides the same quality at lower transmission rates (ITU-T 2023).

1.4.3.7 AU1

An issue associated with compression methods based on MPEG standards pertains to their regulation by *Intellectual Property* rights. Note that a significant number of individual encoding techniques within these standards are patented, requiring specific licenses for their utilization. Addressing this concern, VP8 emerges as an open-source video codec that is devoid of patent restrictions. The primary objective of VP8 is to deliver comparable quality to ITU-T Recommendation H.264 while avoiding the need for licensing. VP8 excels in its incorporation of efficient intra- and interprediction mechanisms, employs 4 × 4 block-oriented DCT compression for all luminance and chrominance error signals, and distinguishes itself by utilizing three reference frames for inter prediction, rather than the conventional two.

VP8 has been extended by VP9 to provide quality comparable to that of ITU-T Recommendation H.265. *AOMedia Video 1* (AU1), in turn, was developed as the successor of VP9 as an open, royalty-free video coding format that supports the following resolutions: 426 × 240, 640 × 360, 854 × 480, 1280 × 720, 1920 × 1080, 3840 × 2160, and 7680 × 4320. For frame rates between 30 fps and 120 fps, these resolutions lead to transmission rates between 1.5 Mbps and 800 Mbps for Open Media (2021).

1.4.3.8 Theora

Theora is an early open source license-free video codec derived from VP3, an ancestor of AU1. It provides similar capabilities to those of ITU-T Recommendation H.262 supporting QCIF, CIF, 4CIF, and HD resolutions of 1440 × 1152 and 1920 × 1152 (xiph.org 2006).

1.4.3.9 MPEG-5 Part 2

MPEG created the MPEG-5 Part 2 standard, also known as *Low Complexity Enhancement Video Coding* (LCEVC), to define a novel enhancement layer that builds on top of the information transmitted over a traditional base codec (i.e., AU1, ITU-T Recommendation H.264, ITU-T Recommendation H.265, ITU-T Recommendation H.266, etc.). This enhancement layer improves QoS by reducing the artifacts introduced by the base codec without significantly increasing the transmission rate (Choi et al. 2020).

1.5 Quality Scores

Due to its significant influence on user experience, the assessment of speech quality is of utmost importance. Specifically, the deterioration in quality caused by factors like packet loss, latency, and jitter is about the need for mechanisms that can evaluate this decline effectively. One such metrics is the SNR, that relies on calculating the ratio between the power associated with the source and error signals, and it is given by

$$\begin{aligned}\mathrm{SNR} &= 10\log_{10}\left(\frac{\sum_n x_n^2}{\sum_n e_n^2}\right)\\ &= 10\log_{10}\left(\frac{\sum_n x_n^2}{\sum_n (x_n - \hat{x}_n)^2}\right)\end{aligned} \tag{1.34}$$

where x_n, e_n, and $\hat{x}_n$ are the input, the error, and the reconstructed samples of the speech under analysis, respectively. Because this method relies on signal characteristics, its precision is confined to waveform codecs exclusively. Additionally, an issue with SNR is that, on average, signals with high amplitudes tend to overshadow those with low amplitudes. This leads to coefficients that are disproportionately larger than anticipated, thus yielding an inaccurate evaluation of the analyzed speech. An alternative approach involves dividing speech streams into frames, allowing for individual SNR calculations within each segment. Then a new metric, *Segmental SNR* (SSNR) is calculated as

$$\mathrm{SSNR} = \frac{1}{N}\sum_{m=1}^{N}\mathrm{SNR}_m \tag{1.35}$$

as an average over all N segments. The idea is that low segmental SNR cannot be masked by high segmental SNR regions anymore.

Considering the nature of SNR and SSNR metrics, their application in evaluating parametric speech codecs yields low outcomes. This is primarily due to these codecs' tendency to distort waveforms while maintaining a level of quality. In such scenarios, an alternative metric becomes necessary, that is, one metric that involves subjective ratings gathered by averaging individuals' opinions during comparisons between source and reconstructed speech waveforms.

From a human standpoint, the perception of speech quality includes various aspects: intelligibility (how easily the speech can be understood), naturalness (the presence of undesirable elements like echoes, or noise and clicking in the reconstructed speech), and recognizability (whether the original speaker can be identified).

Moreover, in the context of parametric codecs, speech quality becomes influenced by several factors. These factors include the speaker's characteristics, the language spoken, signal intensity, background noise levels, the occurrence of channel errors, the presence of non-speech signals such as music, and tandem coding, which refers to processes like decoding, encoding, or transcoding speech across intermediary nodes.

The evaluation of speech can be conducted through various tests, depending on the chosen criteria. One such method is the *Absolute Category Rating* (ACR), where a group of listeners assesses a speech sequence and assigns rankings based on its absolute quality, ranging from excellent (5) to bad (1). The collective average of these ratings yields the *Mean Opinion Score* (MOS).

Conversely, an alternate approach known as *Degradation Category Rating* (DCR) focuses on evaluating the extent of degradation rather than quality. In this process, listeners gauge the degree of degradation experienced in the analyzed speech sequence relative to the original reference. This assessment yields five possible scores: not degraded (5), minimally degraded (4), slightly degraded (3), degraded

(2), and highly degraded (1). Once again, the average scores across all listeners are calculated, resulting in a *Degradation MOS* (DMOS).

Due to its presentation of the reference followed by the degraded speech sequence, the DCR method can introduce bias into the DMOS. An alternative assessment approach, called *Comparison Category Rating* (CCR), involves presenting listeners with both the reference and the degraded speech sequences in a randomized manner. This results in potential scores such as: the first sequence being much better than the second (3), the first sequence being better than the second (2), the first sequence being slightly better than the second (1), both sequences being identical (0), the second sequence being slightly better than the first (−1), the second sequence being better than the first (−2), and the second sequence being much better than the first (−3). As previously, these scores are averaged across all listeners.

As a general trend, increased involvement of listeners tends to yield better and more precise outcomes. Nevertheless, these evaluations demand significant time and resources, making it advantageous to devise an algorithmic solution. This algorithm would take input in the form of speech reference and degraded sequences, generating an objective output score closely aligned with the subjective score that a CCR test would yield.

For this purpose, a multitude of speech reference sequences must be utilized, each subjected to various degradation sequences. This enables the algorithm to generate objective scores, which can then be compared to corresponding subjective scores associated with the sequences.

One of the initial efforts to establish standardized speech quality assessment was undertaken through ITU-T Recommendation P.861. This recommendation outlines the algorithm called *Perceptual Speech Quality Measure* (PSQM). The block diagram framework of PSQM is illustrated in Figure 1.50. In this process, the input speech, acting as the reference, undergoes consecutive encoding and decoding

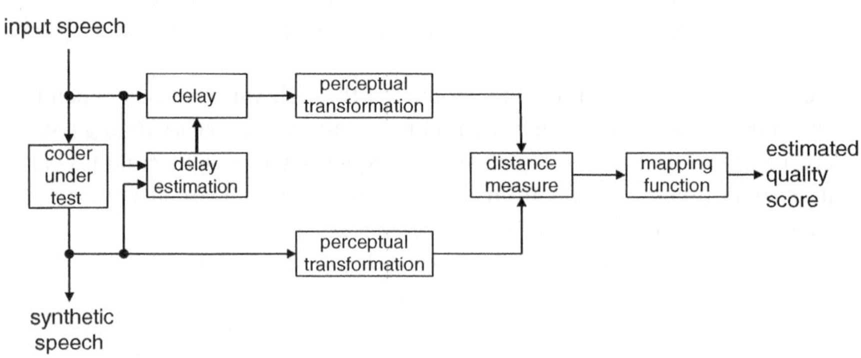

Figure 1.50 Perceptual Speech Quality Measure.

steps to generate synthetic speech. The reference signal is then temporally shifted to account for the latency introduced by codec processing. Utilizing perceptual transformations, parameters are extracted from both the reference and synthetic speech. These parameter differences are then translated into MOS speech quality scores, ranging from 1 (bad) to 5 (excellent).

1.5.1 Network Impairments

Communication networks are made of hosts that can communicate with each other. This communication is performed with the assistance of routers that are in charge of propagating information throughout the network. Both hosts and routers form communication systems with transmitters and receivers connected to channels. Each router supports multiple hosts that, in turn, are connected to sources and sinks. The network provides an alternative to dedicated links between hosts as routers allow for resource sharing in an efficient manner.

Communication networks can be either circuit switching or packet switching-based. Figure 1.51 shows a now obsolete traditional *Plain Old Telephone Service* (POTS) example of circuit switching, where there is a fixed full-duplex path between two hosts. Each telephone acting as a host is connected to a *Digital Loop Carrier* (DLC) that acting as a router allows the speech to enter the *Public Switching Telephone Network* (PSTN) where *Central Offices* (COs) or telephone exchanges

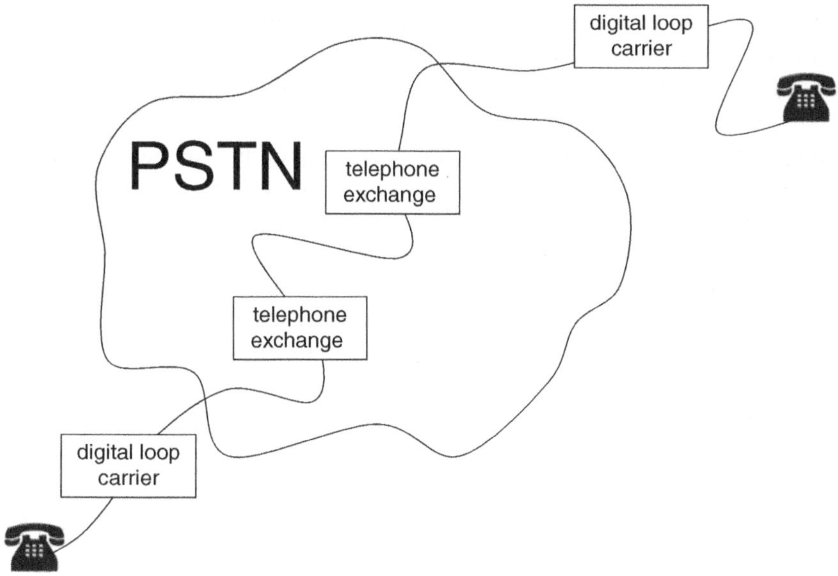

Figure 1.51 Circuit Switching.

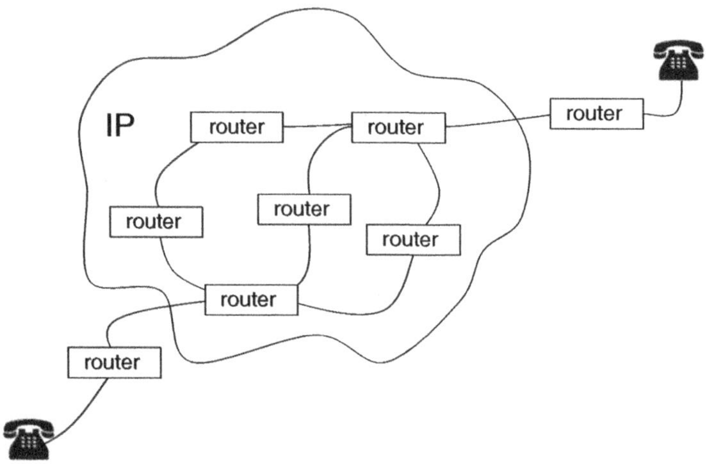

Figure 1.52 Packet Switching.

acting also as routers propagate the signals. A path consists of a common channel that traverses time slots of multiple links that once set up remain uninterrupted. When the path is established, bandwidth and other resources are assigned to the source and destination hosts until it is terminated. Because setup is fast and the path is dedicated, circuit switching is very efficient for transmission of speech.

Figure 1.52 shows an example of telephony based on packet switching to support traditional *Voice over Internet Protocol* (VoIP). Again, phones act as hosts which are now connected to IP routers, switches, and other components like *Session Border Controllers* (SBCs) acting as routers. Speech is chunked, or packetized, and then transmitted throughout the network as packets that follow a specific path from source to sink. Moreover, for each packet, the path is dynamically selected, resulting from network router conditions, congestion, and other factors that determine which set of links are used. Because the path changes on a per-packet basis, source-to-sink delay and other impairments like loss affect each packet differently, imposing quality restrictions on speech. For the same conditions, circuit switching provides much better quality than packet switching, at the cost of less efficient use of available resources. The effect of loss and latency typical of packet switching can be seen in Figure 1.53. Since Host 1 (H1) has some application data that is to be sent to Host 2 (H2), it first divides the data into four packets and adds routing headers (R) to each of them. Router 1 (R1) receives the packets and due to load balancing it forks two of them to Router 2 (R2) and the remaining two to Router 3 (R3). R3 is heavy loaded so it drops packet number 2 and only forwards packet number 4 to Router 4 (R4). Due to a slower link and higher latency, packets 1 and 3 take longer to arrive to R4 than packet 4 such that H2 receives an out of sequence partial number of packets.

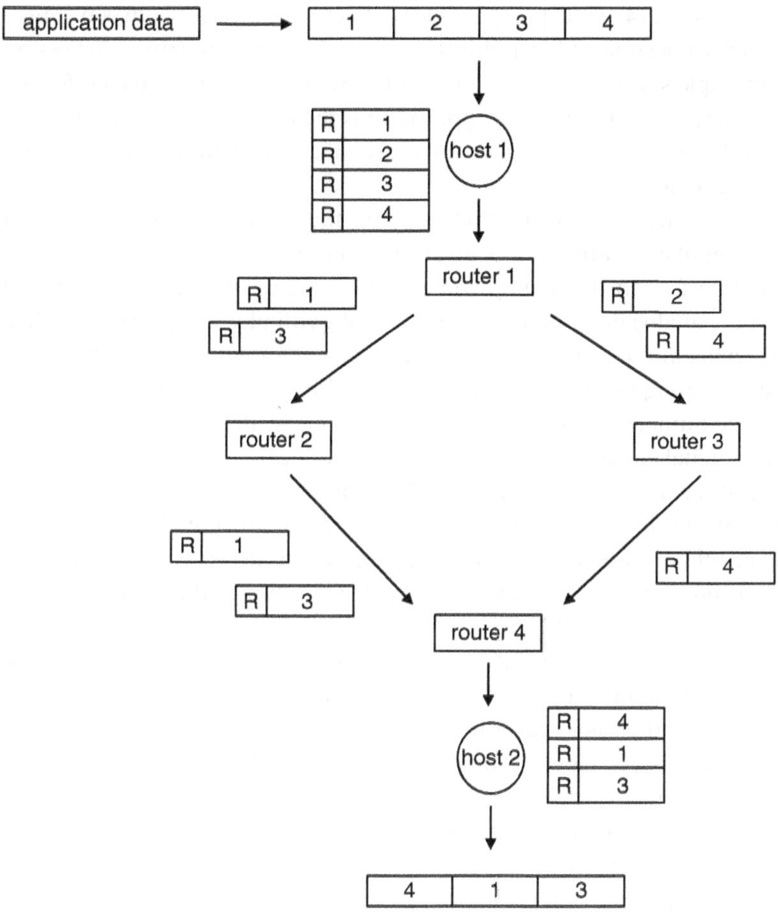

Figure 1.53 Packet Loss and Latency.

Data network communications are based on a layered architecture where each layer is a black box, such that its input is the output of the immediately lower layer, and similarly, its output becomes the input of the immediately higher layer. This exchange of information between layers is performed through each layer's interfaces, where input and output facilities exist. Each layer provides a specific function defined by the *Open Systems Interconnection* (OSI) model. The model is organized into seven layers: (1) Physical Layer handles the transmission and reception of data over the channel, including the conversion of information between the digital and analog domains, involving, therefore, physical phenomena, (2) Data Link Control provides error control mechanisms for reliable transmission of information over the channel, (3) Network is in charge of ensuring packets are delivered from source to sink, (4) Transport provides delivery of messages between source and sink, including support for multiplexing, also known as muxing, (5) Session manages multiple sessions between endpoints, (6) Presentation formats information for further processing, and (7) Application is the layer closest to the end user and is typically involved with services and also conversion of information between the digital and analog domain.

Each layer defines a protocol that governs the conversion of information between the input and output at any given layer interface. Both Layer 1 and Layer 7 have one interface in the analog domain and another interface in the digital domain. In the analog domain, the interface deals with physical phenomena like electrical signals in the channel or acoustic signals coming from a speaker, as in RTC. Layers 2 through 6 are purely digital layers with both interfaces in the digital domain. Since humans and the real world are analog by definition, a conversion between the digital and analog domains is always critical.

The biggest advantage of a layered architecture is the compartmentalization of a communication system, such that when a specific functionality needs to be upgraded, no other components need to be changed. This leads to technologies like *Network Function Virtualization* (NFV) that provide, among other things, shorter deployment times.

The Internet is an example of a data network where different protocols are used to provide services independently at different levels, such that upgrading one layer does not affect another. For scalability reasons, the Internet uses the concept of subnetworks or subnets. Hosts belonging to the same subnet exchange information directly, and routers provide access to hosts in other subnets. Internet does not follow the Open Systems Interconnection (OSI) but instead a simplified 5-layer architecture known as the Internet Engineering Task Force (IETF) architecture. Details of this architecture are presented in Chapter 2.

Although the Internet is an example of packet switching, where each packet takes a different path based on network conditions, protocols at the transport layer like *Transport Control Protocol* (TCP) can provide a circuit switching-like behavior by associating each connection with a stream of information through virtual circuits. Simpler mechanisms like *User Datagram Protocol* (UDP), which preserve the connectionless nature of packet switching, transmit information on a per-packet or datagram basis. Figure 1.54 shows both of these transport layer protocols in the context of the network layer *Internet Protocol* (IP).

Voice over Internet Protocol applications rely on UDP and TCP transport protocols. Unfortunately, the latter involves retransmissions to guarantee data integrity, introducing delays that significantly affect end-user perception. On the other hand, UDP is affected by packet loss as well as delays inherent to packet switching, which degrade quality and affect the overall QoS. Acceptable voice

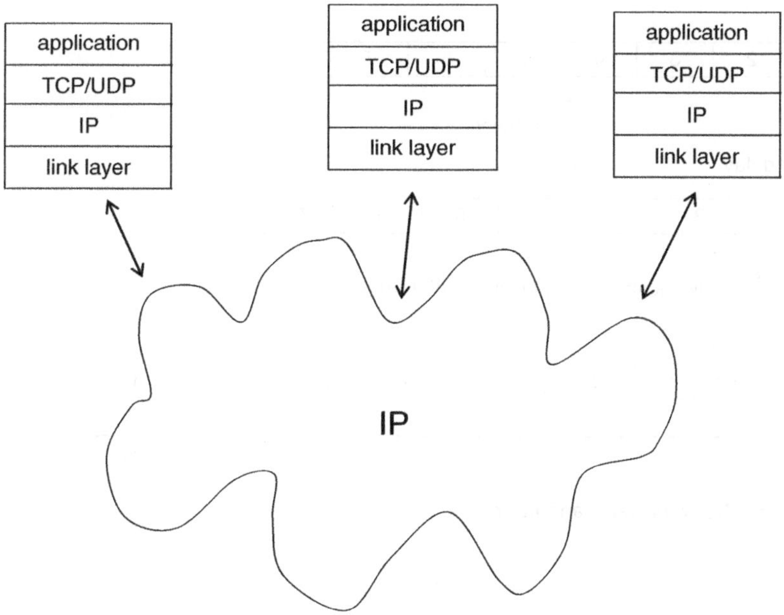

Figure 1.54 IP Protocol Layers.

quality is typically obtained when packet loss is below 3% and latency or delay is 150 milliseconds.

Figure 1.55 shows a time diagram of the effect of network delay in VoIP. If the transmitter sends a sequence of packets (1 through 8), it arrives a bit later due to network delay. Although the interpacket transmission time is constant, the inter-arrival time at the receiver is not always constant, leading to network jitter. To minimize the effect of both delay and jitter, the application layer uses a jitter or playout buffer to ensure that the media sequence can be played smoothly without choppiness or dead air.

1.5.2 PESQ

A refinement to PSQM emerged as *Perceptual Evaluation of Speech Quality* (PESQ), which incorporates more precise perceptual transformations, leading to enhanced accuracy in generating quality MOS. PESQ has been standardized as ITU Recommendation P.862.1 for NB speech and as ITU Recommendation P.862.2 for WB speech. Speech sampled at higher rates like SWB and FB must be downsampled to WB before PESQ can be applied. Note that because downsampling introduces distortion that lowers the scores, scores for SWB and FB sampled signals cannot be accurately obtained.

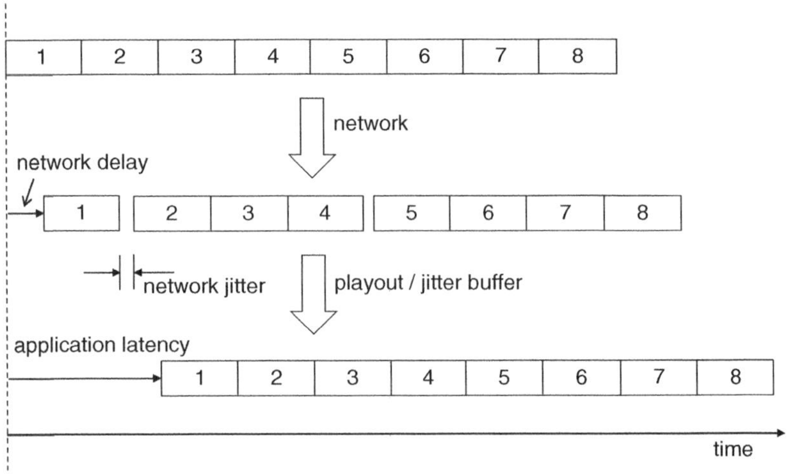

Figure 1.55 Network Delay and Jitter.

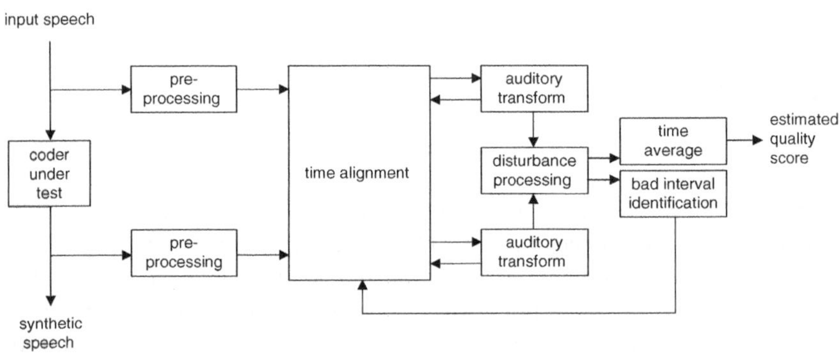

Figure 1.56 PESQ Block Diagram.

Figure 1.56 shows the block diagram of the PESQ scheme. The reference and the signal under test are respectively subject to pre-processing and then time-aligned. These signals are then transformed and auditory-processed. Time averages that lead to the actual PESQ scores are then obtained. PESQ raw scores fall in the −0.5 to 4.5 interval, but they are mapped into *MOS-Listening Quality Objective* (MOS-LQO) values in the 1 to 5 range associated with MOS described in Section 1.5. Specifically, the MOS scale assigns 1 to bad quality, 2 to poor quality, 3 to fair quality, 4 to good quality, and 5 to excellent quality. Typically, different codecs with different grades and levels of compression exhibit different MOS-LQO scores (Herrero and Cadirola 2014; ITU-T Recommendation P.862 2001).

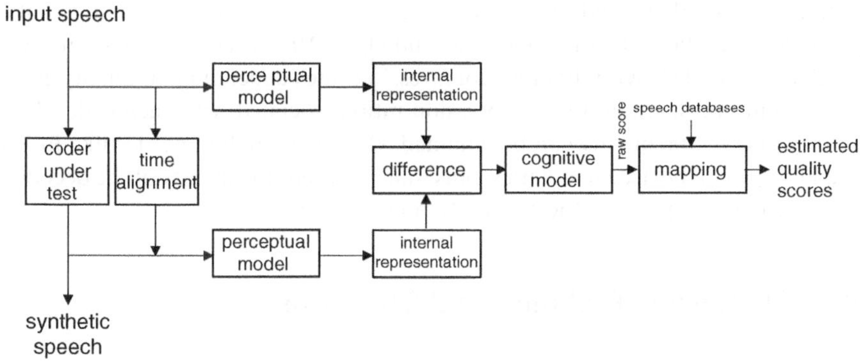

Figure 1.57 POLQA Block Diagram.

1.5.3 POLQA

In 2011, an even more accurate algorithm known as *Perceptual Objective Listening Quality Assessment* (POLQA) was standardized by the ITU through Recommendation P.863 as an evolution of PESQ. Apart from offering improved quality estimation for certain scenarios, POLQA also supports a variety of sampling rates beyond the conventional NB and WB options, including SWB and FB.

Figure 1.57 shows the different blocks involved in POLQA estimations. The reference and the signal under test are time-aligned and subjected to a perceptual model that internally represents critical parameters. The difference between these internal representations is fed to a cognitive model, which produces a POLQA raw score that is then mapped into MOS-LQO scores (ITU-T Recommendation P.863 2018).

1.6 Summary

Media generation and compression are key to RTC. Without understanding how these mechanisms work together, it is not possible to understand RTC in general. The three main types of media (speech, audio, and video) have different characteristics that make them unique from a signal processing perspective and, therefore, compression since compression depends on the nature of the signals. Speech signals are typically compressed through techniques of linear prediction and machine learning in the time domain, while audio and video respectively rely on unidimensional and bidimensional frequency domain analysis. These compression mechanisms lead to the development and use of media codecs.

Speech codecs range from low compression rate ones like the ITU-T Recommendation G.711 to high compression and performance ones like the EVS. Audio codecs, on the other hand, involve lower compression rates to preserve quality. Examples of audio codecs include AAC and Opus. Video codecs have been standardized by the ITU with different compression rates depending on quality goals and resolution. Based on the compression ratio associated with each codec, better or worse media quality can be obtained. Objective quality metrics like PESQ and POLQA can be used to obtain scores that represent the quality of the decoded speech when compared to the originally generated one.

1.7 Homework Problems and Questions

Problem 1.1 The ITU-T Recommendation G.711 codec samples speech at an NB rate. How many 10-millisecond frames are packetized if 240 samples per packet are encapsulated?

Problem 1.2 If the EVS codec is configured at 7.2 Kbps and frames at 20 milliseconds long, how big are 3-frame EVS-encapsulated packets?

Problem 1.3 For an audio signal with bandwidth of 13 kHz, what is the best sampling rate (NB, WB, SWB, or FB)? Why?

Problem 1.4 Under ITU-T Recommendation G.726 with a transmission rate of 16 Kbps, if the sampling rate is NB, how many bits per sample is the raw speech encoded as?

Problem 1.5 Consider a 512 Kbps 1920 × 1080 ITU-T Recommendation H.265 video stream with 4:2:2 frame encoding. If the frame rate is 30, what is the overall compression rate (the ratio between the compressed and non-compressed media rate)?

Problem 1.6 What is the relationship between the raw video transmission rate at 4:2:2 encoding and the width, height, and frame rate of the video stream?

Problem 1.7 What is the trade-off between transmission rate, sampling rate, and quantization level in waveform codecs ?

Problem 1.8 Consider an LPC-based codec, if 16 4-bit coefficients are packed together to encode 10-millisecond frames, what is the transmission rate of the codec?

Problem 1.9 In an NB scenario, if ITU-T Recommendation G.711 exhibits better PESQ/POLQA scores than EVS, why is EVS the standard codec for 5G communications?

Problem 1.10 What are the advantages and disadvantages of circuit switching compared to packet switching in communication networks?

Problem 1.11 What is the purpose of PESQ in voice quality assessment? How does PESQ work?

Problem 1.12 Why do waveform codecs achieve higher PESQ/POLQA scores than LPC ones?

Bibliography

3GPP. (2008a). TS 26.071: Mandatory speech codec speech processing functions; AMR speech codec; general description. Technical Report TS 26.071, 3rd Generation Partnership Project.

3GPP. (2008b). TS 26.190: Speech codec speech processing functions; adaptive multirate – wideband (AMR-WB) speech codec; transcoding functions. Technical Report TS 26.190, 3rd Generation Partnership Project.

3GPP. (2018). TS 126.442: Mandatory speech codec speech processing functions; AMR speech codec; general description. Technical Report TS 126.442, 3rd Generation Partnership Project.

3GPP2. (2004). C.s0014-a: Enhanced variable rate codec, speech service option 3 for wideband spread spectrum digital systems. Technical Report C.S0014-A, 3rd Generation Partnership Project 2.

Alliance for Open Media. (2021). Av1 features. Available from: https://aomedia.org/av1-features/. (Accessed March 15th, 2024)

Benott, H. (2008). *Digital television*. 3rd edition. Boston: Focal Press, pp. vii–viii. ISBN 978-0-240-52081-0. DOI: https://doi.org/10.1016/B978-0-240-52081-0.50002-X. Available from: https://www.sciencedirect.com/science/article/pii/B978024052081050002X. (Accessed March 15th, 2024)

Bluetooth. (2018). An introduction to Bluetooth LE Audio. Available from: https://www.bluetooth.com/blog/a-technical-overview-of-lc3/. (Accessed March 15th, 2024)

Choi, K., Chen, J., Rusanovskyy, D., Choi, K.P., and Jang, E.S. (2020). An overview of the MPEG-5 essential video coding standard [standards in a nutshell]. *IEEE Signal Processing Magazine*, 37(3), pp. 160–167. DOI:10.1109/MSP.2020.2971765.

Chu, W.C. (2003). *Speech coding algorithms: foundation and evolution of standardized coders*. John Wiley & Sons, Ltd. ISBN 978-0-471-66885-5.

ETSI. (1992). GSM full rate speech transcoding. ETSI.

ETSI. (1999). Enhanced full rate (EFR) speech transcoding. ETSI.

FIPS. (1984). FIPS Pub 137, analog to digital conversion of voice by 2,400 bit/second linear predictive coding. FIPS.

Google. (2023). Lyra (codec). Available from: https://github.com/google/lyra. (Accessed March 15th, 2024)

Hagen, R., Andersen, S.V., Linden, J., Astrom, H., Duric, A. and Kleijn, W.B. (2004). Internet Low Bit Rate Codec (iLBC). RFC 3951, December. Available from: https://www.rfc-editor.org/info/rfc3951. (Accessed March 15th, 2024)

Haykin, S. (2009). *Communication systems*. Wiley Publishing. 5th ed. ISBN 0-471-69790-7.

Herrero, R. (2021). Fundamentals of IoT communication technologies. *Textbooks in telecommunication engineering*. Springer International Publishing. ISBN 978-3-030-70079-9. Available from: https://books.google.com/books?id=k70rzgEACAAJ. (Accessed March 15th, 2024)

Herrero, R. (2023). Practical internet of things networking. *Textbooks in telecommunication engineering*. Springer International Publishing. ISBN 978-3-031-28443-4.

Herrero, R. and Cadirola, M. (2014). Effect of FEC mechanisms in the performance of low bit rate codecs in lossy mobile environments. In *Proceedings of the Conference on Principles, Systems and Applications of IP Telecommunications, IPTComm '14*, New York, NY, USA. Association for March 2024 Computing Machinery. ISBN 978-1-450-32124-2. DOI: 10.1145/2670386.2670387. Available from: https://doi.org/10.1145/2670386.2670387. (Accessed March 15th, 2024)

ISO/IEC. (2006). ISO/IEC 13818-7: Advanced audio coding (AAC). Technical Report 13818-7.

ITU-T. (1988a). G.711: Pulse code modulation (PCM) of voice frequencies. ITU-T.

ITU-T. (1988b). G.722: 7 kHz audio-coding within 64 kbit/s. ITU-T.

ITU-T. (1990). G.726: 40, 32, 24, 16 kbit/s adaptive differential pulse code modulation (ADPCM). ITU-T.

ITU-T. (1992). G.728: Coding of speech at 16 kbit/s using low-delay code excited linear prediction. ITU-T.

ITU-T. (1993). H.261: Video codec for audiovisual services at p x 64 kbit/s. ITU-T.

ITU-T. (1995). H.262: Information technology – generic coding of moving pictures and associated audio information: Video. ITU-T.

ITU-T. (1996a). G.723.1: Dual rate speech coder for multimedia communications transmitting at 5.3 and 6.3 kbit/s. ITU-T.

ITU-T. (1996b). G.729: Coding of speech at 8 kbit/s using conjugate-structure algebraic-code-excited linear prediction (CS-ACELP). ITU-T.

ITU-T. (2005). H.263: Video coding for low bit rate communication. ITU-T.

ITU-T. (2008). G.718: Frame error robust narrow-band and wideband embedded variable bit-rate coding of speech and audio from 8-32 kbit/s. ITU-T.

ITU-T. (2009). H.264: Advanced video coding for generic audiovisual services. ITU-T.

ITU-T. (2019). H.265: High efficiency video coding. ITU-T.

ITU-T. (2023). H.266: Versatile video coding. ITU-T.

ITU-T Recommendation P.862. (2001). Perceptual evaluation of speech quality (PESQ): an objective method for end-to-end speech quality assessment of narrow-band telephone networks and speech codecs. Technical Report, International Telecommunication Union, Geneva, Switzerland, November.

ITU-T Recommendation P.863. (2018). Technical Report, International Telecommunication Union, Geneva, Switzerland, September.

Jurafsky, D. and Martin, J.H. (2000). *Speech and language processing: an introduction to natural language processing, computational linguistics, and speech recognition.* 1st ed. USA: Prentice Hall PTR. ISBN 0-130-95069-6.

Proakis, J.G. and Manolakis, D.K. (2006). *Digital signal processing.* 4th ed. Upper Saddle River, New Jersey: Prentice Hall. ISBN 0-131-87374-1. Available from: http://www.amazon.com/Digital-Signal-Processing-John-Proakis/dp/0131873741%3FSubscriptionId%3D192BW6DQ43CK9FN0ZGG2%26tag%3Dws%26linkCode%3Dxm2%26camp%3D2025%26creative%3D165953%26creativeASIN%3D0131873741.

Smith, J.O. (2010). *Physical audio signal processing*, Online Book. 2010 ed. W3K Publishing. ISBN 978-0-974-56072-4.

Valin, J-M. (2016). Speex: A free codec for free speech. CoRR, abs/1602.08668. Available from: http://arxiv.org/abs/1602.08668.

Valin, J-M., Vos, K. and Terriberry, T.B. (2012). Definition of the Opus 115 Audio Codec. RFC 6716, September. Available from: https://www.rfc-editor.org/info/rfc6716. (Accessed March 15th, 2024)

Vos, K., Jensen, S.S. and Soerensen, K.V. (2010). SILK Speech Codec. Internet-Draft draft-vos-silk-02, *Internet Engineering Task Force*, September. Available from: https://datatracker.ietf.org/doc/draft-vos-silk/02/. (Accessed March 15th, 2024).

xiph.org. (2006). Theora video compression. Available from: xiph.org/theora/. (Accessed March 15th, 2024)

2

Media Session Signaling

2.1 Introduction

In the context of packet switching, speech, audio, and video streams must be packetized for transmission over networks. This packetization complies with the layered architecture model detailed in Section 1.5.1, but within IP networks, it involves a streamlined five-layer structure, known as the IETF architecture, as opposed to the traditional seven layers. This architecture consolidates the functionalities of the OSI session, presentation, and application layers into a single IETF application layer, while the lower layers remain functionally equivalent across both architectures, ensuring technical compatibility. The cornerstone of the IETF stack is the network layer protocol, IP, which serves as the foundation for RTC stacks and technologies, as well as this book, given its focus on RTC (Kurose and Ross 2012; Herrero 2021).

This chapter presents the generic physical, link, network, and transport layers that underpin most RTC solutions, serving as foundational building blocks for the application layer protocols that enable media session signaling and facilitate the establishment of RTC sessions. Two key application layer protocols are presented: (1) the *Session Description Protocol* (SDP), utilized by endpoints to negotiate media capabilities within sessions, and (2) the *Session Initiation Protocol* (SIP), which leverages SDP to set up, maintain, and tear down the actual media sessions. Note that even RTC technologies that employ alternatives to SIP for session setup (like WebRTC) still utilize SDP for media capability negotiation (Ahson and Ilyas 2008).

2.2 Lower Layers

Focusing on the application layer as the typical encapsulation point for standard RTC protocols, the remaining layers will be explored, collectively referred to as

Hands-On Real Time Communications: A Practical Guide to RTC Protocols in Non-3GPP, 3GPP, 4G/5G/6G and IoT Networks, First Edition. Rolando Herrero.

Companion website: www.wiley.com/go/herrero/RTCProtocols

lower layers. These include the physical, link, network, and transport layers, each detailed in the following subsections. It is important to note that the technologies associated with the physical and link layers are heavily dependent on the underlying topologies. Since this book addresses three primary topologies, 3GPP, non-3GPP, and IoT, different physical and link layer protocols are typically employed in each case, as discussed in Chapters 5, 6, and 7. It is worth mentioning that the network and transport layers remain common across all physical and link layer technologies.

2.2.1 Physical and Link Layers

The physical layer is intended to support the transmission of frames over a physical medium. This transmission typically involves a modulation scheme that maps a sequence of frame *binary digits* (bits) into symbols that can be propagated as electromagnetic signals over wireline or wireless channels. This mapping is bidirectional, as incoming symbols are also mapped back into bits during demodulation.

The number of symbols transmitted per unit of time represents the signal bandwidth, measured in Hz. The channel bandwidth must comply with the signal bandwidth to enable transmission. The rate of bits input to the physical layer is measured in bps and is known as the transmission rate or network bandwidth. Figure 2.1 illustrates this relationship, where the modulator (M box) converts the incoming digital bit stream into an analog signal, and the demodulator (DM box) converts it back into a digital bit stream.

The rate at which the bit stream arrives at the receiver is called throughput, also measured in bps. Due to channel imperfections, not all transmitted bits reach their destination, leading to frame loss and lower throughput than the actual transmission rate. Throughput is a key indicator of communication channel reliability. Goodput, a related metric, measures the rate of real application data bits arriving at the destination, excluding protocol header bits. It reflects the actual performance experienced by users.

The physical layer transmits data through two routes: guided propagation, confined within human-made channels like waveguides, coaxial cables, and optical fibers (backbone of traditional wireline communication), and free propagation,

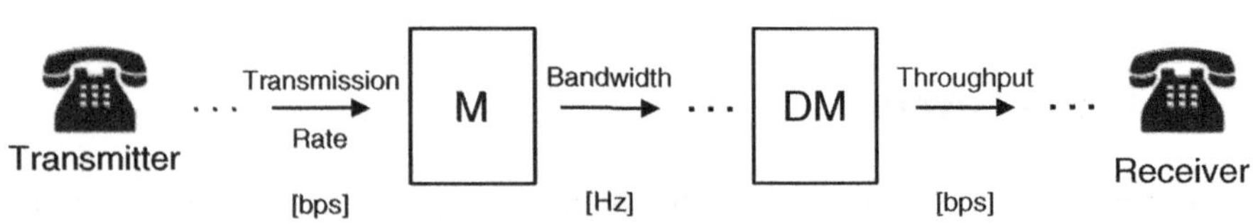

Figure 2.1 Transmission Rate and Bandwidth and Throughput.

carried by electromagnetic waves between antennas across open spaces like the atmosphere and underwater environments (enabling wireless wonders like broadcasting).

In wireline communications, line codes play an important role in frame modulation. Their effectiveness is driven by two key characteristics: minimizing the DC component of the signal to extend its propagation range and reduce repeater needs, and maximizing signal transitions to enhance synchronization between transmitter and receiver. Line codes are baseband signals because their spectral representation includes DC and low-frequency components. This makes them suitable for wireline communications, but they get filtered out when transmitted over passband channels like radio and satellite links. Continuous-wave sinusoidal carriers, on the other hand, are compatible with passband channels, making them the preferred mechanism for digital transmission in wireless environments.

The characteristics of the passband channel, specifically its lowest and highest frequencies, determine the sinusoidal carrier frequency to use in a digital passband modulation scheme. Depending on which carrier parameter is used to carry digital information, modulation can be either *Frequency Shift Keying* (FSK), which changes the carrier frequency, or *Phase Shift Keying* (PSK), which changes the carrier phase (Haykin 2009).

In essence, FSK and PSK establish a one-to-one mapping between the binary digits 0 and 1 and the sinusoidal waveforms illustrated in Figure 2.2.

In PSK, sinusoidal waves are separated by distinct phase shifts to represent digital data. The most basic form, employing a 180° phase separation, is known as *Binary PSK* (BPSK) or 2-PSK. To increase data transmission density, PSK can be expanded to incorporate multiple bits per symbol. *Quadrature PSK* (QPSK), or 4-PSK, achieves this by mapping 2-bit digits onto sinusoidal waves with 90° phase separations (Figure 2.3).

However, QPSK encounters a challenge when transitioning from the symbol 00 to 11, as it requires a 180° phase shift. This abrupt shift can lead to bandwidth

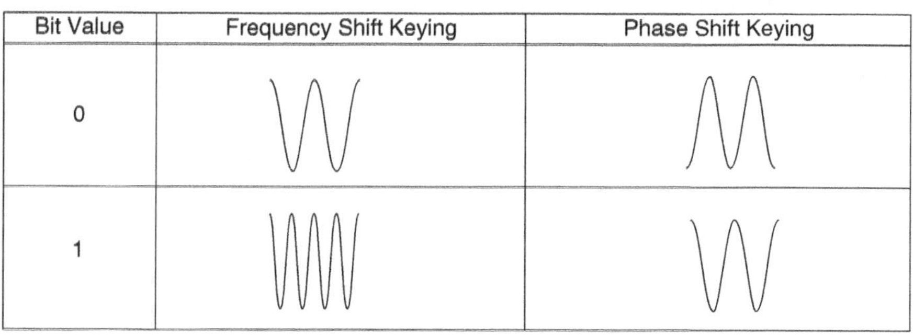

Figure 2.2 FSK and PSK Modulations.

Bit Values	Quadrature Phase Shift Keying
00	
01	
10	
11	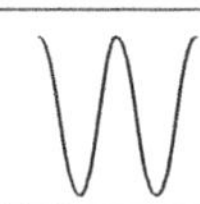

Figure 2.3 Quadrature Phase Shift Keying.

widening and heightened noise. *Offset QPSK* (OQPSK) mitigates this issue by splitting the 180° phase transition into two 90° stages. Another PSK variant, *Differential PSK* (DPSK), bases symbol detection on relative phase differences between consecutive symbols rather than absolute phase values. This enhances receiver performance and is particularly beneficial in scenarios with uncertain phase references. *Differential BPSK* (DBPSK) is a type of DPSK that specifically operates on binary PSK signals. *Quadrature Amplitude Modulation* (QAM) takes digital data transmission a step further by carefully manipulating both the phase and amplitude of the carrier wave. It expands the symbol set, enabling the encoding of multiple bits within a single symbol. For instance, 8-QAM employs eight distinct signals, each representing 3-bit digits.

Example 2.1 Consider a wireless scheme that relies on a QPSK modulation with a symbol duration that is $T = 100$ μs, what are the symbol R_s and transmission R_b rates?

Solution:
Because of QPSK, $M = 4$, the symbol rate is $R_s = \frac{1}{T} = 10000$ symbols per second. The sampling rate is $R_b = R_s \times \log_2 M = 20$ Kbps.

In general, the relationship between the number of symbols (M) and the number of bits per symbol (n) is expressed as $n = \log_2 (M)$. This principle underpins modulation schemes like M-QAM, M-PSK, and M-FSK. However, increasing the number of symbols presents a trade-off. While it elevates the transmission rate, it also heightens the susceptibility to errors during demodulation. This is because a larger symbol set, when transmitted with fixed energy, becomes more vulnerable to channel noise, potentially leading to symbol misinterpretation at the receiver. This interplay between transmission rate and distortion is a fundamental concept in communication systems, known as the *rate–distortion* trade-off. It focuses on the inherent challenge of achieving both high transmission rates and low distortion levels.

To further enhance data transmission capacity, multiple carriers can be used, each independently modulated, to work in unison. This concept, known as multicarrier modulation, involves dividing the available channel bandwidth (W) into a multitude of subchannels, each with a bandwidth of Δf. As shown in Figure 2.4, this division creates approximately $K = \frac{W}{\Delta f}$ subchannels, ready to be individually modulated. When the carrier waves are orthogonal, that is

$$\int_0^T S_i(t)S_j(t)dt = 0 \tag{2.1}$$

where S_n is one M of the symbols associated with the modulation, T is the symbol length, and $i \neq j$, the scheme is called *Orthogonal Frequency Division Multiplexing* (OFDM).

Orthogonality, a key concept in multicarrier modulation, plays an important role in safeguarding data against the detrimental effects of channel noise and multipath interference. By carefully selecting the frequency difference (Δf) between symbol sinusoidal waves such that $T = \frac{1}{\Delta f}$, it can be seen that the carrier waves remain orthogonal, regardless of their phase values. This orthogonality grants two significant advantages: (1) orthogonal carriers exhibit a remarkable ability to withstand packet loss due to channel impairments, ensuring reliable data transmission and (2) orthogonality suppresses interference between subchannels, promoting clean signal separation and maximizing overall system performance. However, orthogonality also introduces a constraint on the symbol period, which in turn limits the transmission rate within each subchannel. To optimize performance,

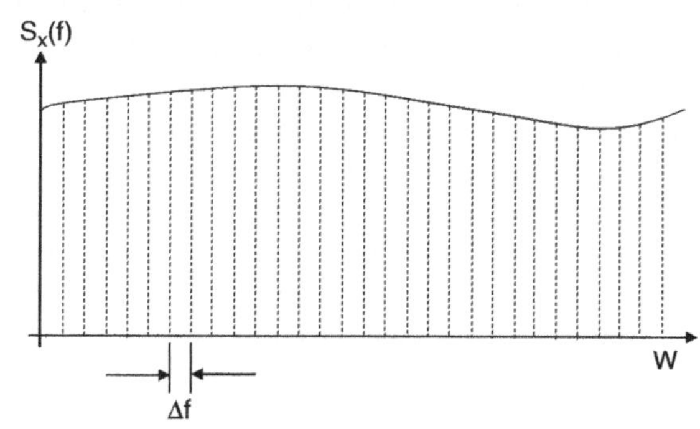

Figure 2.4 Multicarrier Modulation.

different modulation schemes can be tailored to individual subchannels based on their specific SNR characteristics. For instance, consider a scenario with two carriers ($K = 2$). One carrier could employ QPSK modulation, while the other utilizes 64-QAM. This allocation leads to an aggregate transmission rate of $R = \frac{2+6}{T} = \frac{8}{T}$ bps, combining the strengths of both modulation schemes.

While the physical layer handles the conversion of frames into analog signals and vice versa, the task of frame construction falls to the link layer. But this layer's responsibilities extend beyond frame assembly. It serves as the foundation for establishing basic connectivity between adjacent endpoints, enabling them to communicate across a single link. To achieve this goal, the link layer employs a suite of mechanisms:

- Framing: This technique involves adding structured fields and synchronization markers to data received from upper layers. In the context of the link layer, framing specifically refers to the process of attaching headers and delimiters to network layer datagrams, transforming them into frames.
- Channel Coding (Optional): This mechanism seeks to enhance resilience against transmission errors by embedding *Forward Error Correction* (FEC) information within a header. This FEC data enables the receiver to detect and, in some cases, correct errors caused by channel noise, attenuation, or other interference. Traditional FEC methods utilize block and convolutional codes that map frame fields into error-detecting or error-correcting codewords.

Framing and channel coding collectively form the *Link Layer Control* (LLC) sublayer, an important component of the link layer architecture. Beyond framing and channel coding, the link layer also employs two other mechanisms to ensure reliable communication: (1) link access that governs how multiple devices share a common communication link, preventing collisions and ensuring orderly data transmission and reliable delivery that guarantees that frames are successfully delivered to their intended destinations, even in the presence of transmission errors or link disruptions.

In the realm of multi-device communication, order and fairness reign supreme, thanks to a set of regulations known as *Media Access Control* (MAC). This mechanism governs how frames navigate a shared channel, ensuring harmonious transmission and reception. It addresses two key challenges: (1) the link access and (2) the reliable delivery. Link access is about orchestrating shared channels when multiple endpoints compete for access to the same channel and MAC prevents collisions and promotes efficient transmission. To establish clear communication pathways, MAC employs unique MAC addresses to identify source and destination endpoints, as well as any necessary intermediate waypoints. MAC is based on the principle of equitable access, ensuring that all devices have a fair chance to transmit based on their priorities and the

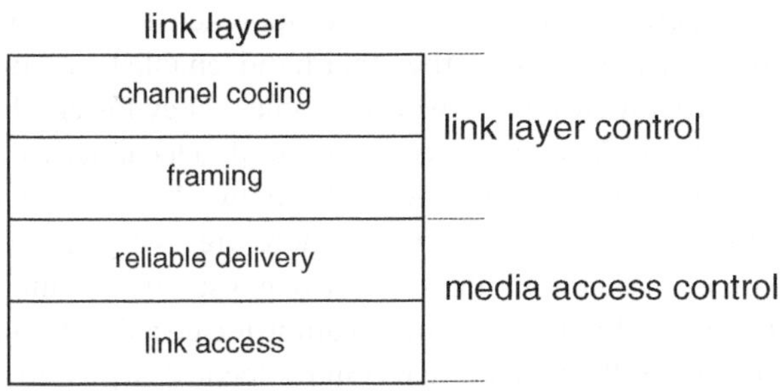

Figure 2.5 Layers of Link Layer.

network's QoS objectives. It balances access times to prevent any single device from dominating the channel. Reliable delivery, on the other hand, supports error resilience by enabling channel coding that helps receivers to detect and discard corrupted frames. To guarantee reliable delivery, MAC offers an optional mechanism that enables retransmission of lost or damaged frames. Receivers provide feedback to senders, typically through acknowledgments (ACKs) or negative acknowledgments (NAKs), indicating the successful or unsuccessful reception of frames. Retransmissions are further safeguarded by timers that trigger retransmission attempts or signal connectivity loss if acknowledgments fail to arrive within a specified timeframe. Together, these two mechanisms, link access and reliable delivery, form the MAC sublayer, operating alongside the LLC sublayer to ensure robust and efficient communication at the link layer. Figure 2.5 illustrates this architecture.

2.2.2 Network Layer

IP reigns as IETF's sole network layer protocol, enabling global end-to-end communication. The network layer acts as a delivery service for datagrams, ensuring their transmission across networks. It encapsulates segments from the transport layer, adds destination addresses, and supports routing and forwarding. Specifically, the network layer wraps segments in datagrams, adding addresses for routing. Routing, in turn, enables the calculation of optimal paths for datagrams, considering network conditions. At each router, the network layer directs incoming datagrams to appropriate output links. Upon arrival, the receiver extracts the original segment for delivery to the transport layer.

Forwarding requires that each host and router possess a forwarding table, which maps input links to output links based on addressing fields within the datagram header. These forwarding tables are populated, either directly or indirectly, by routing algorithms that can be categorized as centralized or distributed. Centralized routing algorithms employ a central entity to construct forwarding tables for the entire network. They distribute these tables to hosts and routers, ensuring a unified approach to datagram forwarding. Distributed routing algorithms, on the other hand, enable hosts and routers to independently construct their own forwarding tables. They foster adaptability and resilience by enabling individual devices to respond to local network conditions.

From the point of view of the network layer's pivotal role in encapsulating and decapsulating datagrams, there are two primary IP versions. The *Internet Protocol Version 4* (IPv4) is a widely established and familiar protocol that has supported Internet communication for decades. Similarly, the *Internet Protocol Version 6* (IPv6) is a more recent protocol, designed to address a critical limitation of IPv4—its finite address space. IPv6 expands the address space exponentially, ensuring ample room for the ever-growing number of devices and services connecting to the Internet.

2.2.2.1 IPv4

IPv4 datagrams encapsulate segments that are either passed down to the link layer or passed up to the transport layer. Specifically, an IPv4 datagram comprises an IPv4 header and a payload containing the transport layer segment. Figure 2.6 illustrates an IPv4 datagram, showcasing its header and payload (RFC 1981a). The header includes the following fields:

- Version (4 bits): Identifies the IP version number, always carrying a value of 4.
- Header Length (4 bits): Specifies the total size of the IPv4 header, including variable-length options.
- Type of Service (ToS) (8 bits): Differentiates datagrams based on Quality of Service (QoS) requirements.
- Datagram Length (16 bits): Indicates the total length of the packet, including the header.

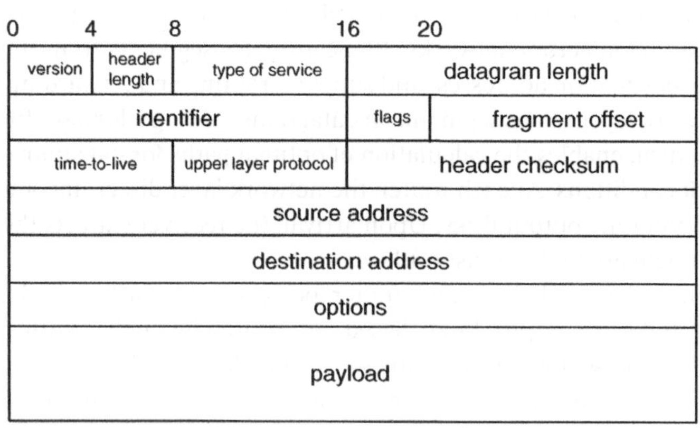

Figure 2.6 IPv4 Header and Payload.

- Identifier (16 bits): Tags the packet as a fragment of a larger datagram.
- Flags (4 bits): Signals the nature of the fragment within the larger datagram.
- Fragment Offset (16 bits): Specifies the position of the fragment within the larger datagram.
- *Time-to-Live* (TTL) (8 bits): Limits the number of times the datagram can be forwarded by decrementing with each retransmission.
- Upper Layer Protocol (8 bits): Indicates the protocol used to encode the payload.
- Header Checksum (16 bits): Ensures the integrity of the IPv4 header, guaranteeing successful transmission.
- Source Address (32 bits): Specifies the origin of the datagram.
- Destination Address (32 bits): Indicates the intended recipient of the datagram.
- Options (variable length): Provides additional information agreed upon between endpoints, such as timestamps. These options are rarely used in order to minimize the datagram size.

Fragmentation serves as a key mechanism within the network layer, ensuring that datagrams can successfully traverse a network from end to end. It becomes necessary when datagrams exceed the maximum size that the physical layer can accommodate. For instance, Ethernet, as a link layer technology, has a frame size limit of 1518 bytes, resulting in a maximum payload size of 1500 bytes. This payload size restriction translates to the *Maximum Transmission Unit* (MTU), which caps the size of datagrams at 1500 bytes. Datagrams that surpass this threshold must undergo fragmentation before transmission.

To illustrate fragmentation, let us examine Figure 2.7, where a sizable segment is partitioned into three distinct chunks. Each chunk then becomes the payload of its own IPv4 segment, ensuring compliance with the MTU size constraints of the underlying link layer technologies. To facilitate reassembly, each fragment bears a shared 16-bit identifier. Additionally, a 4-bit flags field signals whether a given fragment is followed by others. Within a single datagram, all fragments except the final one have the *More Fragments* (MF) bit set within the flags field.

It is worth noting that fragmentation, while essential, is generally considered undesirable due to its potential to amplify network packet loss. For a datagram to be flawlessly reassembled, every fragment must successfully reach its destination.

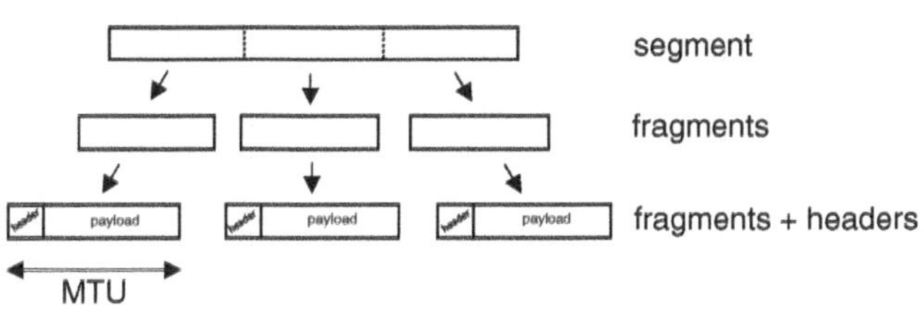

Figure 2.7 IPv4 Fragmentation.

If even a single fragment goes astray, the entire datagram becomes unusable. This is about a fundamental principle in network design: fragmentation, while sometimes necessary, should be avoided whenever possible. And when it becomes unavoidable, the number of fragments per datagram should be minimized.

Within the IPv4 architecture, source and destination addresses, as shown in Figure 2.6, each comprise 32 bits. This design leads to a theoretical capacity of 2^{32} distinct IPv4 addresses, enabling the identification of a vast array of devices within a network. To represent these addresses in a human-readable format, the dotted-decimal notation is employed. It involves the segmentation of each 32-bit address into four 8-bit bytes, with each byte expressed as a decimal number ranging from 0 to 255. These numerical values are then concatenated using dots as delimiters.

Figure 2.8 shows an IPv4 network consisting of three hosts across two distinct subnets. These subnets, designated as *192.168.10.x* and *192.168.20.x*, employ the variable *x* in the final byte of the address space to accommodate a range of hosts from 1 to 255. Within each subnet, hosts are deployed through a shared network segment that interfaces with a router, enabling communication beyond their immediate domain.

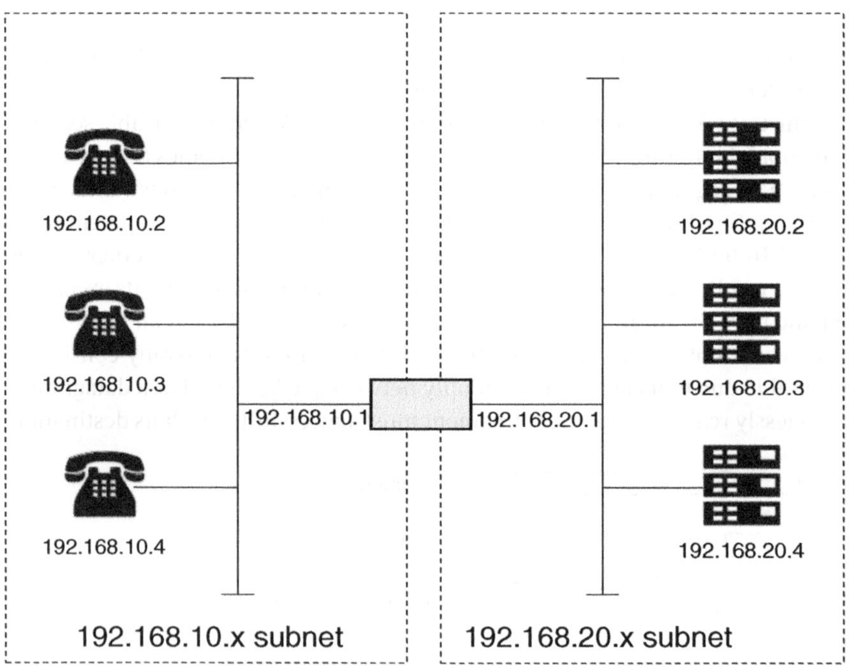

Figure 2.8 Example IPv4 Network.

To optimize address allocation, a strategy known as *Classless Interdomain Routing* (CIDR) is employed. It involves dividing the 32-bit address space into two components: subnet and host identifiers. In this example, three bytes are devoted to identifying the subnet and one byte is left to identify the host. Note that this allocation is concisely expressed in the CIDR notation as *192.168.10.0/24* and *192.168.20.0/24*, where the */24* designates the prefix length, signifying the number of bits dedicated to subnet identification (Kurose and Ross 2012; Herrero 2023).

Classless interdomain routing (CIDR) enables the segmentation of networks into smaller, more efficient subnets. Consider the example of an entity with control over the *192.168.0.0/16* network. By employing CIDR, the 16-bit host space can be subdivided into two distinct components: 2 bits dedicated to subnet identification and 14 bits reserved for host addresses. This allocation yields a collection of $2^2 = 4$ subnets, each capable of accommodating up to $2^{14} = 16384$ hosts within its realm. As indicated in Table 2.1, these subnets are identified by *192.168.0.0/18*, *192.168.64.0/18*, *192.168.128.0/18*, and *192.168.192.0/18*.

Within the realm of IPv4, addresses can be assigned through two primary methods: manual configuration or the dynamic of the *Dynamic Host Configuration Protocol* (DHCP). DHCP uses an exchange of four distinct messages to gracefully automate address allocation: (1) DHCP Server Discovery where the host initiates a broadcast message to discover a DHCP server, (2) DHCP Server Offer where a DHCP server responds with an offering. This communication includes a proposed IPv4 address, a network mask defining the prefix length, and a lease duration, outlining the terms of the address's occupancy, (3) DHCP Request where the host, upon contemplating the server's offer, formally requests the address, signaling its acceptance of the proposed terms, and (4) DHCP Acknowledgment where the server confirms the host's request by means of the address assignment. Note that these DHCP messages are conveyed through the *Bootstrap Protocol* (BOOTP), which in turn relies on the transport services of UDP.

Table 2.1 Subnetting.

Topology	Address Space (binary)	Address Space
Network	192.168.00000000b.00000000b	192.168.0.0/16
Subnet 1	192.168.**00**000000b.00000000b	192.168.0.0/18
Subnet 2	192.168.**01**000000b.00000000b	192.168.64.0/18
Subnet 3	192.168.**10**000000b.00000000b	192.168.128.0/18
Subnet 4	192.168.**11**000000b.00000000b	192.168.192.0/18

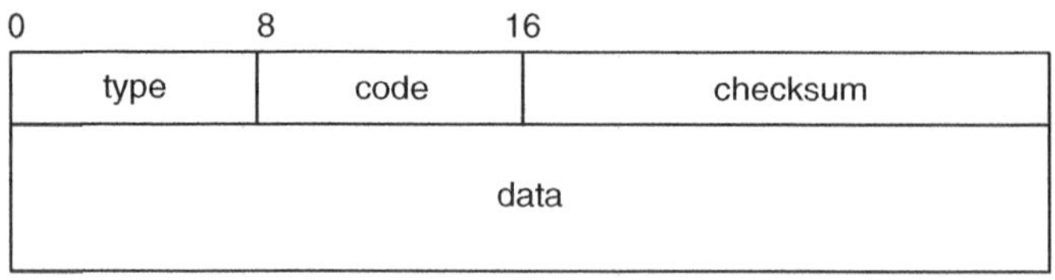

Figure 2.9 ICMP Packet Header and Data.

In the context of IPv4 operations, the *Internet Control Message Protocol* (ICMP) provides essential support services. ICMP transactions support a request-response model, with interactions based on the packet type. A notable example of ICMP is the ping command, a key networking tool for validating connectivity with remote hosts.

Example 2.2 Given a *192.168.10.0/24* network that is divided into 64-host subnets, what are those subnets?

Solution:
Out of the addresses, $\log_2 64 = 6$ bits are used to allocate the hosts, that means that $32 - 6 = 26$ bits are used to represent the networks. This means that subnets are *192.168.10.128/26*, *192.168.10.192/26*, *192.168.10.64/26*, *192 .168.10.0/26*.

The 32-bit structure of IPv4 addresses imposes a finite address space, capable of accommodating approximately $2^{32} \approx 4300 \times 10^6$ unique hosts. This constraint has long since been surpassed, requiring a widespread reliance on *Network Address Translation* (NAT) to extend the reach of IPv4 networks. The main goal is to implement a system that translates combinations of multiple private IPv4 addresses and ports into a single public IPv4 address and its ports. This allows a few public *Wide Area Network* (WAN) IPv4 addresses to map to a virtually unlimited number of private or internal addresses within a *Local Area Network* (LAN). The key factor is utilizing 16-bit transport ports for diverse addressing. However, employing NAT contradicts the fundamental principle of the layered network architecture, which relies on both network and transport layer information for routing. In other words, in a direct connection between two hosts, intermediary routers require transport layer information to route packets. Since routing typically ignores transport layer data, this method is inefficient and deviates considerably from standard behavior. Evidently, using NAT to augment the IPv4 address space is not recommended. A superior solution is to directly increase the size of IP addresses. IPv6 accomplishes this by employing 128-bit addresses instead of 32-bit addresses in IPv4.

2.2.2.2 IPv6

The most significant change introduced by IPv6 is the support for 128-bit addresses, which exponentially increase the potential number of addressable devices. IPv6 address notation is based on eight 16-bit values, each represented by a hexadecimal number with leading zeros removed, and separated by colons. An IPv6 address can be represented in a shortened form by replacing any sequence of consecutive 16-bit values consisting entirely of zeros with a double colon (::). For example, the long form *2001:0:0:0:0:00:41:10* can be represented as the short form *2001::41:10*.

Tables 2.2 and 2.3 display this, as well as other relevant IPv6 address types, alongside their corresponding IPv4 examples. Each IPv6 address possesses a prefix that serves the purpose of network identification. Specifically, the prefix denotes the number of leading bits that establish the network's identity. For instance, if the address is *2001::10:21:10/64*, it signifies that the initial 64 bits, represented as *2001::*, are responsible for identifying the network (Hinden and Deering 1998).

The initial few bits of an IPv6 address consistently reveal its format. Figure 2.10 illustrates the structure of unicast addresses. They start with the binary sequence

Table 2.2 Long vs Short IPv6 Addresses.

Long Form	Short Form	Type
2001:0:0:0:0:0:41:10	2001::41:10	unicast
FF01:0:0:0:0:0:2105:2105	FF01::2105:2105	multicast
FE80:0:0:0:0:3F44:1214:FFCA	FE80::3F44:1214:FFCA	link-local
0:0:0:0:0:0:0:1	::1	loopback
0:0:0:0:0:0:0:0	::	unspecified

Table 2.3 IPv6 vs IPv4 Addresses.

Long Form	IPv4 Equivalent Example
2001:0:0:0:0:0:41:10	192.168.41.10
FF01:0:0:0:0:0:2105:2105	224.0.21.5
FE80:0:0:0:AA31:3F44:1214:FFCA	169.254.10.25
0:0:0:0:0:0:0:1	127.0.0.1
0:0:0:0:0:0:0:0	0.0.0.0

3	45	16	64	bit count
001	global routing prefix	subnet id	interface id	

Figure 2.10 IPv6 Unicast Address Structure.

001, translating to prefixes *2000::/3* and *3000::/3*. This is succeeded by 45-bit and 16-bit fields designated as the global routing prefix and subnet identification, respectively. The global routing prefix is assigned to a site, while the subnet identification is assigned to a subnet within that site. The subsequent 64-bit *Interface Identification* (IID) serves the purpose of identifying the specific device within the subnet. Due to the ability of devices supporting unicast IPv6 addresses to engage in the global IPv6 infrastructure, unicast addresses are referred to as *Global Unicast Addresses* (GUAs). Certain unicast addresses are known as *Unique Local Unicast Addresses* (ULAs) when their routing is confined to a restricted area, such as a site, and they are not intended to be routable within the global Internet. Other types of addresses, include multicast addresses. These addresses are designed to identify a group of devices, with the capability for a single device to associate with multiple groups. The initial 8 bits of a multicast address consist entirely of ones, followed by a flag field and a scope field. The address concludes with a 112-bit group identification field. On the other hand, link-local addresses serve the purpose of bootstrapping and facilitating connectivity within a single link. They start with the 10-bit binary sequence *1111111010*, succeeded by 54 zeros and the 64-bit IID. Consequently, the prefix of a link-local address invariably remains FE80::/10 (Kurose and Ross 2012; Herrero 2023).

In addition to the modified address size, IPv6 introduces a distinct header format, as illustrated in Figure 2.11. It consists of numerous fields, including:

- Version (4 bits): This field designates the IP version number and invariably holds a value of 6.
- Traffic Class (8 bits): This field conveys QoS information.
- Flow Label (20 bits): This field identifies the datagram as a component of a flow of datagrams.
- Payload Length (16 bits): This field indicates the quantity of payload bytes that succeed the header.
- Next Header (8 bits): This field specifies the protocol under which the payload is encoded.
- Hop Limit, also known as the TTL (8 bits): This field functions as a counter that undergoes decrementation as the datagram is forwarded through the network.
- The previously mentioned 128-bit source and destination addresses.

As with IPv4, it is important to note that when the IP TTL reaches zero, the network discards the datagram. Also note that the payload that trails the IPv6

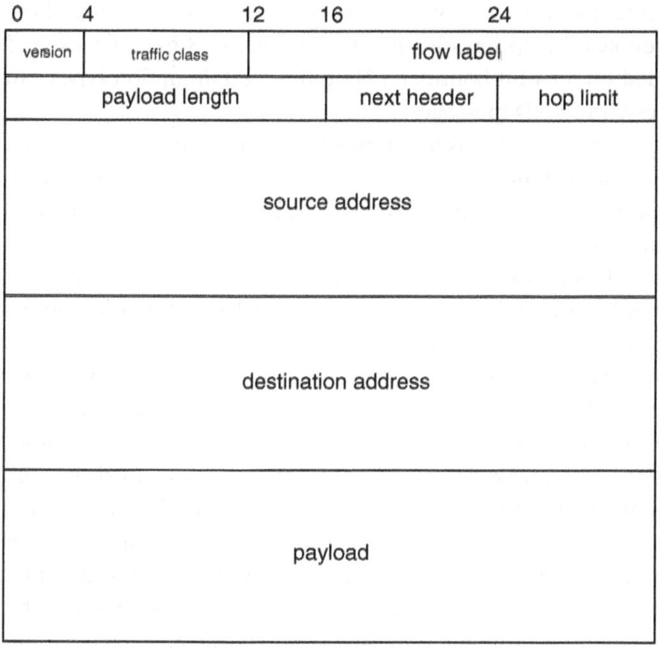

Figure 2.11 IPv6 Header and Payload.

header undergoes processing by the destination device. The IPv6 header excludes numerous fields found within IPv4 headers. Fragmentation/reassembly fields are absent due to IPv6's lack of support for fragmentation and reassembly at intermediate routers. These processes can only be executed at the source and/or destination through the utilization of a specialized header. A header checksum field is also omitted, as it is assumed that error control is performed by either the link-layer or transport layers. Additionally, the IPv6 header lacks an *Options* field, as it is presumed that options can be indicated through the employment of supplementary headers.

IPv6 inherits ICMP from IPv4, a protocol employed by devices and routers to exchange network layer information with one another. ICMPv6, as it is referred to in IPv6, trails the IPv6 header and follows the general format shown in Figure 2.9 and comprises the following fields: (1) an 8-bit type field that dictates the format and interpretation of the message, (2) an 8-bit code field that can be leveraged to support diverse subtypes, and (3) a 16-bit checksum field that provides a basic level of error control.

ICMPv6 plays an important role in facilitating interaction between IPv6 devices through the *Network Discovery* (ND) protocol. ND messages are encoded as ICMPv6, with certain messages incorporating a sequence of *Type-Length-Value*

(TLV) options transmitted within the payload. The type specifies the nature of the option, the length denotes the option's length in multiples of 8-byte blocks, and the values are encoded on a 64-bit boundary. Note that certain option types can appear multiple times within ND messages.

One of the primary functions of ND is to establish communication with routers, enabling connectivity that extends beyond the local link. In fact, on any given link, routers engage in the periodic transmission of multicast (to address FF02::1) *Router Advertisement* (RA) ND messages. These messages serve the purpose of provisioning devices with network parameters such as prefixes and hop limit information. Devices depend on this information to construct a list of potential routers that can be employed to reach other networks.

ICMPv6 RA ND messages contain several fields, including: an 8-bit hop limit field for transmitters on the link, a managed bit that dictates whether addresses are obtained via DHCPv6 or *Stateless Address Autoconfiguration* (SAA), an other bit indicating whether additional configuration information is obtained through DHCPv6, a home agent bit for mobility support, a 2-bit preference field advertising the router's priority relative to others, an 8-bit router lifetime informing devices of the router's availability, a 16-bit reachable timer field specifying the duration for considering a neighbor reachable after reachability confirmation, and a 16-bit retransmission timer field indicating the wait time before retransmitting neighbor solicitation messages. Options, encoded as TLVs, can provide information such as the router's link layer address, network MTU size, and/or prefix information.

The prefix TLV incorporates the following fields: an 8-bit prefix length, an on-link determination bit signifying whether all addresses within the prefix are considered link local, an SAA support bit designating the prefix for use in SAA, a 32-bit valid lifetime field indicating the prefix's validity period for existing connections, a 32-bit preferred lifetime field specifying the prefix's validity period for new connections, and a 128-bit prefix field, padded with zeros to align with the specified length.

In the absence of receiving any RA ND messages, a newly deployed device has the capability to initiate a *Router Solicitation* (RS) ND message to the multicast address (*FF02::2*). This serves the purpose of prompting routers to transmit RAs. RS ND messages are essentially empty ICMPv6 messages, devoid of any additional payload.

The ND protocol additionally addresses device connectivity through the employment of *Neighbor Solicitation* (NS) ND messages, primarily functioning as a replacement for IPv4's *Address Resolution Protocol* (ARP). Specifically, ICMPv6 NS ND messages enable devices to determine the link layer addresses of other devices on the same link. This, in turn, enables devices to ascertain the reachability of other devices on the link by maintaining a *Neighbor Unreachability Detection* (NUD) cache. In response to NS ND messages, a device transmits a *Neighbor Advertisement* (NA) ND message, which incorporates its own link layer address. It is important to mention that NA ND messages can also be sent unsolicitedly to signal alterations in link layer addresses. Another type of ND message is the ICMPv6 Redirect ND message, utilized by routers to inform other IPv6 devices on the link regarding the existence of a more optimal next hop device.

Stateless address autoconfiguration (SAA), as mentioned previously, is a distributed mechanism that grants devices the ability to procure a unique unicast IPv6 address, a necessity for the proper routing of datagrams. A device employing SAA dynamically generates its IPv6 address by combining its 64-bit IID with the prefix supplied by the router. The IID is derived from the device's link layer address, making it contingent upon the link layer technology in use.

For Ethernet interfaces, this entails converting the 48-bit MAC address into a modified IEEE 64-bit EUI-64-based IID. This transformation is achieved by inserting the 16-bit binary sequence *1111111111111110* between the third and fourth bytes of the MAC address, followed by inverting a specific bit known as the universal/local bit within the resulting EUI-64 address.

Each address generated through SAA can reside in one of three distinct states: (1) a Tentative State that is the initial state of an address, during which it cannot be utilized as it undergoes validation through the *Duplicate Address Detection* (DAD) mechanism, (2) a Preferred State that once an address has been validated, it transitions into the preferred state, permitting its use for connectivity, provided it is employed within the designated preferred lifetime, and (3) a Deprecated State where an address enters the deprecated state upon the expiration of its lifetime. Addresses in this state are no longer deemed valid and can only be used for pre-existing flows and connections. The deprecated state persists for the duration of the predefined valid lifetime, enabling devices to transition to new addresses, which must also initially enter the tentative state before becoming preferred.

The rationale behind employing a state machine rather than immediately utilizing IPv6 addresses comes from the potential lack of uniqueness in link layer addresses. While most link layer technologies (like Ethernet) commonly offer unique addresses, this guarantee is not absolute. To address this, a device executes DAD during the tentative state, validating its self-generated IPv6 address by multicasting an NS ND message. This message requests the link layer address associated with the address. If any other device on the link is already employing that IPv6 address, it responds with an NA ND message, indirectly signaling to the querying device that the address is unavailable. In such a scenario, the device must acquire a new address through alternative means. However, if the request times out, the address advances to the preferred state.

The latency introduced during the tentative state can be circumvented by using optimistic DAD instead of plain DAD. Under optimistic DAD, devices start using their generated IPv6 addresses immediately, even while validation is ongoing. To minimize traffic flow in this case, only the transmission of datagrams that do not impact the address translation caches of other devices on the link is permitted. It is important to note that SAA depends on dynamic lifetime values provided by routers within RA ND messages. If a router modifies its lifetime estimations, devices reliant on SAA must correspondingly update their address configuration.

2.2.3 Transport Layer

The transport layer facilitates message transfer between applications running on endpoints, employing multiple transport protocols, including the highly efficient *Quick UDP Internet Connections* (QUIC). However, the two most prominent and important protocols are UDP and TCP. Both TCP and UDP utilize 16-bit port numbers to identify source and destination applications, with port numbers below 49152 designated for well-known and registered applications, and those 49152 or above dynamically allocated. UDP serves as a straightforward wrapper for IP datagrams, incorporating source and destination port information, and its segments are susceptible to the same network impairments as IP datagrams, such as network loss and latency. In contrast, TCP functions by establishing connections, ensuring message delivery, and supporting flow control, enabling rate and congestion control.

2.2.3.1 UDP

In contrast to the connection-oriented nature of TCP, UDP is a connectionless protocol. This means that applications on endpoints can start transmission immediately, as there's no requirement for handshaking. UDP also lacks mechanisms to ensure the successful delivery of messages or guarantee that messages are received in the order they were sent. Due to the absence of flow control and guaranteed delivery, endpoint applications can transmit messages at their maximum speed. UDP's connectionless service makes it well-suited for broadcasting and multicasting scenarios where a single datagram is intended to be received by multiple receivers.

UDP's absence of handshaking for connection establishment and message retransmissions typically translates to low latency. Due to its connectionless nature, there is no need to maintain connection state information. Generally, UDP segments either arrive as quickly as possible or not at all. This characteristic makes UDP ideal for certain real-time scenarios where fast traffic delivery to the destination is very important.

Consider, for instance, RTC technologies like VoIP, where media traffic must reach endpoints promptly to minimize end-to-end delay and prevent degradation of the overall user experience. Another example is real-time IoT scenarios

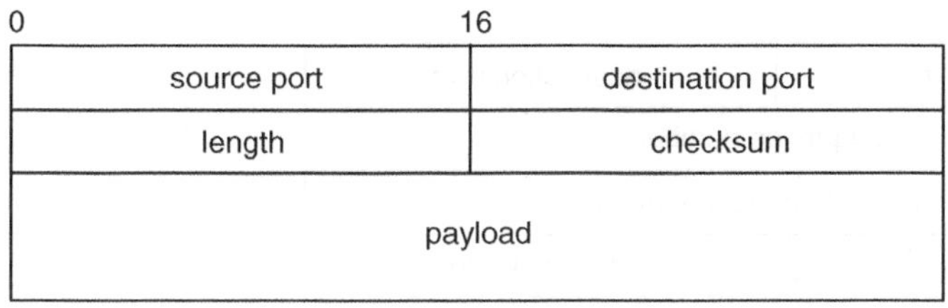

Figure 2.12 UDP Segment Header and Payload.

involving Unmanned Aerial Vehicles (UAVs). In such cases, sensor readouts like position and acceleration are used to establish waypoints that define the UAV's flight path. Any delays in processing these readouts can potentially lead to UAV crashes and physical damage. Within this context, a delayed readout is a lost readout. To mitigate the impact of datagram loss, FEC mechanisms can be employed to introduce controlled redundancy by replicating packet transmissions.

Figure 2.12 shows a UDP segment header (and payload), consisting of a 4-field header and a payload that carries the upper-layer application message. The header includes a 16-bit source port field identifying the transmitting application, a 16-bit destination port field identifying the receiving application, a 16-bit length field indicating the segment size (including both header and payload), and a 16-bit checksum field calculated over the entire segment. It is important to mention that UDP segments are generated and processed exclusively at the endpoints, while intermediate nodes like routers silently forward them. This implies that the checksum is calculated by the transmitter and verified by the receiver, embodying the well-known end-to-end principle and promoting efficiency through the minimization of involvement from non-essential nodes (RFC 1980; Herrero 2023).

2.2.3.2 TCP

Transport control protocol establishes a full-duplex connection-oriented service, distinct from UDP in its point-to-point nature and absence of broadcasting or multiplexing support. It operates within a client/server architecture, where a client initiates a connection with the server. Once established, both endpoints can transmit data streams, which arrive at their destinations uncorrupted and in the correct sequence due to integrated flow and congestion control mechanisms. These mechanisms also avert data duplication. Either endpoint holds the ability to terminate the connection at any point.

Figure 2.13 illustrates a TCP segment that includes a 20-byte header and a payload that carries an application message. The header carries the following fields: (1) a 16-bit source port associated with the sending application, (2) a 16-bit destination port associated with the receiving application, (3) a 32-bit sequence number used to keep track of the transmitted data, (4) a 32-bit acknowledgment

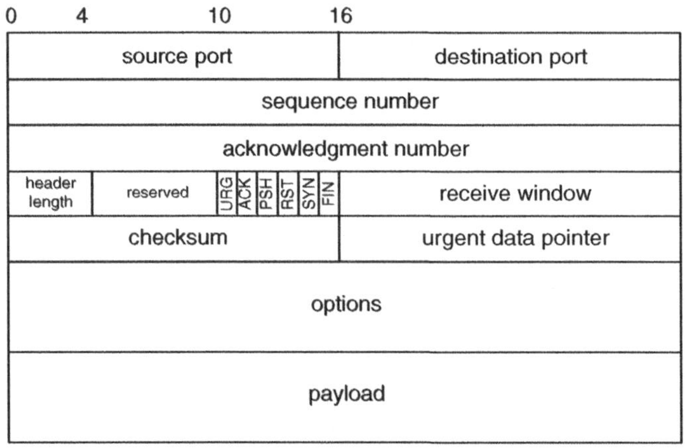

Figure 2.13 TCP Header and Payload.

number used to keep track of the acknowledged data, (5) a 4-bit header length field, (6) a 6-bit flag associated with the state of the connection, (7) a 16-bit receiver window that specifies how much data the receiver is willing to accept, (8) a 16-bit checksum calculated over the whole segment, (9) a 16-bit pointer that signals urgent data within the stream, and (10) a variable length number of options used to support special features.

To ensure efficient transmission and avoid network layer fragmentation, TCP segments messages originating from the application layer into appropriately sized units. These segments, along with their TCP headers, must fit within a single datagram. To achieve this, TCP employs the *Maximum Segment Size* (MSS), which is set to a value equal to or less than the MTU minus the TCP header size.

To set up a connection, a three-way handshake is performed as shown in Figure 2.14. The client initiates by sending an empty segment with the synchronization SYN flag set and a randomly selected sequence number M. The server reciprocates with an empty acknowledgment segment, also bearing the SYN flag set, an acknowledgment number $M+1$, and its own randomly selected sequence number N. The client then acknowledges this exchange with a segment containing sequence number $N+1$, acknowledgment number $M+1$, and, potentially, a payload carrying a segment of an application message.

Figure 2.15 illustrates the graceful termination of a connection through a two-way handshake. Each half-duplex side of the full-duplex connection is dissolved by the transmission of an empty segment with the finalizing FIN flag set, followed by an empty acknowledgment segment in response. To avoid potential packet loss, TCP employs a retransmission mechanism guided by a timeout derived from an estimated *Round Trip Time* (RTT). Single RTT

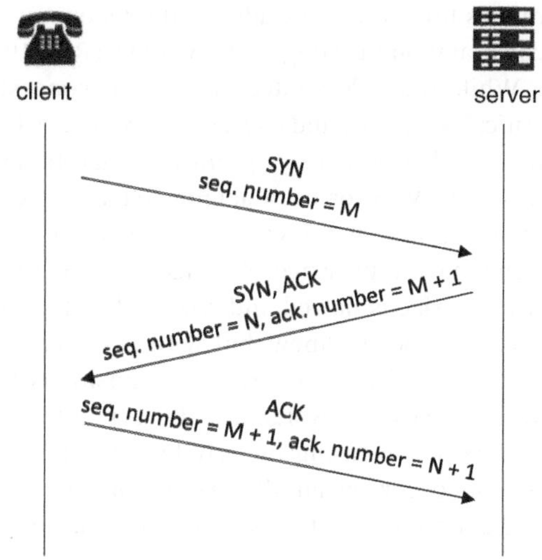

Figure 2.14 Setting up a TCP Connection.

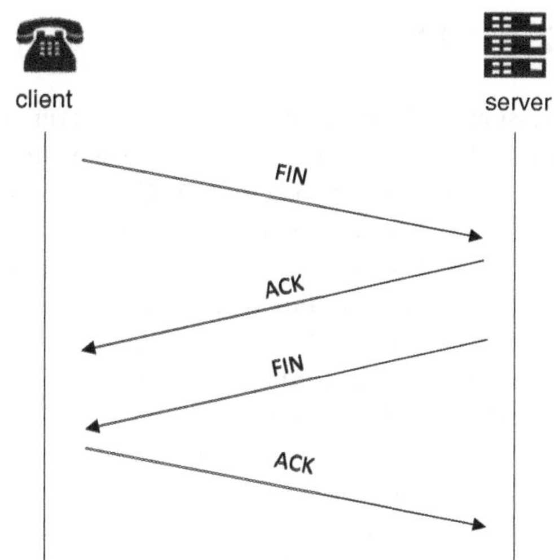

Figure 2.15 Terminating a TCP Connection.

values are calculated as the time interval between segment transmission and acknowledgment reception. These values are then employed to estimate an average RTT through a moving average process. The actual retransmission timeout value is determined by adding four times the standard deviation of the distribution to this estimated average RTT.

TCP flow control enables data transmission through a managed congestion window (cwnd). This window size dictates the number of bytes the sender is permitted to transmit, regulating the transmission rate. The receiver further refines this process by advertising its own window size, imposing an additional constraint on the congestion window (RFC 1981b; Herrero 2023).

While multiple strategies exist for congestion control, two mechanisms are critical: *slow start* and *congestion avoidance*. Under slow start, the transmitter initiates with a cautiously optimistic congestion window size of one MSS, increasing it by one MSS for each acknowledgment received. This exponential growth of the congestion window and transmission rate continues until a timeout occurs, signaling the presence of congestion. At this juncture, the current congestion window size is halved and recorded as the slow start threshold (ssthresh). Slow start then begins anew, but with a subtle shift in strategy once the congestion window reaches the ssthresh—it transitions gracefully into congestion avoidance. With congestion avoidance, the congestion window adopts a more measured approach, increasing linearly by one MSS with each RTT. However, should a timeout occur, the familiar pattern of halving the current congestion window size and entering slow start is once again enacted.

2.3 SIP

Session initiation protocol, a client–server protocol, facilitates communication through requests and responses, each encoded in *American Standard Code for Information Interchange* (ASCII) and adorned with headers and methods. Its initial design favored the agility of UDP transport, a choice to circumvent the latency often associated with TCP connection setup and retransmissions. However, with the introduction of security measures through *Transport Layer Security* (TLS), TCP emerged as a viable alternative for SIP transport. SIP is flexible enough to be the main standard for the establishment of RTC sessions in non-3GPP and 3GPP networks.

Despite this flexibility, however, SIP encounters some challenges within the IoT realm due to its reliance on plain text messages for session establishment, removal, and management. This characteristic can introduce inefficiencies in resource-constrained IoT environments. Furthermore, SIP's deviation from the *Representational State Transfer* (REST) model limits its suitability for seamless IoT device management. REST is an architecture that supports four primitives that are used to interact with the representation of objects. These primitives are associated with *CRUD* functionality where these representations are *Created*, *Read*, *Updated*, and *Deleted*.

Within SIP, endpoints assume distinct roles as *User Agent Clients* (UACs) and *User Agent Servers* (UASs). UACs initiate sessions by generating requests, while UASs respond to these requests. Note that certain topologies allow endpoints to dynamically transition between UAC and UAS functionality, contingent upon whether they initiate or receive a session. To facilitate session establishment, UASs continuously monitor UDP and/or TCP port 5060 (Schooler et al. 2002).

An SIP transaction constitutes a fundamental request/response exchange between a UAC and a UAS. An SIP dialog, in turn, comprises a series of transactions associated with a given session. UASs fulfill diverse roles, including: (1) *Media Servers* that are responsible for generating and delivering audio, speech, and/or video traffic to UACs, (2) *Proxy Servers* that act as intermediaries, relaying SIP messages between clients and other UASs, (3) Redirect Servers that Reroute UAC requests toward alternative UAS destinations, and (4) *Registrars* that maintain registries of UAC locations, enabling efficient routing of incoming sessions.

Deviating from RESTful paradigms, SIP employs a distinct set of request types, or methods, dispatched by UACs to support communication. These methods, unbound by conventional CRUD operations, enable a nuanced approach to session management: (1)INVITE that starts sessions, establishing a foundation for RTC, (2) ACK that enables reliability, ensuring message delivery even amidst the inherent uncertainties of SIP's primary transport protocol, UDP, (3) BYE that is used to terminate sessions, concluding the exchange with finesse, (4) CANCEL that intervenes to rescind pending requests, offering course correction when needed, (5) OPTIONS that unveils UAS capabilities, fostering informed decision-making within the communication landscape, (6) REGISTER that anchors UACs within the SIP realm by enlisting their whereabouts with registrars, ensuring accessibility for incoming sessions, (7) INFO that transcends typical signaling boundaries, enabling out-of-band transmission of supplementary information, including audio tones, for enhanced communication versatility.

Within the realm of SIP, messages rely on ASCII encoding, rendering them readily comprehensible to humans when sniffed by network traffic capture tools. The following request shows an INVITE message sent by a UAC:

```
INVITE sip:userB@callee.com SIP/2.0
Via: SIP/2.0/UDP caller.com;branch=z9hG4bK2105Bg1c2 Max-Forwards: 70
From: <sip:userA@caller.com>; tag=1921051002
To: <sip:userB@callee.com>
Call-ID: 52a19e24b12a2105@caller.com
CSeq: 3012 INVITE
Contact: <sip:userA@caller.com>
Content-Type: application/sdp
Content-Length: 134
```

```
v=0
o=session 5431011 2394312401 IN IP4 userA.caller.com
s=SDP exchange
c=IN IP4 192.168.1.10
t=0 0
m=audio 5200 RTP/AVP 0
a=sendrecv
```

Session initiation protocol requests follow a structured format that resembles that of the *HyperText Transfer Protocol* (HTTP) presented in Section 2.5. The request includes multiple lines that include a request line and multiple header lines. The Request Line declares the method (here, the session-initiating INVITE), the request's *Uniform Resource Identifier* (URI), and the protocol version (typically 2.0). The URI itself unveils its intent through the service identifier **sip:**. The Via header reveals the local address poised to receive the response. It incorporates a *Max-Forwards* field, that prevents routing loops by imposing a maximum hop count on requests. The From and To headers identify the source and destination endpoints, respectively, specified together with unique tags for unambiguous recognition. The Call-ID header, in this instance *52a19e24b12a2105@caller.com*, together with the From and To headers identify the RTC session (known as dialog). The CSeq, here *3012 INVITE*, intertwines an integer with the method. It establishes a sequence for transaction management, initiated with a randomly assigned value and incremented with each subsequent message. This strategy enables endpoints to detect message loss or disorder, crucial when considering unreliable UDP transport. The Contact header provides a direct route to the UAC, by facilitating efficient response delivery. The Content-Type header specifies the nature of the request's content (or body), here specified as *application/sdp*, indicating session characteristics specified by the *Session Description Protocol* (SDP). The Content-Length header indicates the request body's size to enable the successful parsing of the messages at the receiving end.

SIP responses are also encoded as plain ASCII text facilitating their comprehension by both human and machine. The following response, sent by a UAS in response to an INVITE request, is shown as example:

```
SIP/2.0 200 OK
Via: SIP/2.0/UDP callee.com;branch=z9hG4bK110232105;received=192.168.1.20
From: <sip:userA@caller.com>; tag=1921051002
To: <sip:userB@callee.com>; tag=242199212
Call-ID: 52a19e24b12a2105@caller.com
CSeq: 3012 INVITE
```

```
Contact: <sip:userB@callee.com>
Content-Type: application/sdp
Content-Length: 132

v=0
o=session 2841303 91021765 IN IP4 userB.callee.com
s=SDP exchange
c=IN IP4 192.168.1.20
t=0 0
m=audio 5202 RTP/AVP 0
a=sendrecv
```

SIP responses follow a structured format with multiple lines where each line has a specific purpose. The Status Line indicates the protocol version (typically 2.0), the numerical status code, and the corresponding status message (or reason phrase), indicating the outcome of the processing of the request. A list of SIP response codes is in Table 2.4. Note that codes within the 100–199 range signal provisional responses, indicating the arrival of a definitive response bearing a code of 200 or above. The *Via* header delineates the local address designated for response reception, including the precise IP address of the interface that initially received the message. The *From*, *To*, *Call-ID*, and *CSeq* headers mirror those found in the original INVITE request, ensuring continuity within the dialog. The UAS appends a unique tag to the To header to identify the receiving end further. The *Contact* header furnishes a direct route to the UAS, guiding subsequent requests to their intended destination. The *Content-Type* and *Content-Length* headers signal the presence of media negotiation information within the message body. Specifically, they indicate the inclusion of SDP content, which articulates the agreed-upon codec for media exchange.

Table 2.4 SIP Response Codecs.

Range	Meaning
100–199	provisional
200–299	success
300–399	redirect
400–499	client errors
500–599	server errors
600–699	global failure

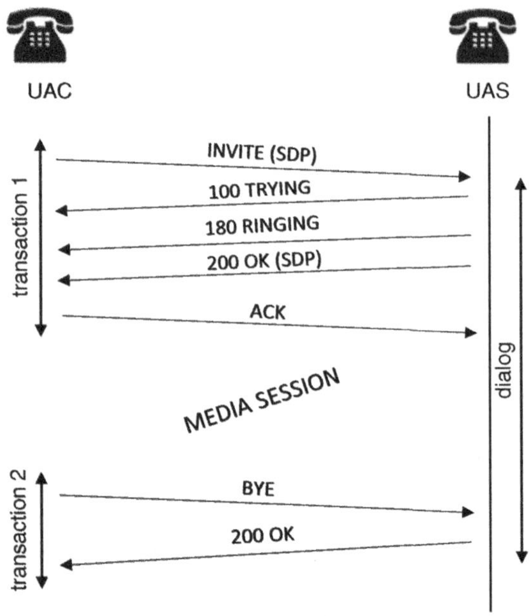

Figure 2.16 SIP Call Flow.

Figure 2.16 shows a dialog between a UAC and a media server UAS, composed of two distinct transactions: one intended to establishing a media session, and its counterpart intended to tear it down. The session establishment transaction starts with the client initiating an INVITE request, carrying an SDP payload that holds the client's media capabilities. The UAS responds with a provisional 100 Trying, letting the client know of the request's reception and ongoing processing. This is followed by another provisional response, 180 Ringing, which prompts the client to generate a ring tone while the INVITE undergoes further consideration. Note that these provisional responses do not carry a body and, thus, do not carry media information.

Example 2.3 Consider a UAS and a UAC connected by a 1 Gigabit Ethernet network and imagine the setup process follows the steps shown in Figure 2.16. Each message sent during this process is around 500 bytes long, and there is no delay in sending the server's responses. How much time elapses between the UAC sending the initial request (SIP INVITE) and the UAC receiving the SDP from the UAS?

Solution:
This corresponds to the RTT time difference between the transmission of the SIP INVITE and the reception of the corresponding 200 OK. This delay is $T = 2 \times \frac{500 \times 8}{1 \times 10^9} = 8\ \mu s$.

The UAS then transmits a definitive 200 OK, to indicate successful negotiation and enclosing the mutually agreed-upon media parameters within its body. To counter for the inherent uncertainties of UDP transport, the client acknowledges this 200 OK with an ACK, completing a robust three-way handshake. Upon identification of the media endpoints, a bidirectional *Real Time Protocol* (RTP) session enables the flow of packetized media datagrams between the participants. To terminate the media session, the client initiates the session termination transaction by transmitting a BYE request, signaling the conclusion of the exchange. The server receives a final 200 OK, with no body as media details are no longer relevant.

2.4 SDP

In the context of media session negotiation, SDP is key. This protocol, supported by media endpoints, specifies media session parameters in a universally comprehensible manner. It facilitates negotiation, enabling endpoints to accept or decline incoming sessions based on their compatibility with specific media types. Although SDP's reliance on ASCII encoding mirrors that of SIP messages, it is acceptable in non-3GPP and 3GPP networks but it introduces inefficiencies in resource-constrained IoT environments. Nevertheless, SDP remains indispensable for outlining media capabilities. It typically enumerates supported media encoding technologies, or codecs, alongside connectivity parameters, such as IP addresses and transport ports. Information is presented in lines following to the format *parameter=value*, where the parameter is denoted by a single character (Begen et al. 2021).

In the INVITE example above, the *v=0* line declares the SDP version, while *c=IN IP4 192.168.1.10* shows endpoint's IP address. The line *m=audio 5200 RTP/AVP 0* shows the media type (audio), the media protocol (RTP), the UDP port (5200), and the codec code (0), signifying ITU G.711 μ-Law coding. The *a=sendrecv* line, an SDP attribute line, signals a full-duplex media session.

It is important to note that SIP, despite its primary role as an application layer protocol for session establishment, it violates the layered architecture by embedding network and transport layer information within the SDP payload. This is an attempt to overcome message traversal challenges in certain NAT scenarios.

2.5 HTTP

HTTP is another important application layer protocol that is relevant to RTC because of its use in WebRTC and in certain IoT scenarios. As SIP, HTTP operates at the session layer, managing client–server interactions for web applications, and relies on TCP for transport, supporting REST principles. Messages are ASCII-encoded, potentially leading to fragmentation and network loss due to lack of TCP header compression. HTTP accesses various web resources, including *HyperText Markup Language* (HTML), scripts, and media, using *Uniform Resource Locators* (URLs), a subset of URIs. HTML documents employ *eXtensible Markup Language* (XML) structure to reference objects via URLs, which consist of service type, hostname, and path, as exemplified by *http://new.nsf.gov/about/history*.

Figure 2.17 shows a client–server interaction that follows to REST principles, characterized by statelessness. Applications acting as clients initiate HTTP requests to retrieve objects from a server. The server responds with HTTP responses containing the requested object, without retaining any state information (Fielding et al. 1999). Figure 2.18 illustrates the TCP connection establishment process, which precedes HTTP traffic transmission. The TCP three-way handshake is employed, consuming one RTT. RTT is defined as the time elapsed for a datagram to traverse from client to server and back, including transmission, propagation, and queuing delays across all nodes. The handshake's final segment is integrated with the HTTP request. Upon receipt, the server transmits an HTTP response containing the actual object. The server response time approximates two RTTs, as evident from the figure.

Configurable connection options impact HTTP traffic: a single persistent TCP connection, handling all requests/responses, contrasts with multiple non-persistent connections, dedicating one per transaction. With a persistent connection, the client initiates a single TCP handshake and subsequently sends requests without requiring additional handshakes. Conversely, non-persistent connections involve repeated handshakes before each request, potentially impacting performance on subsequent readouts.

HTTP transported by means of persistent TCP connections has become more efficient with newer versions of HTTP. The very first version 1.0 of HTTP, formally

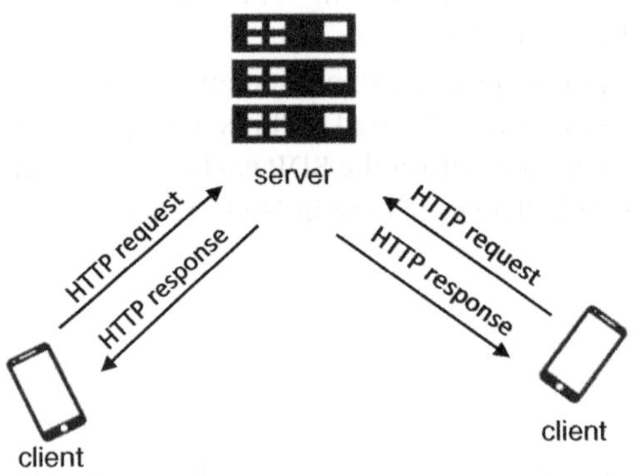

Figure 2.17 HTTP Transactions.

Figure 2.18 HTTP Setup.

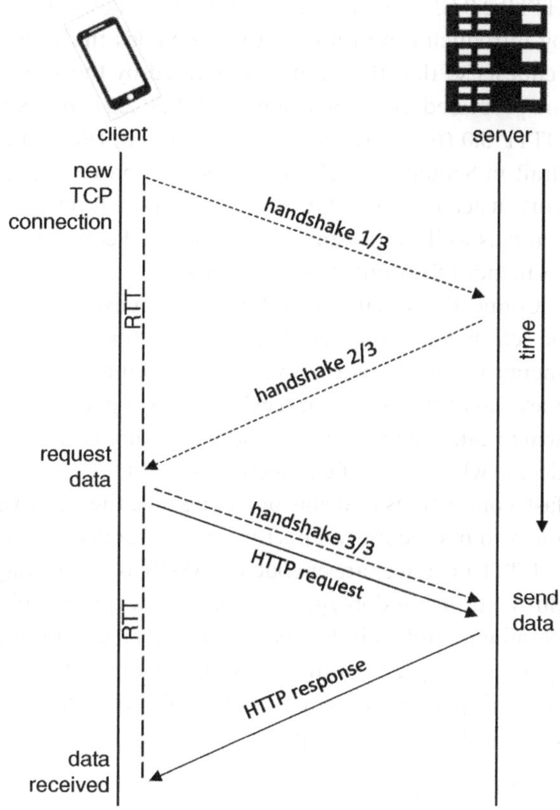

known as HTTP/1.0, only allows clients to send requests in a stop-and-wait fashion. In other words, if a client transmits a request, it must wait for the server response before it can transmit another request. The result of this behavior is excessive latency. HTTP 1.1 (formally known as HTTP/1.1) introduces the concept of pipelining, where the client can transmit many simultaneous requests that are processed and answered by the server in the order in which they are received. HTTP 1.1 exhibits lower latency than HTTP 1.0. Although all the requests may arrive at the server at the same time, some of them may take longer to be processed than the others. For example, a request that retrieves an object that may not be available right away in the server, may take longer to be processed than a request that retrieves an object that is readily available. Therefore, an earlier request may delay the transmission of the response to a later request that may have been already processed. This is known as Head of Line (HOL) blocking because the processing of a request can prevent other responses from being transmitted.

HTTP 2.0 (Belshe et al. 2015) (formally known as HTTP/2.0) addresses this issue by introducing multiplexing that enables the client to simultaneously transmit requests so that they can be answered by the server in the order in which they are processed. As a consequence, HTTP 2.0 exhibits lower latency than HTTP 1.1. HTTP 3.0 (formally known as HTTP/3.0) (Bishop 2020) introduced as an IETF draft in September 2020 attempts to improve latency by relying on QUIC transport instead of TCP. QUIC is a transport layer protocol that lowers connection setup as well as transmission latency and consequently improves congestion performance (Nepomuceno et al. 2018).

Connections that close after each request can slow things down. They need to be set up each time, which takes extra time and resources. They also need extra memory to keep track of everything. Connections that stay open can be faster, but they can also waste resources. They stay open even when they are not being used, which can use up memory and processing power. So, it is a trade-off. One has to decide which type of connection is best for each situation. For things to be really fast, connections that stay open might be the way to go. But if worried about wasting resources, connections that close after each request might be a better choice.

HTTP messages are encoded in ASCII as SIP messages, enabling their inspection during transmission via network traffic capture tools. Following the client–server paradigm, and as indicated above, HTTP communication involves two message types: (1) requests, initiated by clients to elicit server actions, and (2) responses, servers' corresponding replies. The following exemplifies a client-generated HTTP request accessing a webpage:

```
GET / HTTP/1.1
Host: www.l7tr.com
Connection: close
User-Agent: Client
```

HTTP request messages support a specific structure, ensuring consistent communication between clients and servers. Each message comprises a series of lines, each terminated with ASCII carriage return and line feed characters. An additional sequence of these characters marks the message's conclusion. The message's length is variable, accommodating diverse communication needs.

The initial line, termed the request line, plays a pivotal role in defining the request's intent. It includes three essential elements: (1) a method that specifies the desired action to be executed on the server-side resource. Common methods include GET, POST, HEAD, PUT, and DELETE, aligning with CRUD operations prevalent in RESTful architectures, (2) URL that identifies the precise resource being targeted, providing a path for request execution, and (3) Version that indicates the HTTP protocol version employed for communication, ensuring compatibility and adherence to established standards.

Subsequent lines, known as header lines, furnish additional context for the request. These can include diverse information, such as: (1) Host that specifies the server's domain name or IP address, important for routing, particularly when proxy servers mediate communication, (2) Connection that dictates whether the connection should persist for subsequent requests or terminate upon completion, impacting resource utilization and performance, and (3) User-Agent that identifies the client application initiating the request, enabling servers to tailor responses accordingly, enhancing user experience.

Certain method types, like POST requests, accommodate a message body appended to the header lines. This body serves as a vessel for transmitting substantial data payloads from client to server, proving indispensable for scenarios requiring actuation or object creation. Note that GET request responses also leverage the message body to convey object representations. While GET and HEAD methods share commonalities, they diverge in a fundamental aspect: HEAD requests signal the client's intention to examine the object's metadata without retrieving its actual content. This optimization technique minimizes network utilization by averting unnecessary data transfer. In contrast, PUT and DELETE methods support distinct functionalities. PUT typically handles object creation and uploading, while DELETE enables clients to eradicate objects residing on the server.

Figure 2.19 shows a generic HTTP request message, including its structural elements. HTTP responses follow a defined structure, ensuring the seamless

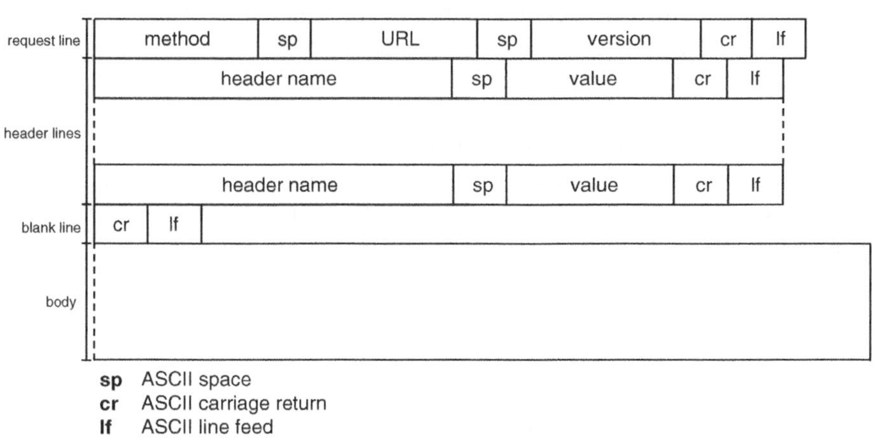

Figure 2.19 HTTP Request.

exchange of information between servers and clients. The example below shows a typical HTTP response generated by a server, transmitting a simple *Hello* message:

```
HTTP/1.1 200 OK
Connection: close
Date: Tue, 23 July 2022 11:24:09 GMT
Server: www.l7tr.com
Last-Modified: Tue, 23 July 2022 10:20:29 GMT
Content-Length: 15
Content-Type: text/xml

<h1>Hello</h1>
```

HTTP responses follow a structured format, mirroring the composition of request messages. This deliberate design ensures clarity and consistency in communication between servers and clients. The Status Line includes a Version field that specifies the HTTP protocol version employed, ensuring compatibility. The Status Code is a numeric code signifying the request's outcome (e.g., 200 for success, 404 for resource not found). The Status Message is a human-readable phrase corresponding to the status code, aiding interpretation. The HTTP headers include (1) the Connection header that dictates connection persistence or closure, impacting resource utilization, (2) the Date header that denotes the response's creation time, providing temporal context, (3) the Server header that identifies the entity that generated the response, enhancing transparency, (4) the Last-Modified header that indicates the timestamp of the object's last modification, facilitating cache management, (5) the Content-Length specifies the message body's byte length, enabling efficient parsing, and (6) the Content-Type header that declares the body's encoding scheme, ensuring accurate interpretation. The message body carries the actual application layer information to be transmitted to the client. Figure 2.20 offers a representation of a generic HTTP response message, further elucidating its structural components.

2.6 Security Considerations

In the realm of SIP communication, security measures emerge as a response to a multitude of threats and attacks. This section presents these challenges and outlines a set of requirements that address network security principles and the development of *Secure SIP* (SIPS).

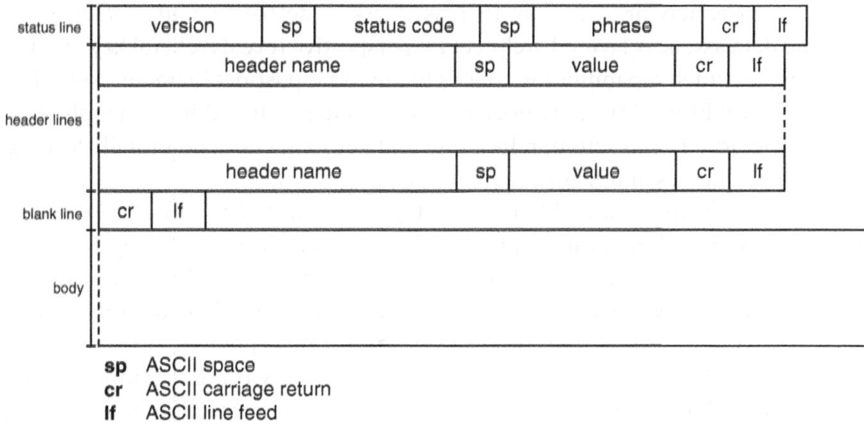

Figure 2.20 HTTP Response.

2.6.1 Challenges and Requirements

Securing SIP traffic faces several unique challenges: (1) endpoints like smartphones and smart devices often rely on batteries, requiring minimal resource consumption for security measures, (2) SIP communication is susceptible to a range of attacks, from passive eavesdropping to active interference, and (3) traditional security mechanisms may not be suitable for the wireless communication used in RTC networks.

Security requirements must be addressed to safeguard SIP traffic: (1) Confidentiality: Information must be encrypted to prevent unauthorized access, ensuring only intended recipients can view sensitive data, (2) Authentication: Verification of sender identities is essential to guarantee that messages originate from trusted sources, preventing impersonation and spoofing, (3) Integrity: Measures must ensure that messages are not tampered with during transmission, maintaining their original content and preventing unauthorized alterations, (4) Freshness: Mechanisms are needed to guarantee that messages are not re-used or replayed, ensuring that communications are timely and valid, (5) Availability: The network must remain functional and accessible even under attack, ensuring that communication continues despite malicious attempts to disrupt it, (6) Resilience: Security measures must remain effective even if some devices or network components are compromised, providing continued protection in the face of partial breaches, (7) Resistance: The network must be designed to withstand attempts by intruders to gain full control, preventing unauthorized takeover and widespread disruption, and (8) Energy efficiency: Security measures should minimize power consumption to accommodate the battery-powered nature of many SIP endpoints, extending device life and reducing energy costs. RTC networks, characterized by wireless access networks, face an increased risk of destructive attacks targeting user and data assets. A layered architecture perspective reveals several key threats, assuming intruder capability for attacks at any time post-deployment. Note that IoT devices within RTC networks present unique vulnerabilities to physical attacks, including destruction, relocation, and key extraction, potentially leading to irreversible losses and firmware manipulation.

RTC networks are vulnerable to multi-layered *Denial of Service* (DoS) attacks that exploit vulnerabilities at the physical, link, and network layers. Physical layer attacks involve interference generation to reduce SNR, while link layer attacks force device energy depletion through media collisions, fake addresses, and sleep deprivation torture, with particularly severe consequences when targeting edge routers. Network layer attacks flood the network with datagrams, causing congestion and throughput reduction. Ironically, replay protection mechanisms, designed to prevent datagram resending, can be leveraged for replay protection attacks, wherein bogus high-sequence datagrams block valid ones from reaching their destination.

RTC networks are also susceptible to a diverse array of network layer attacks, including: (1) Routing Information Spoofing, where attackers manipulate routing information through spoofing, alteration, or replay, creating loops that result in datagram loss, (2) Selective Forwarding, where compromised devices selectively forward only certain datagrams, disrupting connectivity, (3) Sinkhole Attacks, where attackers aim to establish a device as the destination for all traffic within a specific area by broadcasting deceptively low-cost routes to neighboring nodes, (4) Sybil Attacks, where a single device masquerades under multiple identities to become a destination for numerous network nodes, (5) Wormhole Attacks, where datagrams are captured in one location and retransmitted in another, causing routing and connectivity issues, and (6) Neighbor Discovery Attacks, where ND messages, essential for network topology awareness, are either dropped or corrupted, leading to reachability problems (Herrero 2021).

2.6.2 Principles of Network Security

In today's security-critical networks, TCP transport leverages TLS, an application layer protocol (as per IETF definitions) that augments TCP with security features. For UDP transport, *Datagram Transport Layer Security* (DTLS) provides a TLS-like alternative. Both DTLS and TLS foundationally rely on concepts detailed in this Section: (1) Symmetric Encryption by employing shared secret keys for both encryption and decryption, (2) *Public Key Infrastruc ture* (PKI) encryption by utilizing a pair of mathematically linked keys (public and private) for secure communication without prior key exchange, (3) Authentication via Hash Functions by generating unique, fixed-length *fingerprints* of data to ensure integrity and detect alterations, and (4) Digital Signatures by electronically signing data using private keys to establish authenticity and non-repudiation, often in conjunction with *Certification Authorities* (CAs) for trust validation.

Symmetric encryption, a fundamental technique for securing blocks of data like messages, employs a shared secret key (K) to both encrypt and decrypt. It establishes a one-to-one mapping between the original message (m) and its encrypted form ($K(m)$), ensuring reversibility through re-encryption: $K(K(m)) = m$. This process is illustrated in Figure 2.21, where endpoints A and B exchange encrypted messages m_1 and m_2, decrypted upon reception as $K(K(m_1))$ and $K(K(m_2))$. Common symmetric encryption algorithms include the *Data Encryption Standard* (DES) and the *Advanced Encryption Standard* (AES), with AES offering varying key sizes and robustness levels (AES-128, AES-192, and AES-256).

To safeguard message confidentiality and prevent pattern analysis, large messages are divided into smaller blocks, each encrypted individually. However, the one-to-one mapping inherent in symmetric encryption poses a challenge: If multiple messages share similar blocks, their encrypted versions will also exhibit identical blocks, potentially revealing sensitive information to eavesdroppers. To mitigate this, controlled randomization techniques, such as *Cipher Block Chaining* (CBC) and *Electronic Codebook* (ECB), are employed to introduce variability and make it significantly more difficult for attackers to discern patterns or relationships between messages.

PKI addresses a key challenge in symmetric encryption: the inherent insecurity of key distribution. In symmetric encryption, endpoints must share a secret key, often transmitted over potentially vulnerable communication channels. PKI offers a more secure alternative by employing mathematically linked public–private key pairs. Under PKI, a message (m) encrypted with a recipient's public key (K_+) can only be decrypted using their corresponding private key (K_-). This

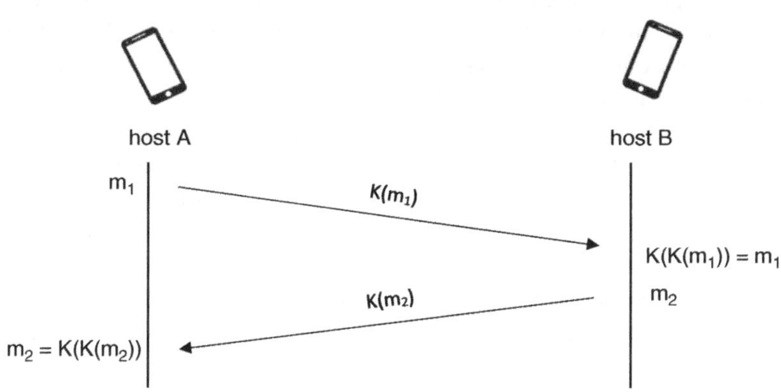

Figure 2.21 Symmetric Encryption Flow.

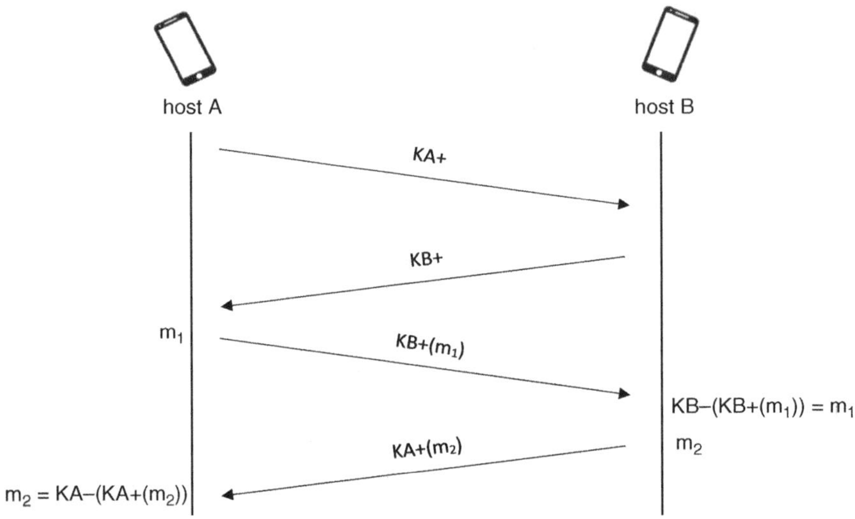

Figure 2.22 PKI-based Encryption Flow.

ensures confidentiality even without pre-shared secrets. In opposition, a message digitally signed with a private key can be verified using the associated public key, guaranteeing authenticity and non-repudiation.

Figure 2.22 illustrates a PKI-based interaction between endpoints A and B. A generates a public-private key pair (KA−, KA+), while B generates (KB−, KB+). A sends its public key KA+ to B, and B sends its public key KB+ to A. A encrypts a message m_1 with KB+ and sends it to B. B decrypts m_1 with its private key KB−. B encrypts a message m_2 with KA+ and sends it to A. A decrypts m_2 with its private key KA−.

Common PKI algorithms include (1) *Rivest-Shamir-Adleman* (RSA) that is used for both encryption and decryption, (2) *Digital Signature Algorithm* (DSA) that is used primarily for digital signatures and authentication, and (3) *Elliptic Curve Cryptography* (ECC) that offers efficient key generation and smaller key sizes.

In addition to encryption, message authentication stands as a cornerstone mechanism within both TLS and DTLS. While encryption safeguards message confidentiality from unauthorized access, authentication ensures message integrity throughout network transmission, preventing any attempts at tampering.

Message authentication hinges upon the utilization of hash functions. Formally defined as a many-to-one mapping, a hash function, denoted as $H(\ldots)$, maps an input message m to a shorter, fixed-length output sequence known as the hash value $H(m)$. This hash value exhibits an important property: even minute alterations to the original message yield a vastly different hash value, rendering any undetected modifications highly improbable.

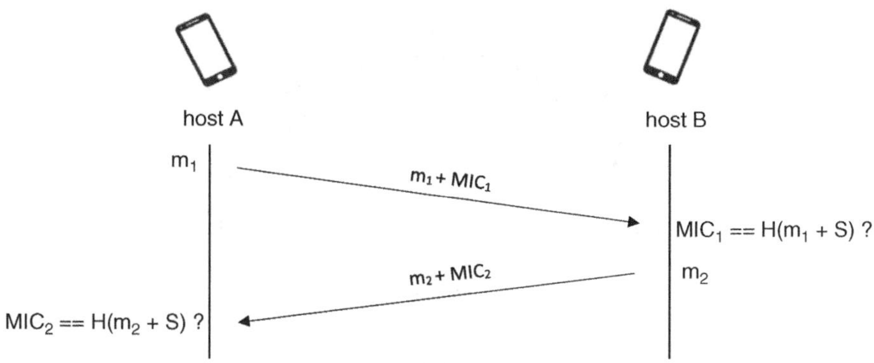

Figure 2.23 Message Authentication Flow.

Prevalent hash functions include families such as the *Secure Hash Algorithm* (SHA) and the *Message Digest* (MD) algorithms. These include SHA-1, SHA-128, SHA-256, MD4, and MD5, with varying hash lengths. Figure 2.23 shows the message authentication process. Entity A, initiating communication with entity B, transmits message m_1 alongside a *Message Integrity Code* (MIC), denoted as MIC_1, for authentication purposes. MIC_1 is computed as $H(m_1 + S)$, encapsulating the hash of the message m_1 concatenated with a shared secret S. Upon receiving m_1 and MIC_1, entity B initiates integrity validation by replicating the MIC calculation locally. It appends the shared secret S to the received m_1, generates the hash $H(m_1 + S)$, and compares it with the received MIC_1. Identical values signify successful message authentication. Reciprocally, when entity B transmits message m_2 to entity A, it generates MIC_2 for authentication, validated by entity A upon receiving m_2 and MIC_2.

While *Message Authentication Code* (MAC) is a synonymous term for MIC, the latter is often used to circumvent ambiguity with Media Access Control (also abbreviated as MAC).

An important challenge within PKI lies in establishing unequivocal trust in the public keys exchanged between communicating endpoints. While public key transmission is integral for enabling secure encryption, a mechanism to unequivocally bind a public key to its genuine owner is important to prevent interception and substitution attacks. CAs address this challenge by leveraging digital signatures to incontrovertibly associate public keys with their rightful owners (Kurose and Ross 2012).

Figure 2.24 shows the digital signature process. Endpoints A and B independently compute hashes of their respective public keys, H(KA+) and H(KB+). These hashes are transmitted to the trusted CA. The CA, employing its private key KCA−, encrypts each hash, generating digital signatures Sign(A) and Sign(B),

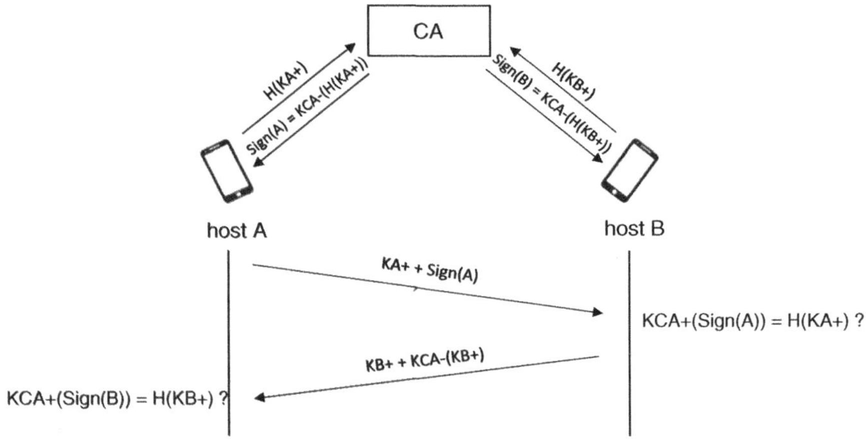

Figure 2.24 Certification Authority Flow.

which are subsequently transmitted back to the respective endpoints. Endpoint A transmits its public key KA+ and the corresponding signature Sign(A) to endpoint B. Endpoint B calculates the hash H(KA+) of the received public key KA+. It employs the CA's widely-known public key KCA+ to decrypt the signature Sign(A), yielding: $\text{KCA} + (\text{Sign}(A)) = \text{KCA} + (\text{KCA}-(H(\text{KA}+))) = H(\text{KA}+)$ identicality between the calculated hash H(KA+) and the decrypted signature validates the authenticity of the received public key KA+, indicating that it originates from endpoint A and has not been tampered with. Endpoint B follows an analogous process to transmit its public key KB+ and signature Sign(B), which are validated by endpoint A. In essence, CA-issued digital signatures establish a bedrock of trust within PKI, ensuring the integrity and authenticity of public keys, thereby fortifying secure communication.

2.6.3 TLS and DTLS

In adherence to the security principles outlined in Section 2.6.2, both TLS and DTLS protocols support a traffic flow between client and server entities to establish secure communication sessions. This section looks at the intricacies of this exchange.

Figure 2.25 shows the sequential establishment of a TCP connection, followed by the exchange of security parameters and certificates. The process starts with the client transmitting a TCP SYN segment to initiate the connection. The server responds with a SYN-ACK segment, acknowledging the client's request and

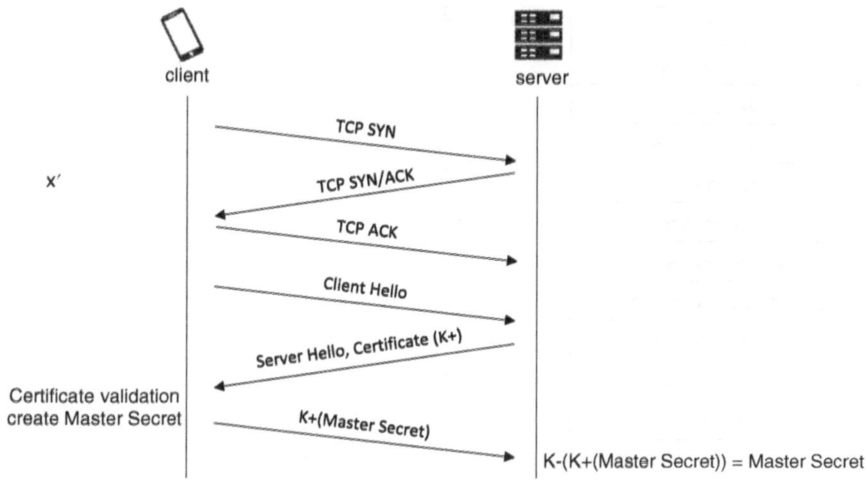

Figure 2.25 TLS Session Setup Flow.

simultaneously initiating its own connection establishment. The client completes the TCP handshake by transmitting an ACK segment. Upon successful establishment of the TCP connection, the client transmits a Hello message, delineating the authentication and encryption security parameters it supports. In a server authentication scenario, the server responds by transmitting a certificate, including its public key and the signature generated by the CA (as detailed in Section 2.6.2), alongside a Hello message that corroborates the selected security parameters (Rescorla 2018).

The certificate format, shown in Figure 2.26, also specifies other information like its validity period. Once the client validates the certificate signature against the CA public key, it creates a Master Secret that includes four keys: (1) a symmetric encryption key for traffic from client to server, (2) a symmetric encryption key for traffic from server to client, (3) a secret for authentication of traffic from client to server, and (4) a secret for authentication of traffic from server to client.

The Master Secret is encrypted by the client using the public key in the certificate and transmitted to the server, which holds the private key. The server decrypts the Master Secret and becomes aware of the symmetric keys and secrets. At this point, both client and server start sending encrypted and authenticated messages to each other. Note that PKI is only used to exchange symmetric keys and secrets that are used in the long term for the encryption and authentication of messages. This makes sense since PKI is computationally complex and its use must be ideally minimized. This makes it possible for both TLS and DTLS to run on low-end constrained embedded processors and devices.

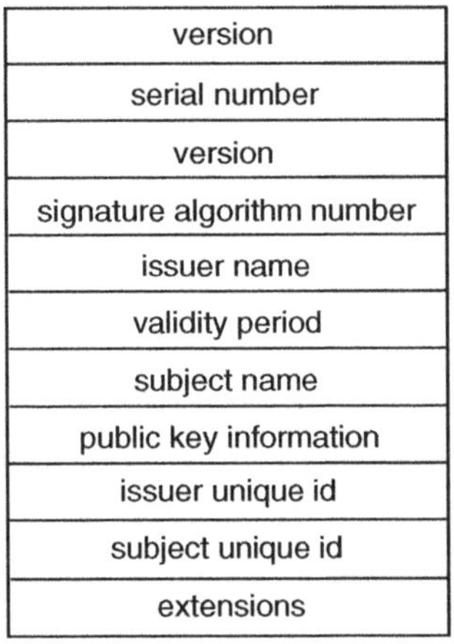

Figure 2.26 ITU X.509 Format.

For enhanced security, mutual authentication is also supported. In this case, the client sends its own CA-signed certificate along with the Hello message for the server to validate it.

Within TLS and DTLS, data is encapsulated within structures known as records, which comply with a well-defined format shown in Figure 2.27. These records exhibit a strategic blend of unencrypted and encrypted elements, enabling partial visibility for network analysis tools while safeguarding sensitive information. Each record comprises several distinct fields: (1) a 1-byte unencrypted type field that serves as an important indicator, revealing whether the record encapsulates a handshake packet, critical for establishing secure communication, or a payload-bearing message, (2) a 2-byte unencrypted version field that explicitly specifies the TLS or DTLS version employed within the communication, ensuring compatibility and adherence to established protocols, (3) a 3-byte unencrypted length field that communicates the precise length of the record, enabling accurate parsing and interpretation by receiving entities, (4) a variable-length data field that typically carries the encrypted message and (5) a variable-length MIC field, also encrypted, that safeguards the integrity of the record's contents. Note that its calculation includes both the payload and a sequence number, although the latter is not explicitly transmitted within the record itself (Rescorla and Modadugu 2012).

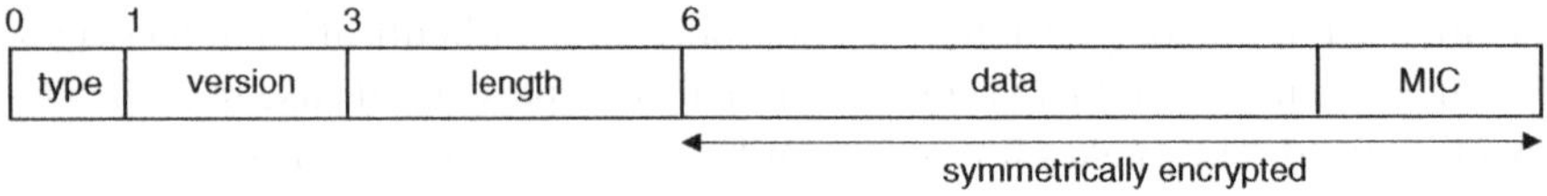

Figure 2.27 TLS Record Format.

The aforementioned sequence number, while absent from the transmitted record, plays a pivotal role in maintaining order and integrity. Both client and server track this sequence number, enabling them to reconstruct the correct sequence of records, independent of potential reordering or loss that might occur at the transport layer. This mechanism proves particularly invaluable within DTLS, as it mitigates inherent limitations associated with the connectionless nature of UDP transport. Note that both TLS and DTLS offer a specialized record type known as alerts. These alerts predominantly serve the purpose of gracefully terminating sessions, contributing to orderly communication closure. Additionally, they can function as a mechanism for error control, as each alert type carries a distinct semantic meaning, signaling the occurrence of errors or other critical conditions that require attention.

2.6.4 SIP over TLS

To safeguard the confidentiality and integrity of SIP communications, SIP messages are typically encrypted and authenticated via transmission over TLS. This protocol offers a choice between two authentication mechanisms: (1) Server authentication where the client just verifies the server's identity, ensuring that clients are communicating with the intended server and not an imposter and (2) Mutual authentication that extends security further by requiring both client and server to authenticate themselves, enhancing trust and mitigating potential impersonation attacks.

While the theoretical possibility exists to transmit SIP messages over DTLS, SIP has been standardized exclusively for TLS transport. This decision reflects TLS's suitability for reliable, connection-oriented communication, aligning with SIP's inherent characteristics. In contrast to unsecured SIP, which typically utilizes UDP or TCP over port 5060, secure SIP employs TLS over TCP, operating on port 5061. This distinct port assignment facilitates clear differentiation between secure and unsecured SIP traffic.

The establishment of a secure SIP session follows a sequential process. The UAC initiates a TCP connection with the UAS, forming the foundation for subsequent secure communication. Upon successful TCP connection, a TLS session is negotiated and established, ensuring confidentiality and integrity of subsequent communications. Only after the secure TLS session is in place can the actual SIP session be configured, enabling the exchange of SIP messages within a protected environment.

This layered approach safeguards SIP communications from eavesdropping, tampering, and impersonation, contributing to the overall security and trustworthiness of SIP-based applications.

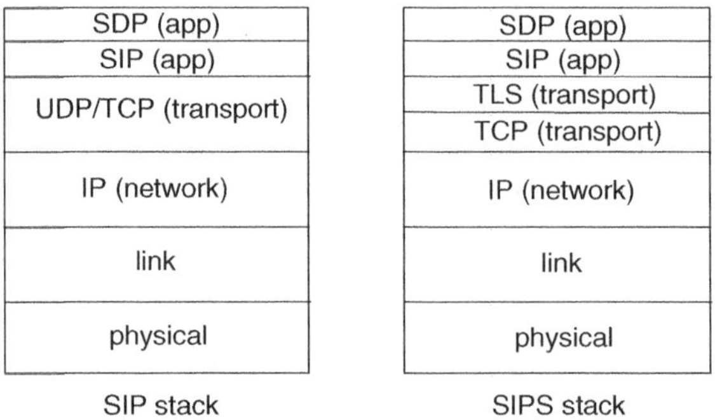

Figure 2.28 Secure and Unsecure SIP Protocol Stacks.

Figure 2.28 illustrates the architectural differences between unsecured SIP and its secure counterpart, SIPS. The main distinction resides at the transport layer, where SIPS incorporates an additional layer of security in the form of TLS, operating exclusively over TCP transport.

Figure 2.29 shows the message flow exchanged between a UAC and a UAS during the establishment and termination of a secure SIP session. The UAC initiates a TCP connection by transmitting a SYN segment. The UAS responds with a SYN-ACK segment, acknowledging the connection request and initiating its own connection establishment. The UAC completes the TCP handshake by transmitting an ACK segment. This three-way handshake typically consumes a single RTT. The UAC then embeds a TLS client *Hello* message within a TCP ACK segment. The UAS responds with a server *Hello* message, presenting its server certificate for authentication purposes. The UAC, in the case of mutual authentication, transmits a Master Secret containing its client certificate. Both UAC and UAS exchange *Finished* messages to signal completion of the TLS handshake. The establishment of the secure session typically spans at least two RTTs. With encryption and authentication in place, the UAC initiates an SIP session by transmitting an INVITE request to negotiate media session parameters. The UAS responds with a sequence of SIP responses (100, 180, and 200), indicating progress and eventual acceptance of the session establishment. The UAC confirms receipt of the final response with a SIP ACK message, marking the successful establishment of the media session.

For the SIP session termination, the UAC initiates session termination by transmitting a SIP BYE message. Both UAC and UAS exchange *Close Notify* alerts to gracefully conclude the TLS session. Then both endpoints transmit TCP FIN segments to close the underlying TCP connection. Note that these two sequences of steps ensure the confidentiality, integrity, and authenticity of SIP communications, safeguarding sensitive information from unauthorized access or manipulation.

When implementing secure SIP, two key modifications occur. The initial service identifier, typically **sip:**, transitions to **sips:** to signal the activation of secure

Figure 2.29 SIPS Message Flow.

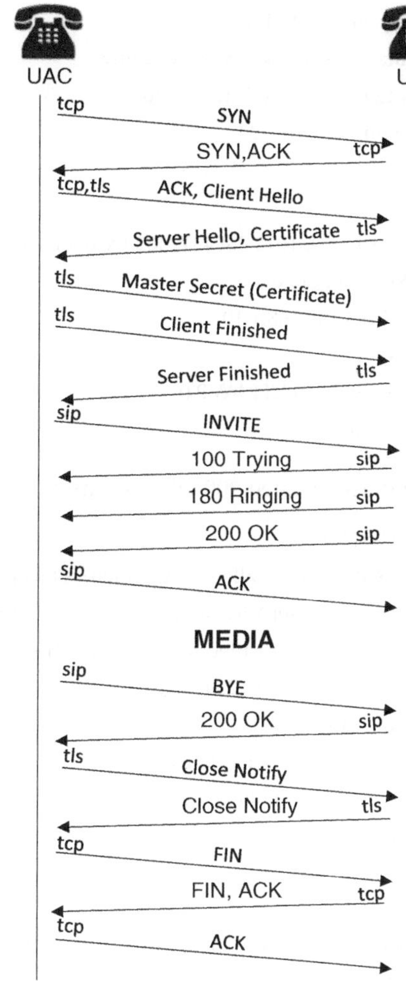

communication protocols and the protocol responsible for data transmission switches from either UDP or TCP to TLS. This ensures the confidentiality and integrity of SIP messages through encryption (Audet 2009). Consider the following example of an SIPS INVITE:

INVITE sips:userB@callee.com SIP/2.0

Via: SIP/2.0/TLS caller.com;branch=z9hG4bK2105Bg1dd Max-Forwards: 70

From: <sips:userA@caller.com>; tag=1921051090

To: <sips:userB@callee.com>

Call-ID: 53a19e23b12a4105@caller.com

```
CSeq: 3013 INVITE
Contact: <sips:userA@caller.com>
Content-Type: application/sdp
Content-Length: 134

v=0
o=session 5431012 2394312402 IN IP4 userA.caller.com
s=SDP exchange
c=IN IP4 192.168.1.10
t=0 0
m=audio 5200 RTP/AVP 0
a=sendrecv
```

and the associated 200 OK response:

```
SIP/2.0 200 OK
Via: SIP/2.0/TLS callee.com;branch=z9hG4bK110232105;received=192.168.1.20
From: <sips:userA@caller.com>; tag=1921051090
To: <sips:userB@callee.com>; tag=242199220
Call-ID: 53a19e23b12a4105@caller.com
CSeq: 3013 INVITE
Contact: <sip:userB@callee.com>
Content-Type: application/sdp
Content-Length: 132

v=0
o=session 2841304 91021775 IN IP4 userB.callee.com
s=SDP exchange
c=IN IP4 192.168.1.20
t=0 0
m=audio 5202 RTP/AVP 0
a=sendrecv
```

More importantly, despite the security-related adjustments, the request line and headers within SIP messages remain identical regardless of whether secure or unsecure SIP is employed.

2.7 Summary

Media session establishment requires setting up sessions where media is transmitted between endpoints. This requires a communication network. The Internet Protocol Suite defines mechanisms deployed on an architecture that divides a communication system into interdependent layers. The lowest layer, the physical layer, interfaces between the analog and digital worlds. Signal modulation plays a crucial role in this layer. Together with the physical layer, the link layer enables access connectivity. Then, the network layer introduces IP, which supports end-to-end packet delivery through the network. Its main variations are IPv4 and IPv6.

On top of the network layer, the transport layer supports multiplexing, allowing multiple applications to share the network. This layer represents a trade-off between minimizing latency and minimizing packet loss. UDP prioritizes low latency but may experience more packet loss, while the TCP prioritizes reliable delivery (low loss) even if it increases latency. Finally, the application layer enables encoding and transmission of application data and sessions. Two main protocols, SIP and HTTP, support media traffic. SIP is specifically designed to establish, manage, and terminate media sessions, and it uses Session Description Protocol (SDP) for media negotiation. In all cases, security protocols like TLS or DTLS are necessary for secure communication.

2.8 Homework Problems and Questions

Problem 2.1 Consider a wireless scheme that relies on a 16-QAM modulation with a symbol duration that is $T = 100$ μs, what are the symbol R_s and transmission R_b rates?

Problem 2.2 What is the main difference between centralized and distributed forwarding tables in computer networking?

Problem 2.3 Given a 192.168.10.0/24 network that is divided into 8-host subnets, what are those subnets?

Problem 2.4 What are the key differences between the IPv4 and IPv6 headers, particularly in terms of fields and functionality? How does IPv6 handle fragmentation and reassembly compared to IPv4?

Problem 2.5 Consider the transmission of an IPv4 datagram that is 1800 bytes long. Then, assuming a 10% network packet loss and a 1500-byte MTU size, what is the actual datagram loss probability?

Problem 2.6 Why do the UDP and TCP headers include the source and destination port numbers at the same offset locations? What is the advantage of this?

Problem 2.7 Imagine that ten files (100 bytes each) are downloaded from a server using HTTP 1.0. How much faster would it be to download them all at once if the connection between the client and the server stayed open (persistent connection) compared to opening a new connection (non-persistent connection) for each file? Consider that it takes 100 milliseconds on average for data to travel back and forth (RTT).

Problem 2.8 Consider a UAS and a UAC connected by a 1 Gigabit Ethernet network and imagine the setup process follows the steps shown in Figure 2.16. Each message sent during this process is around 500 bytes long, and there is no delay in sending the server's responses. How much time elapses between the UAC sending the initial request (SIP INVITE) and the UAS receiving the confirmation (SIP ACK)?

Problem 2.9 Why, as opposed to HTTP, the interaction between a UAC and a UAS, requires the client to send an SIP ACK message to acknowledge the SIP 200 OK sent by the UAS?

Problem 2.10 Can the SDP of an INVITE include two media lines for audio and video associated with the same port number? Justify.

Problem 2.11 How does symmetric encryption work, and what are some common symmetric encryption algorithms?

Problem 2.12 What are the advantages and disadvantages of using SIPS as opposed to plain SIP (consider all factors like latency and network impairments in general, not just security)?

Problem 2.13 Why is it often believed that the SDP protocol violates the principles of the layered architecture?

Bibliography

RFC. (1980). User datagram protocol. RFC 768, August. Available from: https://www.rfc-editor.org/info/rfc768.

RFC. (1981a). Internet protocol. RFC 791, September. Available from: https://www.rfc-editor.org/info/rfc791. (Accessed March 15th, 2024).

RFC. (1981b). Transmission control protocol. RFC 793, September. Available from: https://www.rfc-editor.org/info/rfc793. (Accessed March 15th, 2024).

Ahson, S.A. and Ilyas, M. (2008). *SIP handbook: services, technologies, and security of session initiation protocol*. 1st ed. USA: CRC Press, Inc. ISBN 1-420-06603-X.

Audet, F. (2009). The use of the SIPS URI scheme in the session initiation protocol (SIP). RFC 5630, October. Available from: https://www.rfc-editor.org/info/rfc5630. (Accessed March 15th, 2024).

Begen, A.C., Kyzivat, P., Perkins, C., and Handley, M.J. (2021). SDP: Session description protocol. RFC 8866, January. Available from: https://www.rfc-editor.org/info/rfc8866. (Accessed March 15th, 2024).

Belshe, M., Peon, R., and Thomson, M. (2015). Hypertext transfer protocol version 2 (HTTP/2). RFC 7540, May. Available from: https://rfc-editor.org/rfc/rfc7540.txt. (Accessed March 15th, 2024).

Bishop, M. (2020). Hypertext transfer protocol version 3 (HTTP/3). Internet-Draft draft-ietf-quic-http-29, *Internet Engineering Task Force*, June. Available from: https://datatracker.ietf.org/doc/html/draft-ietf-quic-http-29. (Accessed March 15th, 2024).

Fielding, R., Gettys, J., Mogul, J., Frystyk, H., Masinter, L., Leach, P., and Berners-Lee, T. (1999). RFC 2616, Hypertext transfer protocol – http/1.1. Available from: http://www.rfc.net/rfc2616.html. (Accessed March 15th, 2024).

Haykin, S. (2009). *Communication systems*. 5th ed. Wiley Publishing. ISBN 0-471-69790-7.

Herrero, R. (2021). Fundamentals of IoT communication technologies. *Textbooks in telecommunication engineering*. Springer International Publishing. ISBN 978-3-030-70079-9. Available from: https://books.google.com/books?id=k70rzgEACAAJ. (Accessed March 15th, 2024).

Herrero, R. (2023). Practical internet of things networking. *Textbooks in telecommunication engineering*. Springer International Publishing. ISBN 978-3-031-28443-4.

Hinden, B. and Dr., Deering, S.E. (1998). Internet protocol, version 6 (IPv6) specification. RFC 2460, December. Available from: https://www.rfc-editor.org/info/rfc2460. (Accessed March 15th, 2024).

Kurose, J.F. and Ross, K.W. (2012). *Computer networking: a top-down approach*. 6th ed. Pearson. ISBN 0-132-85620-4.

Nepomuceno, K., Oliveira, I.N.d., Aschoff, R.R., Bezerra, D., Ito, M.S., Melo, W., Sadok, D., and Szabo, G. (2018). QUIC and TCP: a performance evaluation. In 2018 IEEE Symposium on Computers and Communications (ISCC), pp. 00045–00051.

Rescorla, E. (2018). The transport layer security (TLS) protocol version 1.3. RFC 8446, August. Available from: https://rfc-editor.org/rfc/rfc8446.txt. (Accessed March 15th, 2024).

Rescorla, E. and Modadugu, N. (2012). Datagram transport layer security version 1.2. RFC 6347, January. Available from: https://rfc-editor.org/rfc/rfc6347.txt. (Accessed March 15th, 2024).

Schooler, E., Rosenberg, J., Schulzrinne, H., Johnston, A., Camarillo, G., Peterson, J., Sparks, R., and Handley, M.J. (2002). SIP: Session initiation protocol. RFC 3261, July. Available from: https://rfc-editor.org/rfc/rfc3261.txt. (Accessed March 15th, 2024).

3

Media Packetization

3.1 Introduction

While Chapter 2 focused on the establishment, management, and termination of media sessions, this chapter focuses on the process of breaking media into packets for transmission (or media packetization) in non-3GPP, 3GPP, and IoT networks. Once a media session is up and running, the actual media data needs to be sent across the network, and that is where RTP (briefly mentioned in Chapter 2) and *Real Time Control Protocol* (RTCP) come in (Schulzrinne et al. 2003; Herrero 2023).

These protocols form a stack, similar to the ones shown in Figure 2.28, with a few key differences indicated in Figure 3.1. In most cases, the lower layers (physical, link, and network) are the same for both media session signaling and media packetization. However, in certain scenarios involving media servers or other RTC edge/cloud devices, separate network layers with different IP addresses might be used for signaling and media. The transport layer for RTP and RTCP typically relies on UDP, with port numbers dynamically negotiated through SIP, as discussed in Chapter 2.

Because security is key in all RTC applications, this chapter also introduces *Secure RTP* (SRTP) and *Secure RTCP* (SRTCP), which respectively provide efficient encryption and authentication of RTP and RTCP packets (Carrara et al. 2004; Ahson and Ilyas 2008).

3.2 RTP

Real Time Protocol (RTP) is responsible for delivering audio, speech, and video sessions in real time, ensuring smooth and timely communication. It is a dedicated delivery service for media content. To set up an RTP session, and as described in

Hands-On Real Time Communications: A Practical Guide to RTC Protocols in Non-3GPP, 3GPP, 4G/5G/6G and IoT Networks, First Edition. Rolando Herrero.

Companion website: www.wiley.com/go/herrero/RTCProtocols

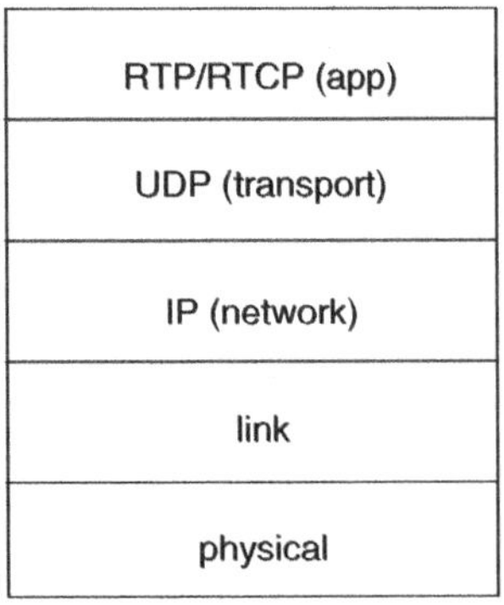

Figure 3.1 RTP/RTCP Stack.

Chapter 2, SIP and SDP work together, exchanging information to establish the connection. While SIP enables negotiation, SDP encodes the media parameters needed for endpoints to understand each other (Perkins 2003; Schulzrinne et al. 2003).

RTP is transported over UDP (RFC 1980), which prioritizes low latency over reliability. This is important for media, as delayed packets often become useless. To smooth out any minor inconsistencies in packet arrival times, RTP applications typically use playout buffers which temporarily store packets before playing them in order. This helps prevent choppy playback.

The focus of UDP on speed also makes RTP ideal for multicast transmissions, where a single sender can reach multiple receivers efficiently. This is particularly useful for multicast conferencing, where multiple participants can join a single session without overwhelming the network. RTP does not limit itself to one media type at a time. It can handle multiple simultaneous streams, allowing a single SIP dialog to coordinate synchronized audio and video.

Beyond its primary role in real-time media delivery, RTP also supports transcoding (the process of converting media streams from one codec to another). This can be useful for compatibility reasons, but it requires complex signal processing techniques and can be computationally demanding.

Before media streams can travel across the network, they are divided into packets and become the payload of RTP headers (shown in Figure 3.2). As with any other header, the fields in the RTP header are overhead needed to support media packetization. The header contains several fields: (1) Version that is a 2-bit field that always states 2, indicating the current RTP version, (2) Padding that is a flag that signals whether extra padding bytes are added to the end of the packet, often needed for certain codecs to function correctly. The last byte of the padding clarifies its length,(3) Extension field that is a flag that indicates whether an additional header section follows the main RTP header, providing extra information, (4) CSRC Count 4-bit field that specifies the number of

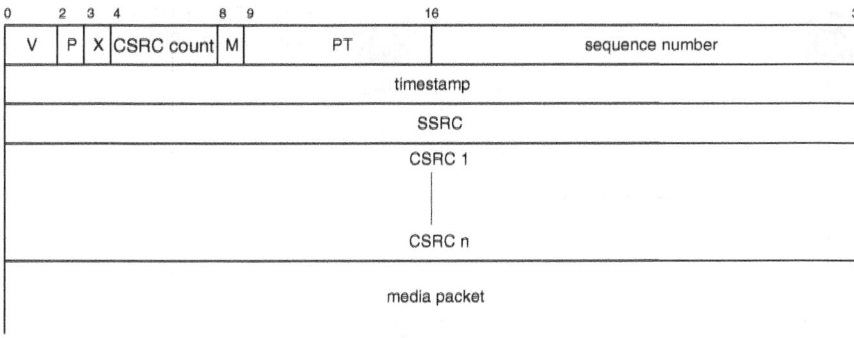

V version
P padding
X extension
M marker
PT payload type
SSRC synchronization source
CSRC contributing sources

Figure 3.2 RTP Header Format and Payload.

Contributing Source (CSRC) fields included in the header, (5) Marker that is a multi-purpose flag can signal important events or boundaries within the media stream, (6) Payload Type that is a 7-bit field that identifies the specific codec used to encode the media data, ensuring proper decoding at the receiving end, (7) Sequence Number that is a 16-bit field that helps receivers track the order of packets and detect any that might have gotten lost along the way, (8) Timestamp that is an important 32-bit field that pinpoints the exact sampling instant of the first byte in the payload, essential for synchronization and estimating any timing irregularities (jitter), (9) *Synchronization Source* (SSRC) that is a 32-bit random number that uniquely identifies the particular media stream, preventing confusion when multiple streams are in play and (10) the CSRC List that is a list of 32-bit CSRC values specifying the original sources of the media data in a packet, commonly used when multiple streams are combined by a mixer application (i.e., conference application) (Schulzrinne et al. 2003).

RTP packets travel back and forth between the UAC and UAS, as shown in Figure 3.3. As already indicated, the media is divided into smaller packets (or packetized), which are sent at regular intervals by both parties. Each RTP packet carries a unique sequence number that increases with every transmission. However, due to the nature of UDP transport, some packets may inevitably go missing during transmission. The figure demonstrates this with both the UAC and UAS sending eight packets, but not all of them successfully reaching their destination. Despite these potential losses, humans are remarkably resilient to missing audio or video information, so the overall impact on quality is often

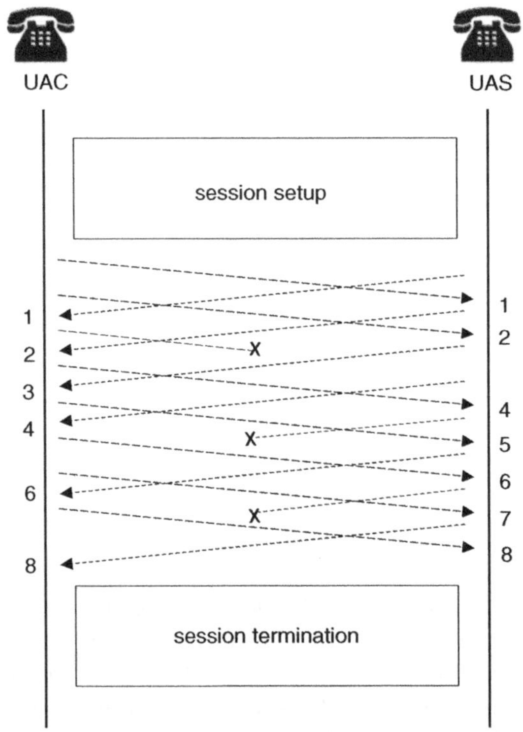

Figure 3.3 RTP Media Transmission Flow.

minimal. Additionally, some codecs have built-in mechanisms to conceal missing packets, further contributing to a positive QoS experience.

Media streams undergo segmentation into frames, which are subsequently compressed using codecs. Multiple speech and audio frames are packetized in accordance with specific packetization rules, as specified in Section 3.5. For a given default codec frame size, possessing a frame duration of f, the RTP packet duration is calculated as $T = n \times f$, where n denotes the number of frames encapsulated within a single RTP packet. Subsequently, for a given sampling rate F_s, the number of speech and audio samples conveyed per RTP packet is determined by $F_s \times n \times f$. To illustrate, consider ITU-T Recommendation G.711, operating with narrowband speech sampled at 8000 sps and two frames per packet ($n = 2$). As the codec frame size equates to 10 milliseconds, the number of speech samples per RTP packet is computed as $8000 \times 2 \times 0.010 = 160$ samples. In the context of video captured at a specified frame rate, a video frame typically undergoes compression and partitioning into smaller chunks, carefully sized to preclude fragmentation. These individual chunks are then transmitted independently over RTP packets.

For synchronization purposes, receivers leverage SSRCs to distinguish individual media streams and sequence numbers to ascertain the precise order in which to render media packets. Specifically, the sequential increment of the sequence number with each transmitted packet enables receivers to estimate packet loss and subsequently endeavor to reconstruct the original packet sequence. In conjunction with the sequence number, the timestamp field provides an additional piece of information, instructing the receiver on the exact timing for media packet playback and providing synchronization cues. Note that certain applications, such as video, exhibit a characteristic where multiple consecutive RTP packets bear identical timestamps, as they collectively constitute a single video frame. Conversely, in the majority of audio and speech scenarios, timestamps exhibit a change in tandem with sequence numbers.

Example 3.1 A microphone is sampled at a WB rate. Samples are compressed as 16-bit values and buffered into 320-byte frames that are transmitted as RTP packets. Assume the link layer header is 14 bytes long. Assume IPv4. What is the overall transmission rate?

Solution:
320 bytes corresponds to 160 samples per RTP packet. At 16000 sps, this implies 100 RTP packets are transmitted per second. Each packet is $320 + 12 + 8 + 20 + 14 = 374$ bytes, leading to $R = 374 \times 8 \times 100 = 299.2$ Kbps. Note that the UDP and IPv4 headers are respectively 8 and 20 bytes long.

RTP stands as a versatile protocol, offering functionality that bears variable interpretations contingent upon the specific codec in question. For instance, the extension bit serves to signal the presence of an extension header following the RTP header. This extension header, in turn, may include additional fields that hold relevance to the codec's operations. In a similar fashion, the marker bit typically finds use, depending on the codec, to delineate the start of a stream. This proves particularly significant in scenarios where a codec extends support for DTX. Under DTX, *Voice Activity Detection* (VAD) may initiate a period of silence during which the codec refrains from producing any frames. Upon subsequent detection of speech, the media stream resumes, and this resumption is signaled by setting the marker bit within the RTP header.

3.3 RTCP

The RTP suite incorporates an additional protocol, named RTCP, which serves the purpose of quality control over media streams. Essentially, for each RTP stream,

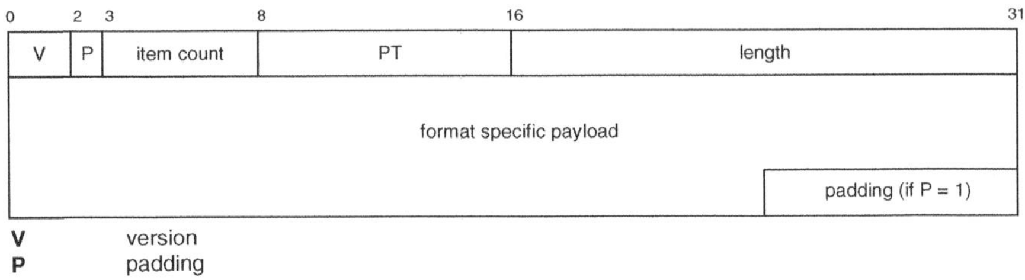

Figure 3.4 RTCP Header Format and Payload.

there exists an optional, associated RTCP stream that enables receivers to furnish feedback regarding the quality of the incoming media packets. Analogous to RTP streams, RTCP streams are transported over UDP, typically employing contiguous ports. As such, RTCP holds particular significance within the context of conferencing multimedia applications.

RTCP establishes an association with an RTP stream through its *Canonical Name* (CNAME) rather than its SSRC, due to the potential for SSRC modification during a session. RTCP also possesses the capability to include supplementary information for reference purposes, such as email addresses or phone numbers. Figure 3.4 shows the format of a generic RTCP header. It starts with a 2-bit version field, invariably set to *2*, followed by a padding bit that signals the presence of additional bytes subsequent to the format-specific information payload. The header additionally incorporates a 5-bit item count, indicating the enumerated items within the payload, an 8-bit packet type specifying the RTCP message type, and a 16-bit length delineating the size of the payload that trails the RTCP header (Schulzrinne et al. 2003).

RTCP employs a diverse array of message types to convey distinct categories of information, including (1) the *Sender Report* (SR) message type that communicates transmission and reception statistics originating from active senders, (2) the *Receiver Report* (RR) message type that conveys reception statistics from passive senders, (3) the *SDP Security Description for Media Streams* (SDES) message type that furnishes a description of a sender, including the CNAME, (4) the *BYE* message type that serves as an explicit indicator of the termination of participation within a session, and (5) the *APP* message type that caters to application-specific functionalities.

3.4 SRTP and SRTCP

Analogous to SIPS, introduced in Section 2.6, security within the realm of RTP traffic is a consequence of requirements that address a multitude of challenges associated with attacks and threats. These requirements have directly spurred the development of SRTP and SRTCP. SRTP offers two distinct approaches for negotiating the security of RTP packets: (1) SDES-Based Mechanism that hinges upon SDES to activate encryption and authentication functionalities and (2) DTLS-Based Mechanism that, in contrast, leverages DTLS to establish the cryptographic keys upon which SRTP depends.

Prior to focus on the specifics of these mechanisms, a comprehensive exploration of generic SRTP transport principles will be undertaken. IETF RTC 3711, formally known as the *Secure Real-Time Transport Protocol* (SRTP), establishes a standardized approach for the generation of SRTP and SRTCP packets. To execute encryption and authentication within both SRTP and SRTCP, session encryption keys, authentication keys, and salting keys are derived from a master key and a master salt. In the realm of cryptography, salting introduces an additional layer of randomization, preventing table-based attacks. The exchange of the master key and master salt transpires through signaling mechanisms such as the SDP within SIP or WebRTC sessions.

IETF RFC 3711 outlines two distinct AES-based CBC mechanisms for encryption within SRTP, each designed to facilitate controlled randomization of the encrypted data: (1) *Counter Mode* (CM) Encryption and (2) f8-Mode Encryption, founded upon the f8 algorithm. Analogously, authentication within SRTP enables the generation of *HASH-based Message Authentication Codecs* (HMAC), which are embedded within every message when enabled. The hash function employed in this context is SHA-1, and the authentication tag can be configured as either a short 32-bit tag or a lengthier 80-bit tag (Carrara et al. 2004).

The most prevalent SRTP suites are outlined within two key IETF RFCs: IETF RFC 3711 and IETF RFC 6188, titled *The Use of AES-192 and AES-256 in Secure RTP*. Key SRTP suites from IETF RFC 3711 include (1) AES_CM_128 _HMAC_SHA1_32 that employs AES-128 in CM mode with a 32-bit SHA-1 authentication tag, (2) AES_CM_128_HMAC_SHA1_80 that utilizes AES-128 in CM mode with an 80-bit SHA-1 authentication tag, and (3) F8_128_HMAC _SHA1_32 that leverages AES-128 in f8-Mode with a 32-bit SHA-1 authentication tag.

Similarly, key SRTP suites from IETF RFC 6188 include (1) AES_192_CM _HMAC_SHA1_32 that employs AES-192 in CM mode with a 32-bit SHA-1 authentication tag, (2) AES_192_CM_HMAC_SHA1_80 that utilizes AES-192 in CM mode with an 80-bit SHA-1 authentication tag, (3) AES_256_CM_HMAC _SHA1_32 that employs AES-256 in CM mode with a 32-bit SHA-1 authentication tag, and (4) AES_256_CM_HMAC_SHA1_80 that utilizes AES-256 in CM mode with an 80-bit SHA-1 authentication tag.

The SRTP header, illustrated in Figure 3.5, expands upon the RTP header to provide enhanced security features. It incorporates padding to guarantee that the

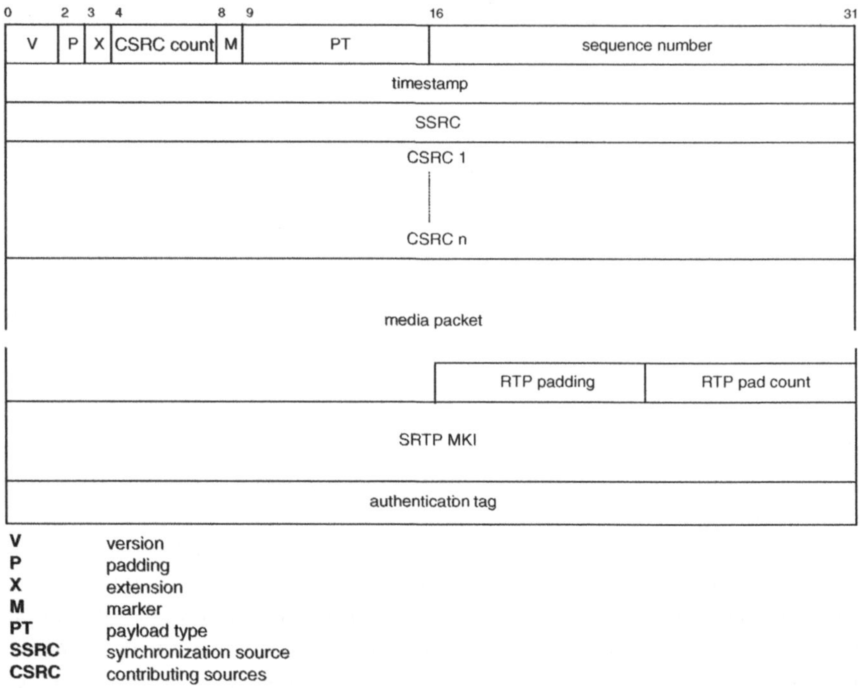

Figure 3.5 SRTP Header Format and Payload.

payload aligns seamlessly with byte boundaries, a necessity for compatibility with the cipher's block size. It also offers the option to include a *Master Key Identifier* (MKI) field, a variable-length component that signals the specific master key employed for encryption and authentication within the RTP packet. Additionally, it integrates an HMAC authentication tag, computed over the original RTP header and payload to ensure integrity. It is important to note that the MKI field itself remains outside the scope of authentication.

Figure 3.6 shows the structure of the SRTCP header, which mirrors the SRTP header's approach to security enhancements. It builds upon the foundation of the RTCP header, as illustrated in Figure 3.8, by incorporating an optional variable-length MKI field and an HMAC authentication tag. These additions enhance both confidentiality and integrity within RTCP communications.

The upcoming sections will discuss two pivotal mechanisms that support this process: SDES and DTLS-SRTP. What follow are comprehensive explanations of their respective roles and functionalities.

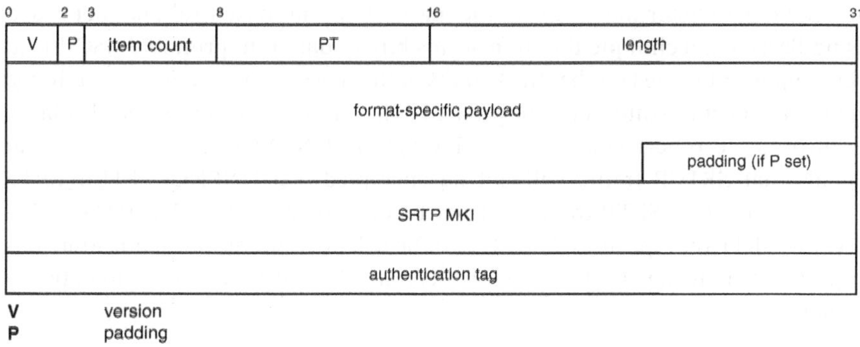

Figure 3.6 SRTCP Header Format and Payload.

3.4.1 SDES

IETF RFC 4568, titled *SDP Security Description for Media Streams* outlines a method for effectively communicating the master key and master salt, as introduced in Section 3.4. These critical parameters are conveyed within the SDP of SIP messages. To achieve this, they are encoded as an SDP attribute line of the crypto type (Andreasen et al. 2006). To understand how this works, let us introduce a practical example. Consider the following SDP transmitted within an SIP INVITE:

```
v=0
o=session 5432105 2394210529 IN IP4 userA.caller.com
s=SDP exchange
c=IN IP4 192.168.1.10
t=0 0
m=audio 5200 RTP/AVP 0
a=sendrecv
a=crypto:1 AES_CM_128_HMAC_SHA1_80 inline:LUXAiAmCt+
s80cCUPz128i2V4ERaoc/PoaL+9ue/
```

Note that the crypto attribute includes the following components: (1) a tag that is a decimal number (1 in this example) that serves as a unique identifier, important for correlating offers and answers within the negotiation process, (2) the SRTP suite that is a string that represents the chosen SRTP suite, AES_CM_128_HMAC_SHA1_80, in this specific case. Refer back to Section 3.4 for a comprehensive list of commonly used crypto suites that can be negotiated through SDES, (3) the Inline parameter that encloses the master key and master salt, both encoded using Base64, a technique that transforms binary data into printable ASCII text. Very important is the fact that the lengths of the master key and master salt hinge upon the specific suite being negotiated. The crypto line offers the flexibility to incorporate other parameters, including: (1) UNENCRYPTED_SRTCP that signals that SRTCP traffic will not be encrypted, (2) UNENCRYPTED_SRTP that indicates that SRTP traffic will not be encrypted, and (3) UNAUTHENTICATED_SRTP that specifies that SRTP traffic will not undergo authentication. It is important to note that, by default, both SRTP and SRTCP packets are automatically authenticated.

3.4.2 DTLS-SRTP

While SRTP itself handles the encryption and authentication of RTP packets, DTLS-SRTP emerges as a powerful mechanism for negotiating the master key and master salt, as defined in Section 3.4. This process unfolds through a secure exchange between endpoints, specified by IETF RFC 5764 DTLS. In this context, DTLS possesses the inherent ability to accommodate extensions, effectively expanding its functionality. In this context, IETF RFC 5764 introduces the *use_srtp* extension, tasked with specifying protection profiles associated with specific encryption and authentication suites. Once the use_srtp extension is in play and a DTLS session has been successfully established, the master key and salt can be seamlessly exported from the underlying session information. This addresses a key concept: the DTLS session itself acts as the foundation for determining the keys that will subsequently govern the SRTP session. Note that the actual RTP packets are not encrypted or authenticated directly using DTLS. Instead, this responsibility falls to the mechanisms introduced in Section 3.4. This is because SRTP is designed for efficiency and effectiveness in handling RTP packets, surpassing DTLS in these specific areas. In essence, DTLS-SRTP combines the strengths of both protocols: DTLS supports a secure key exchange, while SRTP excels in the application of those keys to safeguard media traffic (McGrew and Rescorla 2010).

The DTLS-based SRTP security mechanism establishes a robust correlation with the actual media session path (as described in Chapter 2) through the exchange of fingerprints between the participating endpoints. A fingerprint, in this context, refers to the uniquely derived value obtained by applying a cryptographic hash function, such as the venerable SHA-256, to the endpoint's DTLS certificate.

In the fingerprint generation process, the endpoint starts by applying the aforementioned SHA-256 hash function to its SRTP-DTLS certificate, leading to the fingerprint. This generated fingerprint is then sent to the remote endpoint. Upon receipt of the fingerprint, the remote endpoint mirrors the process,

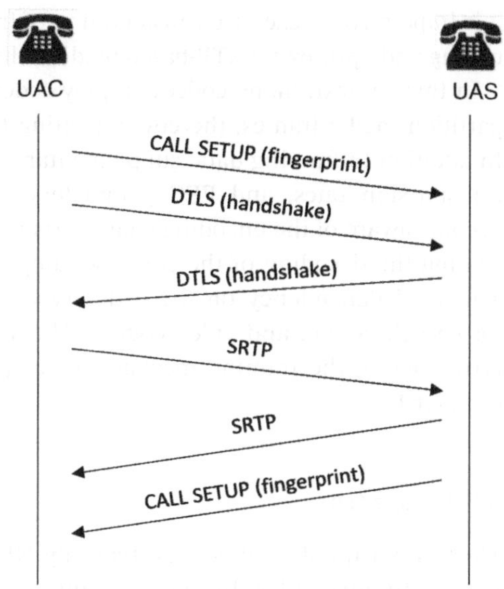

Figure 3.7 DTLS-SRTP Flow.

independently calculating the fingerprint of the received certificate. It then engages in a comparison between the locally generated fingerprint and the one received over the control path. Any discrepancy between the two fingerprints implies that the sending side cannot be authenticated. The fingerprint information is transported within SDP payloads, embedded within either SIP signaling in traditional telephony applications or HTTP exchanges in the context of WebRTC. Figure 3.7 shows the certificate and fingerprint exchange between two endpoints.

3.5 Framing

Due to the nature of media packetization over UDP transport, which lacks any inherent segmentation mechanisms, RTP packets become susceptible to the negative effects of IP fragmentation. As indicated in Section 2.2.2.1, fragmentation introduces a considerable amplification of the actual datagram loss probability, far surpassing the nominal network layer packet loss probability.

Several factors are responsible for this effect. First, the magnitude of fragmentation-induced loss escalates in direct proportion to the size of the datagram relative to the MTU. In the context of RTP traffic, this correlation translates to a dependency on the media frame count encapsulated within individual RTP packets. Moreover, the loss of a single datagram causes a cascading loss of multiple media frames, further amplifying the adverse impact on media quality. Therefore, it is imperative to exercise consideration of fragmentation dynamics when architecting and optimizing RTP-based media delivery systems.

In this context, many codecs employ codec-specific framing rules that aim to partition media frames, thereby mitigating the occurrence of IP fragmentation. In addition to framing data, supplementary information such as sampling rates, transmission rates, and FEC parameters is conveyed to enable endpoints to become aware of the encoding characteristics of the actual media without necessitating the decoding of the corresponding media frame. Framing rules exhibit a marked dependency on the codec type, with distinct approaches employed for speech, audio, and video codecs. The subsequent subsections focus on the intricacies of the framing mechanisms employed by the codecs introduced in Chapter 1.

3.5.1 Speech

There are a number of codecs, typically NB codecs, that do not adhere to any specific framing rules. Instead, multiple compressed media frames are sequentially packaged within the payload of a single RTP packet for these codecs. Some codecs that have been standardized by ITU-T Recommendations, like G.711, G.726, G.723.1, G.729, G.728, and G.722, fall under this category. In addition, GSM 6.10, Speex, iLBC, GSM-EFR, RTAudio, Lyra, and LPC-10 also do not specify RTP-friendly framing mechanisms (Schulzrinne et al. 2003).

Consider Figure 3.8, where a media stream is processed by a codec to generate nine media frames. If three media frames are encapsulated into a single RTP packet, then the media stream results in the transmission of three RTP packets that are sequentially encoded and decoded upon their reception. If the codec supports a transmission rate of R bps and n d-millisecond frames are encapsulated in each RTP packet, then each packet carries $R \times n \times \frac{d}{8000}$ bytes. For an ITU-T Recommendation G.711, where each frame is $d = 10$ milliseconds long and two frames are packetized, each RTP packet carries $64000 \times 2 \times \frac{10}{8000} = 160$ bytes. If the MTU size, measured in bytes, is M, then $R \times n \times \frac{d}{8000} < M$ and, therefore, $n < \frac{M \times 8000}{d \times R}$. For the same ITU-T Recommendation G.711 scenario and assuming an MTU that

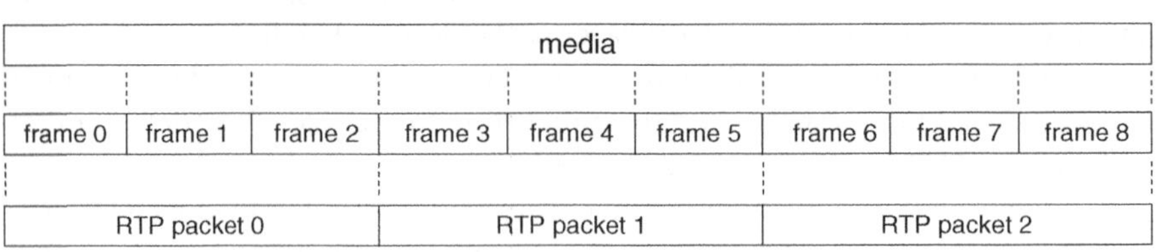

Figure 3.8 Simple Framing Example.

is $M = 1500$ bytes long, then $n < \frac{1500\times8000}{10\times64000} = 18.75$. This means that up to 18 frames can be encapsulated in every RTP packet without causing IP fragmentation.

For all these codecs, the SDP payload signals the codec name and sampling rate. Consider the following snippet of an SDP payload associated with the ITU-T Recommendation G.711:

```
...
m=audio 5100 RTP/AVP 0
a=rtpmap:0 pcmu/8000
...
```

The payload type 0 is reserved for ITU-T Recommendation G.711 μ-Law, and the sampling rate as with any NB codec is 8000. Table 3.1 lists codec names and their corresponding payload types. Note that ITU-T Recommendation G.726 includes a mode that originates on the *ATM Adaptation Layer 2* (AAL2). *Asynchronous Transfer Mode* (ATM) is a legacy networking technology that has been largely superseded by IP networking. Also note that payload types 96 (and above) are dynamically negotiated and they are not assigned to one specific codec. The following subsections address the non-standard framing of other relevant speech codecs.

Table 3.1 Codec Name and Payload Type.

Name	Payload Type	Description
pcmu	0	ITU-T Recommendation G.711 μ-Law
pcma	8	ITU-T Recommendation G.711 A-Law
g726	96	ITU-T Recommendation G.726
aal2-g726	96	ITU-T Recommendation G.726 AAL2
g723	4	ITU-T Recommendation G.723
g729	18	ITU-T Recommendation G.729
g728	15	ITU-T Recommendation G.728
g722	9	ITU-T Recommendation G.722
gsm	3	GSM 6.10
speex	96	Speex
ilbc	96	iLBC
gsm-efr	96	GSM-EFR
x-msrta	96	RTAudio
lyra	96	Lyra
lpc10	96	LPC-10

Example 3.2 If 20 milliseconds of raw speech are compressed by means of the ITU-T Recommendation G.711 codec and transmitted over RTP packets. Assume IPv4 and a 14-byte link layer header. How big are these packets?

Solution:
The ITU-T Recommendation G.711 codec samples at 8000 sps leading to $8000 \times 0.02 = 160$ samples per packet. Since each sample is encoded with one byte, the RTP packet size is $160 + 12 + 8 + 20 + 14 = 214$ bytes where 8 and 20 are respectively the UDP and the IPv4 header sizes.

3.5.1.1 AMR and AMR-WB

RTP Packetization of AMR and AMR-WB codec frames is standardized by IETF RFC 4867 *RTP Payload Format and File Storage Format for Adaptive Multi-Rate (AMR) and Adaptive Multi-Rate Wideband (AMR-WB) Audio Codecs.* Because AMR-WB is, in turn, standardized as ITU-T Recommendation G.722.2, this RFC covers its packetization as well (Sjoberg et al. 2007).

The RFC lays out a specific blueprint for RTP packet construction, as illustrated in Figure 3.9. Each packet is organized into three distinct sections: (1) a payload header that serves as a dynamic control center, allowing for adaptability in selecting between AMR and AMR-WB rates, (2) a Table of Contents (ToC) that outlines the types of frames that reside within the speech payload, and (3) a speech payload that carries the actual speech frames themselves.

The payload header is shown in Figure 3.10. It contains a 4-bit *Codec Mode Request* (CMR) field that specifies the actual AMR or AMR-WB rate being requested to the receiver. The field value corresponds to the index shown in Table 3.2. If the CMR is set to 15 (all ones), then the transmitter requests any rate from the receiver. In octet-aligned mode, four reserved bits (set to zero) follow the CMR field. In bandwidth-efficient mode, only the 4-bit CMR field is transmitted.

Figure 3.11 shows the structure of the ToC fields. It starts with a 1-bit continuation flag that it is either true (1) or false (0) and determines whether subsequent

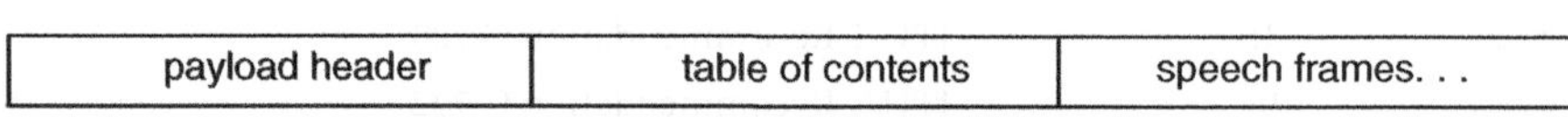

Figure 3.9 IETF RFC 4867 Format.

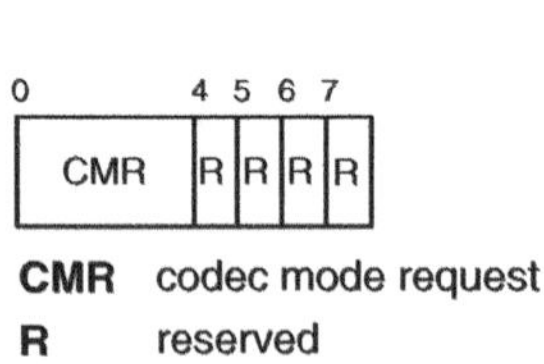

Figure 3.10 Payload Header.

Table 3.2 AMR and AMR-WB Rates (Kbps).

Index	AMR	AMR-WB
0	4.75	6.60
1	5.15	8.85
2	5.90	12.65
3	6.70	14.25
4	7.40	15.85
5	7.95	18.25
6	10.20	19.85
7	12.20	23.05
8		23.85

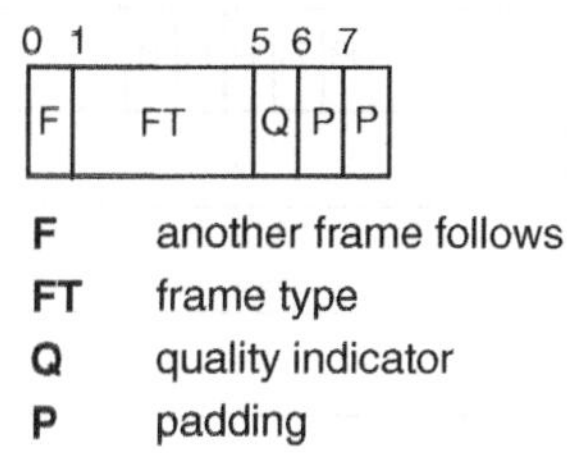

Figure 3.11 Payload Header.

sets of ToC fields are to follow. It continues with a 4-bit *Field Type* (FT) that indicates the rate of the speech frame associated with the ToC, as encoded in Table 3.2. The final 1-bit *Frame Quality* (Q) flag, also indicating true (1) or false (0), communicates whether the corresponding frame is intact or has sustained damage. Octet-aligned mode incorporates two padding bits, set to zero, following the Q field. In contrast, bandwidth-efficient mode omits these padding bits entirely.

Figures 3.12 and 3.13 offer a representation of the encoding process for a 148-bit speech frame, showcasing both bandwidth-efficient and octet-aligned modes, respectively. In bandwidth-efficient mode, the 10-bit payload header and ToC initiate the frame, followed by the speech frame itself. This combination yields an RTP payload size of 158 bits. To achieve octet alignment, 2 padding bits are added, resulting in a total payload size of 20 bytes. In octet-aligned mode, the payload header and ToC, now including 16 bits, precede the 148-bit speech frame. The resulting 164-bit payload receives 4 padding bits to ensure octet alignment, leading to a final payload size of 21 bytes. Note that the difference in size between the two modes is a mere single byte.

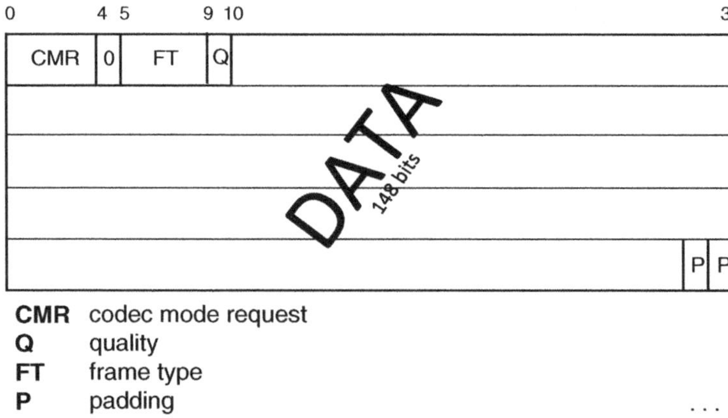

Figure 3.12 Bandwidth Efficient Mode Encoding.

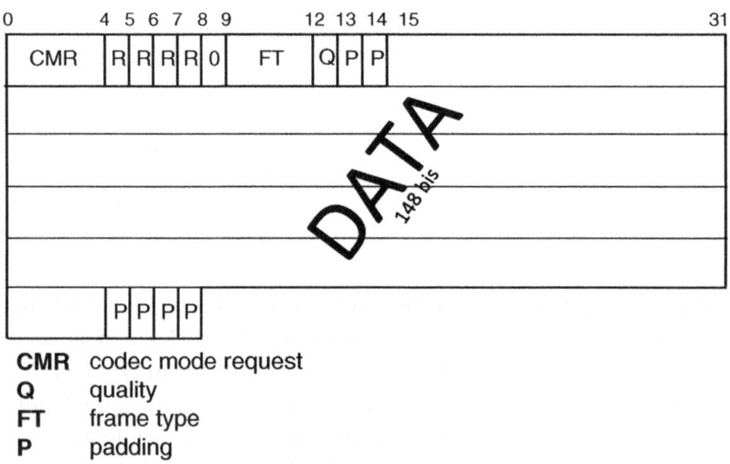

Figure 3.13 Octet-aligned Mode Encoding.

Consider the following SDP payload associated with AMR negotiation:

```
...
m=audio 5100 RTP/AVP 96
a=rtpmap:96 amr/8000
a=fmtp:96 octet-align=0; robust-sorting=1; mode-set=0
...
```

The codec name is amr, and the NB sampling rate is 8000. In addition, the *format specific parameter* (fmtp) attribute is used to specify codec-specific parameters like transmission rates. The octet-aligned parameter indicates whether octet-aligned mode or bandwidth-efficient mode is being requested. The mode-set parameter signals the actual transmission rate. On the other hand, the robust-sorting parameter indicates that the frame payload content must be rearranged to minimize the effect of loss and other impairments. For AMR-WB, the codec name is amr-wb, and the WB sampling rate is naturally 16000.

3.5.1.2 EVRC and EVRC-B

IETF RFC 3558, titled *RTP Payload Format for Enhanced Variable Rate Codecs (EVRC)* establishes a way to structure EVRC speech frames within RTP packets (Li 2003). IETF RFC 4788, titled *Enhancements to RTP Payload Formats for EVRC Family Codecs* builds upon this framework to accommodate the framing of EVRC-B speech frames as well (Xie and Kapoor 2007).

IETF RFC 3558 defines two distinct EVRC speech frame transmission formats over RTP: the EVRC Header-Free Packet Format (Figure 3.14), carrying a single codec frame directly after the RTP header for efficiency, and the EVRC Interleaved Packet Format (Figure 3.15), interleaving multiple codec frames within the RTP payload to mitigate packet loss by dispersing frames and minimizing contiguous frame loss.

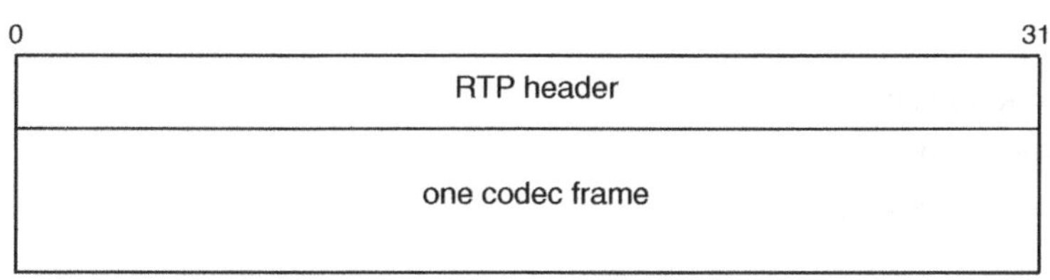

Figure 3.14 EVRC Header-Free Packet Format.

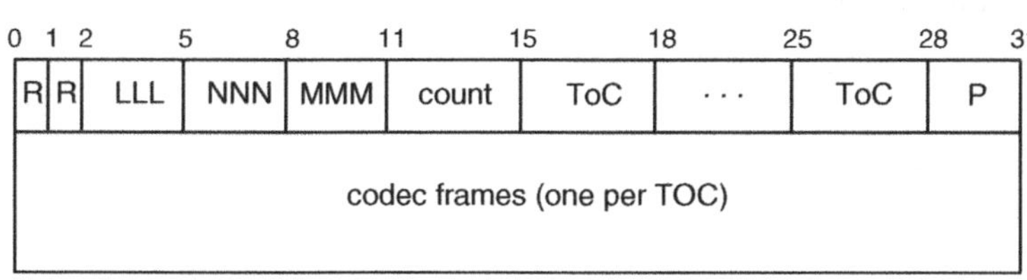

R reserved
LLL interleave length
NNN interleave index
MMM mode request
ToC table of content
P padding

Figure 3.15 EVRC Interleaved Packet Format.

This EVRC Interleaved Packet Format combats packet loss by interleaving multiple codec frames within the RTP payload. This format starts with two reserved bits set to zero, followed by a 3-bit interleave length field specifying the number of interleaved frame groups, a 3-bit mode request field (encoded as per Table 3.3) to communicate the desired incoming transmission rate, a 4-bit ToC count indicating the number of ToC entries, and then a count-specified number of 3-bit ToC entries. Padding bits (set to zero) complete the sequence to ensure 8-bit boundary packing. Note that when the interleave length is set to zero, bundling is enabled, allowing a single group to carry a sequence of multiple speech frames.

Consider the example of an interleave length of four (LLL = 3) and three frames per RTP packet. In this case, the following RTP packets would be sent:

1. LLL = 3, NNN = 0, Frame 0, Frame 4, Frame 8
2. LLL = 3, NNN = 1, Frame 1, Frame 5, Frame 9
3. LLL = 3, NNN = 2, Frame 2, Frame 6, Frame 10
4. LLL = 3, NNN = 3, Frame 3, Frame 7, Frame 11

In addition, IETF RFC 4788 introduces a compact bundled format that supports the packing of multiple same-type EVRC and EVRC-B speech frames as indicated in Figure 3.16. The following is an SDP payload associated with the negotiation of EVRC parameters:

```
...
m=audio 5100 RTP/AVP 96
a=rtpmap:96 evrc/8000
a=fmtp:96 maxinterleave=2
...
```

where the maxinterleave parameter under the fmtp attribute specifies the maximum value for interleaving.

Table 3.3 EVRC and EVRC-B Rates (Kbps).

Index	Rate
0	blank
1	0.80
2	2.00
3	4.00
4	8.55
5	erasure

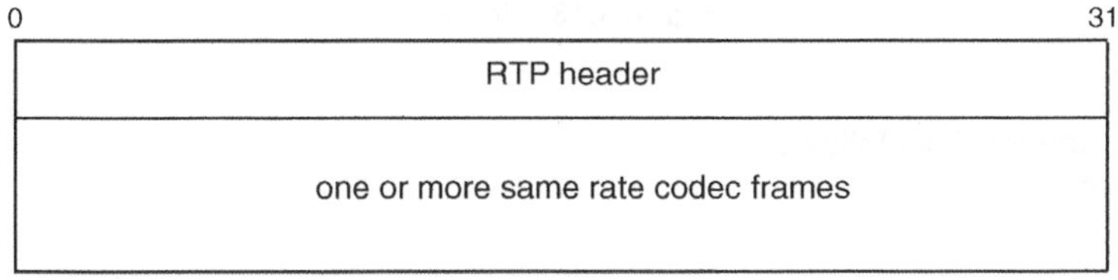

Figure 3.16 Compact Bundled Format.

3.5.1.3 EVS

To ensure a consistent and standardized approach to transmitting EVS codec traffic over RTP networks, the *European Telecommunications Standards Institute* (ETSI) has established a specific framing format, formally defined in TS 126 445 *Codec for Enhanced Voice Services (EVS)*. This format provides two distinct modes of operation, each tailored to address specific transmission needs. It prioritizes efficiency by minimizing overhead within RTP packets, optimizing bandwidth utilization and it emphasizes robustness and flexibility by incorporating additional information within the RTP header, enhancing error resilience and enabling more granular control over transmission parameters (3GPP 2015).

These modes are designed to effectively handle both primary and AMR-WB IO EVS speech frames, ensuring reliable and efficient delivery of EVS codec traffic across RTP networks. Compact mode mirrors the efficient frame packing of EVRC Header-Free Packet Format (Figure 3.14), carrying a single EVS speech frame per RTP packet. Header full mode, akin to AMR framing, enables multiple EVS frames per RTP packet, optionally including a CMR field for receiver rate requests. Both modes invariably feature a ToC entry list within each header full packet.

In the EVS RTP framing format shown in Figure 3.17, a single zero-value bit signals the presence of a ToC. Each ToC entry subsequently indicates whether additional ToCs are packed within the packet. The corresponding sequence of speech frames directly follows the ToC entries, ensuring a clear and ordered structure for packet transmission.

Within the EVS RTP framing scheme, ToC entries follow a structured format, shown in Figure 3.18. Each ToC begins with a fixed zero-value bit, followed by a single bit that signals whether another ToC is present. The remaining six

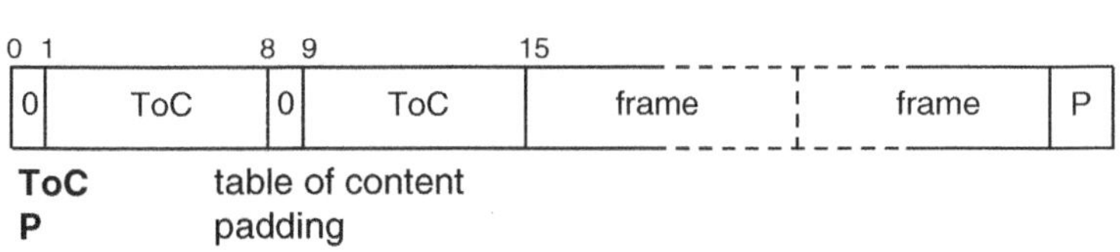

Figure 3.17 ToC-only Format.

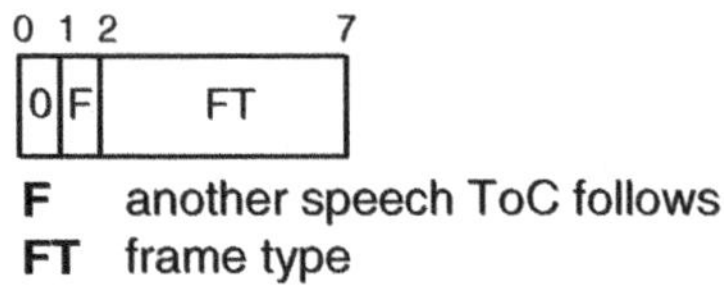

Figure 3.18 ToC Format.

Table 3.4 EVS Frame Type.

EVS mode	AMR-WB Q bit?	EVS Bitrate	Description
0	0	0	primary 2.80
0	0	1	primary 7.20
0	0	2	primary 8.00
0	0	3	primary 9.60
0	0	4	primary 13.20
0	0	5	primary 16.40
0	0	6	primary 24.40
0	0	7	primary 32.00
0	0	8	primary 48.00
0	0	9	primary 64.00
0	0	10	primary 96.00
0	0	11	primary 128.00
1	Q	0	AMR-WB IO 6.60
1	Q	1	AMR-WB IO 8.85
1	Q	2	AMR-WB IO 12.65
1	Q	3	AMR-WB IO 14.25
1	Q	4	AMR-WB IO 15.85
1	Q	5	AMR-WB IO 18.25
1	Q	6	AMR-WB IO 19.85
1	Q	7	AMR-WB IO 23.05
1	Q	8	AMR-WB IO 23.85

bits encode the frame type, adhering to the encoding specifications outlined in Table 3.4.

To accommodate CMR fields, the format in Figure 3.17 can be extended as illustrated in Figure 3.19. The presence of a CMR field is signaled by setting the first bit to one, directly followed by the CMR field itself. The subsequent structure remains consistent, including the ToC list and corresponding speech frames.

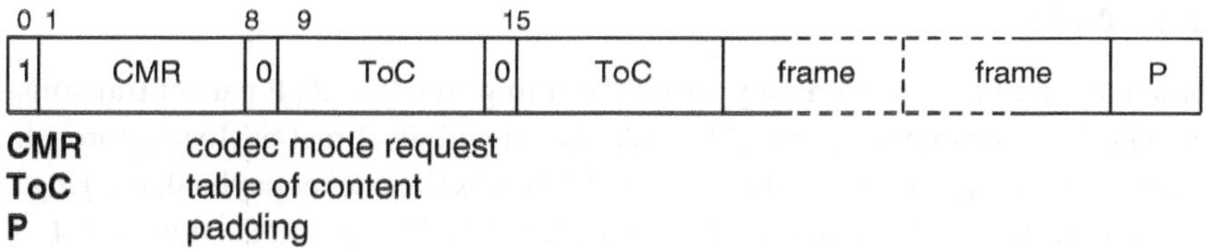

Figure 3.19 CMR and ToC Format.

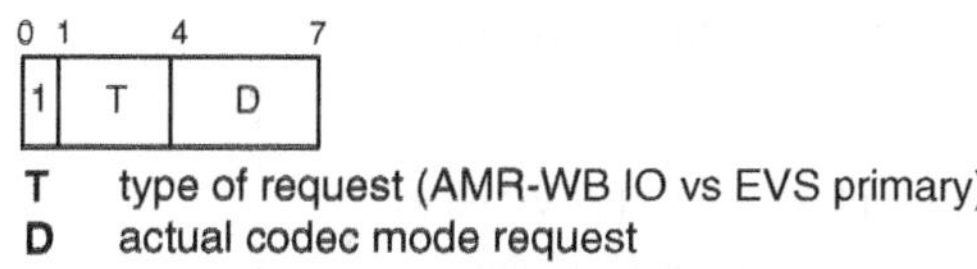

Figure 3.20 CMR Format.

Figure 3.20 shows the format of the CMR field. After the first set bit, the CMR continues with a 3-bit field that specifies whether the request is a primary or an AMR-WB IO one. This field also specifies the sampling rate associated with mode: NB, WB, SWB, and FB. For primary rates, if the field is 0, the sampling rate is NB, if it is 2, the sampling rate is WB, if it is 3, the sampling rate is SWB, if it is 4, the sampling rate is FB. For AMR-WB IO, if the field is 1, then the sampling rate is, of course, WB. The last four bits of the CMR specify the actual requested transmission rate for each of the modes and sampling rates.

The CMR field, illustrated in Figure 3.20, has a structured format that includes (1) a single bit flag set to 1, indicating the presence of the CMR field, (2) a 3-bit mode and sampling rate field that encodes both the request type (primary or AMR-WB IO) and the associated sampling rate (NB, WB, SWB, or FB) and a 4-bit requested transmission rate field that explicitly specifies the requested transmission rate. The following is an example of an SDP payload for EVS:

```
...
m=audio 5100 RTP/AVP 96
a=rtpmap:96 evs/8000
a=fmtp:96 cmr=1;hf-only=1;evs-mode-switch=1;mode-set=0
...
```

The fmtp attribute comprises a minimum of four parameters, each serving a distinct purpose. The cmr parameter indicates whether the CMR field is requested, enabling transmission rate negotiation. The hf-only parameter facilitates support for header full mode, a robust framing format for EVS codec traffic. The evs-mode-switch parameter specifies whether primary or AMR-WB IO modes are in operation, determining the codec mode for voice transmission. The mode-set parameter is used to negotiate the specific transmission rate to be employed within the chosen codec mode, ensuring efficient bandwidth utilization.

3.5.2 Audio

While most audio codecs employ specific framing rules for RTP packet transmission, ITU-T Recommendation G.722.1 stands out with its direct payload approach. It seamlessly integrates one or more audio frames as the packet payload, offering a streamlined encapsulation method. IETF RFC 5577 *RTP Payload Format for ITU-T Recommendation G.722.1* formally outlines this distinctive approach along with other pertinent details. The following is an illustrative snippet of an SDP payload associated with this codec:

```
...
m=audio 5100 RTP/AVP 96 97
a=rtpmap:96 g7221/16000
a=fmtp:96 bitrate=24000
a=rtpmap:97 g7221/32000
a=fmtp:97 bitrate=48000
...
```

IETF RFC 5577 mandates the inclusion of a bitrate parameter within SDP payloads, complementing standard parameters like codec name and sampling rates. This parameter explicitly designates the actual transmission rate employed for dynamic payload types 96 and 97, ensuring consistency in network communications (Luthi and Even 2009). In contrast to codecs with defined framing specifications, iSAC, ITU-T Recommendation G.718, and LC3 adopt a more direct approach. They forgo complex framing structures and instead simply position one or more audio frames as the payload within RTP packets, similar to the efficient method illustrated in Figure 3.16. While IETF drafts were proposed to standardize framing for iSAC and G.718, they ultimately expired without becoming official RFCs, leaving these codecs to operate with this straightforward payload-based approach.

3.5.2.1 AAC

IETF RFC 6416 *RTP Payload Format for MPEG-4 Streams* establishes the standardized framing for AAC over RTP, including both audio and video streams. Within this specification, AAC specifically complies with the *Low-overhead MPEG-4 Audio Transport Multiplex* (LATM) format, a structure designed to minimize overhead while ensuring efficient transmission of AAC audio data (Doehla et al. 2011).

As illustrated in Figure 3.21, the LATM format arranges a sequence of audioMuxElements, each potentially containing multiple AAC frames, as the payload within RTP packets. This efficient encapsulation strategy may optionally

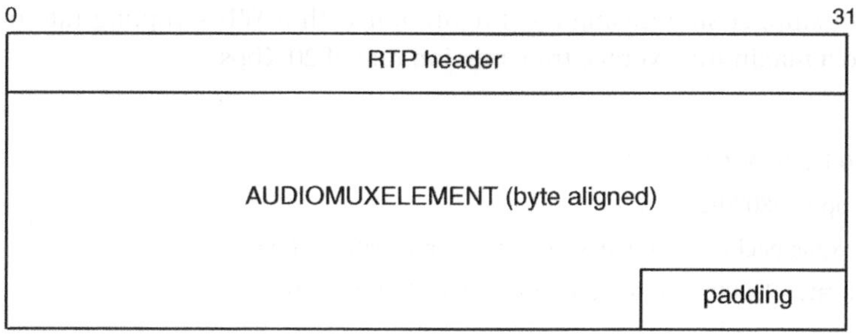

Figure 3.21 LATM Format.

incorporate padding bits (set to zero) to achieve complete byte alignment within the RTP payload. Consider the following snippet of an SDP payload signaling a 64 Kbps AAC stereo bitstream sampled at FB (48000 sps):

```
...
m=audio 5100 RTP/AVP 96
a=rtpmap:96 mp4a-latm/48000/2
a=fmtp:96 profile-level-id=44; bitrate=64000; cpresent=0; config=40005623101fe0;
SBR-enabled=1
...
```

The codec name, specified as mp4a-latm, aligns with the LATM format for AAC audio, while the trailing 2 in the rtpmap line designates stereo audio. Within the fmtp attribute line, five parameters define the audio configuration: (1) profile-level-id unambiguously identifies the AAC profile in use, (2) bitrate explicitly conveys the negotiated transmission rate in bps, (3) cpresent, operating as a Boolean flag (1 for true, 0 for false), indicates whether codec configuration information is embedded within the payload, (4) config furnishes a string encoding the comprehensive codec configuration details, and (5) SBR-enabled signals the presence or absence of *Spectral Band Replication* (SBR), a technique designed to augment coding efficiency at lower bitrates.

3.5.2.2 Opus

IETF RFC 7587 *RTP Payload Format for the Opus Speech and Audio Codec* establishes the standardized framing for Opus over RTP. Mirroring the direct payload approach employed by ITU-T Recommendation G.722.1, multiple Opus audio frames are seamlessly integrated as the payload within RTP packets, as shown in Figure 3.16. Note that the RFC does not mandate any specific SDP parameters for negotiation during session establishment, offering flexibility in configuration (Spittka et al. 2015). As an illustrative example, consider the following SDP snippet showcasing a configuration with a WB sampling rate of 16000 sps and a maximum average transmission rate of 20 Kbps:

```
...
m=audio 5100 RTP/AVP 96
a=rtpmap:96 opus/48000/2
a=fmtp:96 maxplaybackrate=16000; sprop-maxcapturerate=16000;
maxaveragebitrate=20000; stereo=1; useinbandfec=1; usedtx=0
...
```

The fmtp attribute line within the SDP snippet outlines the following key parameters: (1) maxplaybackrate specifies the highest sampling rate that the receiver can effectively render, (2) sprop-maxcapturerate specifies the maximum sampling rate that the sender is capable of generating, (3) maxaveragebitrate, conveying the average rate anticipated by the receiver, ensures alignment between transmission and reception capabilities, (4) stereo, when set, designates stereo audio, otherwise defaulting to mono, (5) useinbandfec, when present, signals support for FEC, enhancing resilience against packet loss, and (6) usedtx, when set, denotes support for DTX, a technique for conserving bandwidth during periods of silence. It is important to note that the rtpmap attribute consistently indicates 2-channel stereo support at FB 48000 sps.

3.5.3 Video

Due to the substantial size of video frames, which often surpass the capacity of individual RTP packets, stringent framing and packetization guidelines are essential to avert IP fragmentation and the associated heightened risk of packet loss. Most of the standardization efforts for video codecs has predominantly been carried out by the ITU.

3.5.3.1 ITU-T Recommendation H.261

IETF RFC 4587 *RTP Payload Format for H.261 Video Streams* establishes the standardized framing for ITU-T Recommendation H.261 over RTP. To circumvent IP fragmentation, each video frame undergoes partitioning into blocks that align with the capacity of individual IP datagrams. The encapsulation process involves positioning the H.261 stream directly after the H.261 header, as illustrated in Figure 3.22 (Even 2006).

To safeguard against truncation of *Macroblocks* (MBs) and other image elements during RTP transmission, each packet starts with two important 3-bit fields: *Start Bit Position* (SBIT) and *End Bit Position* (EBIT). These fields collectively ensure the integrity of image data within packets. The header then proceeds with two

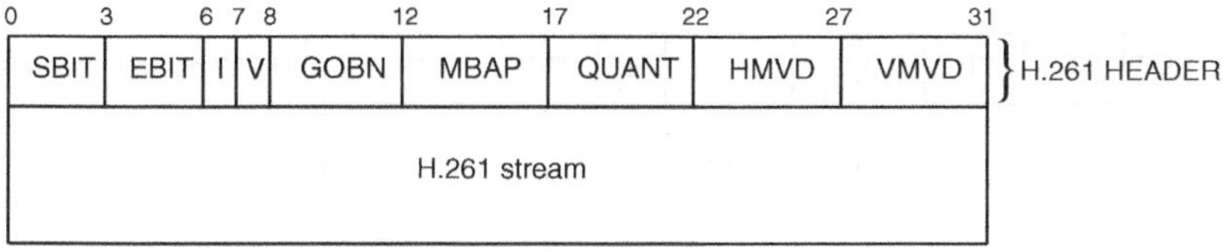

SBIT start bit position
EBIT end bit position
I intra frame data
V motion vector data
GOBN GOB number
MBAP MB address predictor
QUANT quantizer
HMVD horizontal MB data
VMVD vertical MB data

Figure 3.22 H.261 Header Format.

flags, *I* and *V*, which respectively indicate the presence of intraframe encoding and motion vector encoding. A subsequent 4-bit field, *GOBN*, specifies the *Group of Blocks* (GOB) number, followed by a 5-bit *MB Address Predictor* (MBAP) that designates the final *MB Address* (MBA) encoded within the preceding packet. Next, a 5-bit *QUANT* quantizer field communicates the quantizer value that was active prior to the start of the current RTP packet. The header concludes with two 5-bit fields harboring horizontal and vertical motion vector reference data, providing essential information for motion estimation and compensation.

3.5.3.2 ITU-T Recommendation H.263

While IETF RFC 2190 *RTP Payload Format for H.263 Video Streams* initially standardized the framing for ITU-T Recommendation H.263, it is now considered a legacy format. Newer implementations adhere to IETF RFC 4629 *RTP Payload Format for ITU-T Recommendation H.263 Video*. This updated format offers three distinct modes of operation (A, B, and C), each tailored for specific packet size requirements. Mode A employs fragmentation based on GOB boundaries, ensuring alignment with larger packet sizes, Mode B and Mode C facilitate fragmentation at the more granular level of MB boundaries, enabling finer control over packet size and adaptability to varying network conditions (Zhu 1997).

Figure 3.23 shows the structure of the Mode A header, which bears similarities to the framing in IETF RFC 4587. It starts with the SBIT and EBIT fields, each spanning 3 bits, ensuring the integrity of image elements within packets. The 3-bit *Source Format* (SRC) field maps to the resolutions supported by ITU-T Recommendation H.263. Four flags convey essential encoding details: I (intraframe mode), U (unrestricted motion vector mode), S (arithmetic coding mode), and A (advanced prediction mode). The header further incorporates a 2-bit *Differential*

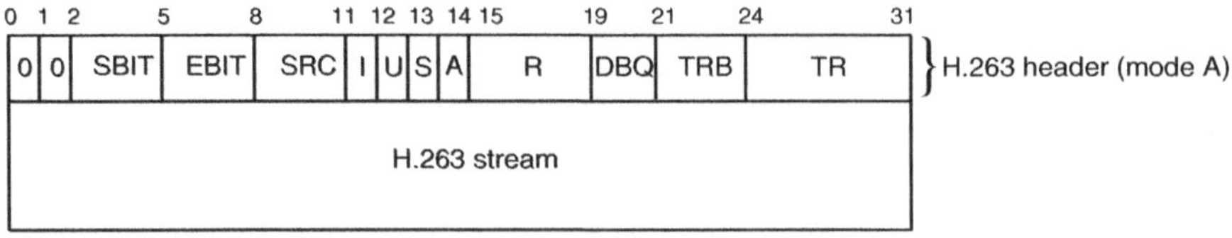

SBIT start bit position
EBIT end bit position
SRC source format (resolution)
I inter vs intra coded
U unrestricted MV
S syntax-based arithmetic coding
A advanced prediction
R reserved (all zeros)
DBQ differential quantization parameter
TRB temporal reference for the B frame
TR temporal reference for the P frame

Figure 3.23 H.263 Header Format (Mode A).

Quantization parameter (DBQ), a 3-bit *temporal reference for B frames* (TRB) field, and an 8-bit *temporal reference for P frames* (TR) field, each aligning with the specifications defined within ITU-T Recommendation H.263.

The Mode B header, shown in Figure 3.24, includes all parameters found within the Mode A header, but further introduces several additions. The QUANT (5 bits)

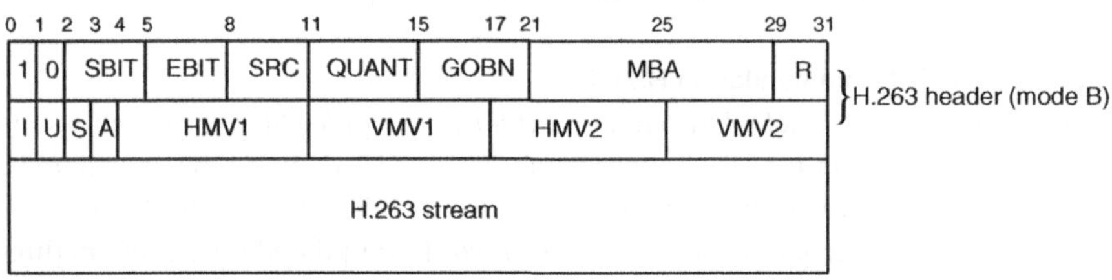

SBIT start bit position
EBIT end bit position
SRC source format (resolution)
QUANT quantization value for first MB
GOBN GOB number at the start of packet
MBA MB address
R reserved (all zeros)
I picture coding type
U unrestricted MV
S syntax based arithmetic coding
A advanced prediction
HMV1/2 horizontal motion vector predictor
VMV1/2 vertical motion vector predictor

Figure 3.24 H.263 Header Format (Mode B).

and GOBN (5 bits) mirror those in IETF RFC 4587 and convey the quantizer value and GOB number, respectively. The MBA (9 bits) field designates the MB address within the GOB of the initial MB positioned within the RTP packet. The HMV1 (7 bits) and HMH1 (7 bits) fields furnish horizontal and vertical motion vectors, respectively, for the foremost MB residing within the packet. The HMV2 (7 bits) and HMH2 (7 bits) fields, in parallel with HMV1 and HMH1, provide horizontal and vertical motion vectors for block 3 within the first MB situated in the packet.

Figure 3.25 presents the framing structure employing a Mode C header, which supports the entire set of parameters included within both Mode A and Mode B headers, including temporal references. IETF RFC 4629 further introduces a mechanism specifically designed to accommodate framing for both ITU-T Recommendation H.263 and ITU-T Recommendation H.263v2 (also known as H.263+) traffic (Wenger et al. 2007). Figure 3.26 shows the header format devised in accordance with this RFC. It is characterized by a notable simplification and a reduced number of fields. Key elements include the (1) *picture start* (P) flag that designates the start of either a complete picture or a segment within the RTP packet, the (2) *video redundancy coding* (VRC) flag that signals the presence or

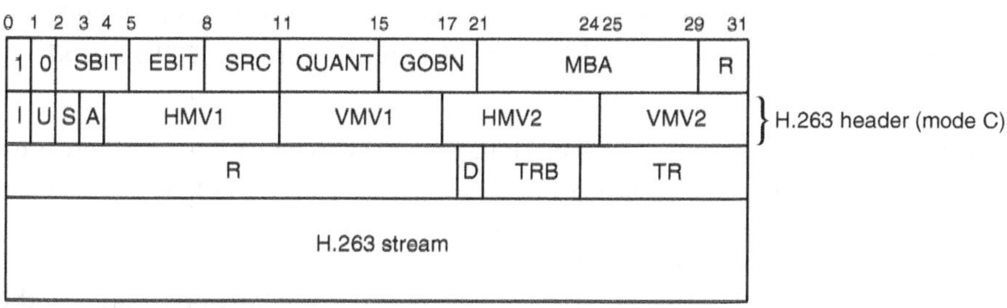

SBIT	start bit position
EBIT	end bit position
SRC	source format (resolution)
QUANT	quantization value for first MB
GOBN	GOB number at the start of packet
MBA	MB address
R	reserved (all zeros)
I	picture coding type
U	unrestricted MV
S	syntax based arithmetic coding
A	advanced prediction
HMV1/2	horizontal motion vector predictor
VMV1/2	vertical motion vector predictor
D	differential quantization parameter
TRB	temporal reference for the B frame
TR	temporal reference for the P frame

Figure 3.25 H.263 Header Format (Mode B).

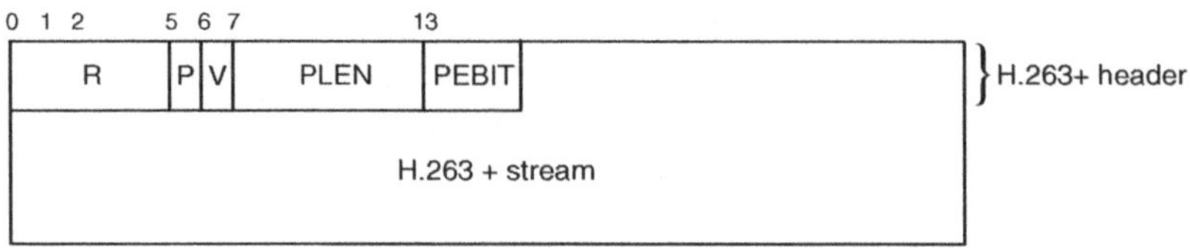

R	reserved (all zeros)
P	picture start
V	video redundancy coding present
PLEN	length of extra picture header
PEBIT	bits to be ignored in the last byte of picture header

Figure 3.26 H.263+ Header Format.

absence of (VRC) information subsequent to the header, the (3) 6-bit PLEN field that facilitates the signaling of additional picture header information beyond the core header elements, and the (4) 3-bit PEBIT field that pinpoints the precise number of bits that should be disregarded within the final byte of the picture header, ensuring accurate data interpretation.

Example 3.3 If a simple 10 fps video raw stream is compressed at 64 Kbps by relying on ITU-T Recommendation H.263. Assume an MTU size of 1500 bytes. Assuming IPv4 and a 14-byte link layer header, how big are the RTP packets?

Solution:
At 10 fps, each video frame is 6400 bits long (or 800 bytes long). Because there is no fragmentation, each RTP packet is $800 + 12 + 8 + 20 + 14 = 854$ bytes on average.

3.5.3.3 ITU-T Recommendation H.264

IETF RFC 6184 *RTP Payload Format for H.261 Video Streams* outlines the standardized framing approach for ITU-T Recommendation H.264. It provides a robust framework for encoding one or more *Network Abstraction Layer Units* (NALUs), fundamental building blocks within H.264 video streams. The RFC delineates several distinct types of NALUs that can be efficiently transported as payloads within RTP packets. Single NALU Packets directly encapsulate a solitary NALU within their payload, offering a straightforward transmission scheme. Aggregation packets, designed for enhanced efficiency, aggregate multiple NALUs into a single RTP packet, minimizing overhead. Two distinct types of Single Time Aggregation Packets (STAP-A and STAP-B) and two types of Multi Time Aggregation Packets (MTAP16 and MTAP24) are available, catering to diverse aggregation requirements. Fragmentation Unit packets, when a single NALU surpasses the capacity of an individual RTP packet, enable fragmentation, distributing the

Table 3.5 NALU Types and Packet Types.

NALU Type	Packet Type
0	reserved
1–23	NALU
24	STAP-A
25	STAP-B
26	MTAP16
27	MTAP24
28	FU-A
29	FU-B
30 and 31	reserved

NALU across multiple packets to ensure successful transmission. Two versions, FU-A and FU-B, offer flexibility in fragmentation strategies (Jesup et al. 2011).

Table 3.5 enumerates the distinct NALU type values that are intrinsically associated with each NALU type within the H.264 framework. Furthermore, IETF RFC 6184 designates three distinct packetization modes, each permissible for every NALU type, furnishing versatility in transmission strategies. The single NALU Mode favors simplicity, encapsulating a single NALU within each RTP packet, ensuring clarity and straightforwardness in transport. In the non-interleaved node: multiple NALUs can reside within a single RTP packet, but they must adhere to a strict sequential order, fostering efficiency while maintaining a clear structure. In interleaved mode, that offers enhanced flexibility, NALUs from distinct video frames coexist within a single RTP packet, unlocking potential for improved resilience against packet loss and reduced latency.

Figure 3.27 illustrates the generic format employed for an RTP payload that dutifully carries a single NALU. It is important to note that all other NALU types, regardless of their specific functionalities, invariably start with the same fundamental trio of fields: (1) 1-bit *Forbidden Zero Bit* that serves as a key indicator of potential errors or syntax violations within the payload, safeguarding against erroneous data propagation, (2) 2-bit *NAL Reference Id* field that designates whether the NALU in question holds references to other NALUs. If it does, it can be discarded without triggering error propagation, optimizing transmission efficiency, and (3) 5-bit NALU Type field that unambiguously identifies the precise type of NALU encapsulated within the payload, ensuring correct interpretation and processing by receiving entities.

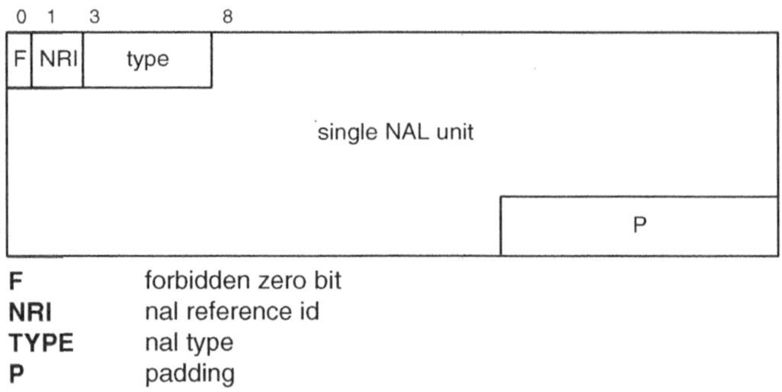

Figure 3.27 Single NAL Unit Packet (H.264).

3.5.3.4 ITU-T Recommendation H.265

IETF RFC 7798 *RTP Payload Format for High Efficiency Video Coding (HEVC)* establishes the standardized framing approach for ITU-T Recommendation H.265. Note that the H.265 codec preserves the utilization of NALUs, which were first introduced within the predecessor codec, ITU-T Recommendation H.264, as detailed in Section 3.5.3.3. This continuity in architectural design facilitates a smoother transition for both developers and implementers (Wang et al. 2016).

Figure 3.28 exemplifies an RTP payload structure that encapsulates a solitary NALU, showing the evolution in header design compared to its predecessor. While IETF RFC 6184 mandated an 8-bit NALU header for ITU-T Recommendation H.264, IETF RFC 7798, governing H.265 framing, prescribes a more expansive 16-bit header. This expansion accommodates enhanced capabilities within the codec.

More importantly, all packets, regardless of their type start with a standardized NALU header, ensuring a consistent framework for interpretation. This header comprises four essential fields, the (1) 1-bit *Forbidden Zero Bit* flag that signals the

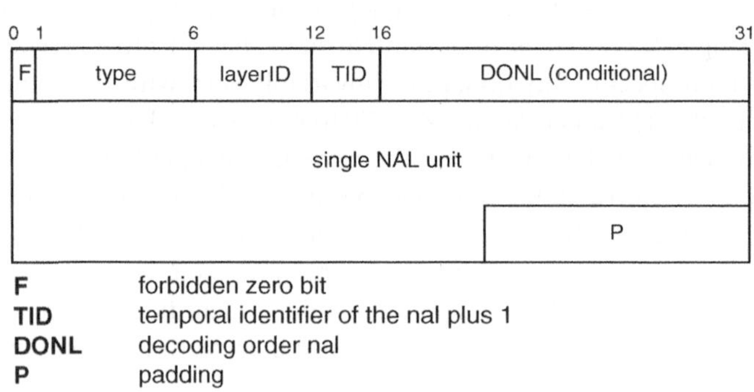

Figure 3.28 Single NAL Unit Packet (H.265).

potential presence of errors or syntax violations within the payload, safeguarding against the propagation of erroneous data, the (2) 5-bit NALU Type field that designates the specific type of NALU encapsulated within the payload, guiding its correct interpretation and processing, the (3) 6-bit Layer Id field, currently set to zero, that holds potential for future utilization in 3D video, offering a foundation for expanded codec capabilities, and the (4) 3-bit Temporal Identifier of the NAL plus 1 field that conveys temporal information pertaining to the NALU, promoting efficient decoding and rendering processes. Note that certain packets, including Single NALU packets, further incorporate a 16-bit Decoding Order NAL (DONL) field. When present, this field explicitly communicates the 16 least significant bits of the NALU's designated decoding order, ensuring precise reconstruction of the video stream.

3.5.3.5 ITU-T Recommendation H.266

IETF RFC 9328 *RTP Payload Format for Versatile Video Coding (VVC)* outlines the framing mechanism for ITU-T Recommendation H.266 packets. Note that H.266, in alignment with its predecessors H.264 and H.265, employs NALUs as fundamental building blocks for video packetization. Furthermore, IETF RFC 9328 mirrors the structural scheme established in IETF RFC 7798, as shown in Figure 3.28, where a solitary NALU resides within the RTP payload (Zhao et al. 2022).

The primary distinction between these two RFCs surfaces in the designated NALU types. For instance, Aggregation Packets bear type values of 48 in IETF RFC 7798, while IETF RFC 9328 assigns them a type of 28. Similarly, Fragmentation Packets hold type values of 49 under IETF RFC 7798, while IETF RFC 9328 designates them with type 29. These distinctions ensure compatibility and prevent potential conflicts during packet interpretation.

3.6 Summary

Codecs, configured through negotiation relying on information exchanged via the SDP protocol, encode raw media into frames. These frames are then encapsulated and transmitted using the RTP protocol to the endpoints. RTP packets are sent as UDP payloads and include sequence numbers and timestamps to guarantee efficient playback with minimal distortion. The RTCP protocol provides a way to carry metrics relevant to the quality of the media streams, including application layer message loss and latency.

When security is critical, RTP and RTCP packets can be respectively encrypted and authenticated through the SRTP and SRTCP protocols. As a way to exchange security parameters, the SDES protocol enables the specification of master and salting keys used to derive encryption and authentication keys. DTLS-SRTP can also be used as an alternative to SDES for exchanging these keys. To mitigate the negative effects of IP fragmentation, most codecs introduce framing schemes that allow for application-controlled fragmentation. Additionally, framing supports the signaling of information needed by the receiving devices, such as transmission rates, sampling rates, and general quality parameters.

3.7 Homework Problems and Questions

Problem 3.1 Why do RTP headers include both sequence numbers and timestamps? Shouldn't just one of them be enough?

Problem 3.2 Why do video codecs fragment video frames at the application layer instead of relying on standard IP fragmentation mechanisms?

Problem 3.3 What is the advantage of using DTLS for key negotiation when compared to SDES?

Problem 3.4 What do SDP signaling ITU-T Recommendation G.711 and EVS speech capabilities look like?

Problem 3.5 A microphone is sampled at an NB rate. Samples are compressed as 8-bit values and buffered into 160-byte frames that are transmitted as RTP packets. Assume the link layer header is 14 bytes long. Assume IPv6. What is the overall transmission rate?

Problem 3.6 Why is RTP transported over UDP, and how do playout buffers help improve media playback quality in RTP applications?

Problem 3.7 How does RTP packetization work, and what considerations are involved when encapsulating media frames into RTP packets?

Problem 3.8 30 milliseconds of raw speech are compressed by means of the ITU-T Recommendation G.711 codec and transmitted over RTP packets. Assuming IPv4 and a 14-byte link layer header, how big are these packets?

Problem 3.9 If 20 milliseconds of raw speech are compressed by means of the 6.6 Kbps AMR-WB codec and transmitted over RTP packets.

Assuming IPv4 and a 14-byte link layer header, how big are these packets?

Problem 3.10 If a simple 30 fps 4:2:2 CIF video raw stream is compressed at 245 Kbps by relying on ITU-T Recommendation H.265, What is the compression rate (compressed video transmission rate divided by raw video transmission rate)? Assuming the link layer header is 14 bytes long, what is the overall transmission rate of the traffic when taking into account all headers?

Problem 3.11 What does an RTP packet carrying two 16.4 Kbps Primary Rate EVS frames look like?

Problem 3.12 What does an RTP packet carrying three octet-align 7.4 Kbps AMR frames look like?

Bibliography

3GPP. (2015). TS 126.445: Codec for enhanced voice services (EVS). Technical Report TS 126.445, 3rd Generation Partnership Project.

Ahson, S.A. and Ilyas, M. (2008). *SIP handbook: services, technologies, and security of session initiation protocol.* 1st ed. USA: CRC Press, Inc. ISBN 1-420-06603-X.

Andreasen, F., Wing, D., and Baugher, M. (2006). Session description protocol (SDP) security descriptions for media streams. RFC 4568, July. Available from: https://www.rfc-editor.org/info/rfc4568.

Carrara, E., Norrman, K., McGrew, D., Naslund, M., and Baugher, M. (2004). The secure real-time transport protocol (SRTP). RFC 3711, March. Available from: https://www.rfc-editor.org/info/rfc3711. (Accessed March 15th, 2024)

Doehla, S., de Bont, F., Kim, J., and Schmidt, M. (2011). RTP payload format for MPEG-4 audio/visual streams. RFC 6416, October. Available from: https://www.rfc-editor.org/info/rfc6416. (Accessed March 15th, 2024)

Even, R. (2006). RTP payload format for H.261 video streams. RFC 4587, August. Available from: https://www.rfc-editor.org/info/rfc4587. (Accessed March 15th, 2024)

Herrero, R. (2023). Practical internet of things networking. *Textbooks in telecommunication engineering*. Springer International Publishing. ISBN 978-3-031-28443-4.

Jesup, R., Kristensen, T., Wang, Y., and Even, R. (2011). RTP payload format for H.264 video. RFC 6184, May. Available from: https://www.rfc-editor.org/info/rfc6184. (Accessed March 15th, 2024)

Li, A.H. (2003). RTP payload format for enhanced variable rate codecs (EVRC) and selectable mode vocoders (SMV). RFC 3558, July. Available from: https://www.rfc-editor.org/info/rfc3558. (Accessed March 15th, 2024)

Luthi, P. and Even, R. (2009). RTP payload format for ITU-T Recommendation G.722.1. RFC 5577, July. Available from: https://www.rfc-editor.org/info/rfc5577. (Accessed March 15th, 2024)

McGrew, D. and Rescorla, E. (2010) Datagram transport layer security (DTLS) extension to establish keys for the secure real-time transport protocol (SRTP). RFC 5764, May. Available from: https://www.rfc-editor.org/info/rfc5764. (Accessed March 15th, 2024)

Perkins, C. (2003). *RTP: audio and video for the internet*. 1st ed. Addison-Wesley Professional. ISBN 0-672-32249-8.

RFC. (1980). User datagram protocol. RFC 768, August. Available from: https://www.rfc-editor.org/info/rfc768. (Accessed March 15th, 2024)

Schulzrinne, H., Casner, S.L., Frederick, R., and Jacobson, V. (2003). RTP: a transport protocol for real-time applications. RFC 3550, July. Available from: https://rfc-editor.org/rfc/rfc3550.txt. (Accessed March 15th, 2024)

Sjoberg, J., Lakaniemi, A., Westerlund, M., and Xie, Q. (2007). RTP payload format and file storage format for the adaptive multi-rate (AMR) and adaptive multi-rate wideband (AMR-WB) audio codecs. RFC 4867, April. Available from: https://www.rfc-editor.org/info/rfc4867. (Accessed March 15th, 2024)

Spittka, J., Vos, K., and Valin, J-M. (2015). RTP payload format for the Opus speech and audio codec. RFC 7587, June. Available from: https://www.rfc-editor.org/info/rfc7587. (Accessed March 15th, 2024)

Wang, Y.-K., Sanchez, Y., Schierl, T., Wenger, S., and Hannuksela, M.M. (2016). RTP payload format for high efficiency video coding (HEVC). RFC 7798, March. Available from: https://www.rfc-editor.org/info/rfc7798. (Accessed March 15th, 2024)

Wenger, S., Bormann, C., Even, R., Sullivan, G., and Ott, J. (2007). RTP payload format for ITU-T Rec. H.263 video. RFC 4629, January. Available from: https://www.rfc-editor.org/info/rfc4629. (Accessed March 15th, 2024)

Xie, Q. and Kapoor, R. (2007). Enhancements to RTP payload formats for EVRC family codecs. RFC 4788, January. Available from: https://www.rfc-editor.org/info/rfc4788. (Accessed March 15th, 2024)

Zhao, S., Wenger, S., Sanchez, Y., Wang, Y.-K., and Hannuksela, M.M. (2022). RTP payload format for versatile video coding (VVC). RFC 9328, December. Available from: https://www.rfc-editor.org/info/rfc9328. (Accessed March 15th, 2024)

Zhu, C. (1997). RTP payload format for H.263 video streams. RFC 2190, September. Available from: https://www.rfc-editor.org/info/rfc2190. (Accessed March 15th, 2024)

4

Media over Network

4.1 Introduction

The opening chapters of this book, Chapters 1 to 3, explain the intricacies of media in modern communication systems, focusing on its transmission and signaling. Chapter 1 lays the groundwork, exploring the mechanisms behind signal generation, processing, compression, and encoding for speech, audio, and video. Chapter 2 then focuses on the protocols and technologies that establish media sessions between endpoints. Finally, Chapter 3 tackles the nuts and bolts of transmission protocols, enabling the delivery of packetized media.

This chapter explores the tools and techniques that enable media sessions across diverse networks, including 3GPP, non-3GPP, and IoT. It focuses on unifying technologies like IP-based media packetization (e.g., RTP) and signaling (e.g., SIP) that facilitate a standardized approach for RTC traffic transmission. While network architectures may vary in their lower layers, the chapter shows the common ground found in upper-layer protocols within the IETF layered architecture. This commonality enables a smooth integration across diverse communication networks. Subsequent chapters will detail the specific protocol stacks and topologies that underpin this integration. Table 4.1 summarizes the main characteristics of the technologies discussed in this Chapter.

4.2 Non-3GPP Networks

Non-3GPP networks operate outside the 3GPP standards commonly used for mobile communications. They leverage traditional wired and wireless physical and link layers to transmit data. This section explores two such networks, Ethernet and Wi-Fi, which support IPv4 encapsulation directly, without requiring any type of adaptation.

Hands-On Real Time Communications: A Practical Guide to RTC Protocols in Non-3GPP, 3GPP, 4G/5G/6G and IoT Networks, First Edition. Rolando Herrero.

Companion website: www.wiley.com/go/herrero/RTCProtocols

Table 4.1 Technology Comparison.

Technology	Licensed	Power Efficient	Coverage
3GPP	yes	no	long
Non-3GPP	no	no	short
IoT (WPAN)	no	yes	short
IoT (LPWAN)	both	yes	long

4.2.1 Ethernet

As a physical and link-layer protocol, Ethernet is very good at connecting devices through rapid packet-switching. It takes data from higher layers, packs it into frames, and sends them across wired links. Standardized by IEEE 802.3, Ethernet transmits data at high speeds, reaching up to 400 Gbps. It offers a variety of signals and cabling options and while it is typically used with higher-layer protocols like IP, UDP, and TCP, Ethernet proves surprisingly good for high-end RTC.

Ethernet has supported various media for data transmission since its invention. It started by supporting coaxial cables, later expanding to support twisted pair and fiber optics. This adaptability mirrored technological advancements, propelling its initial 10–100 Mbps speeds into Fiber-optic options that opened the door to gigabit (1 Gbps), ten-gigabit (10 Gbps), forty-gigabit (40 Gbps), and even hundred-gigabit (100 Gbps) data streams. In 2017, Ethernet was standardized to support 200 Gbps and 400 Gbps rates under IEEE 802.3bs-2017. Pushing the boundaries further, 800 Gbps and 1.6 Tbps nominal transmission rates are in the process of standardization as part of IEEE 802.3df.

At its core, Ethernet relies on line codes for efficient data transmission across these media. These codes, ranging from Manchester encoding to multi-level *Pulse Amplitude Modulation* (PAM), enable modulation and demodulation to ensure seamless communication between endpoints. To maintain perfect synchronization, Ethernet relies on a specific bit transition pattern. The following section focuses on this intricate synchronization protocol (Haykin 2009).

Figure 4.1 shows an Ethernet frame, revealing the structure that carries network data over the payload. The initial bytes, a sequence of alternating ones and zeros called the preamble, act as a handshake between stations, ensuring perfect bit synchronization (explained further in Section 2.2.1). Following this signal, the *Start Frame Delimiter* (SFD), a distinct *10101011* sequence, marks the true beginning of the frame, guiding the receiver to navigate the subsequent fields containing destination, source, and the precious data payload itself. This scheme of bits enables the flawless flow of information within the Ethernet realm.

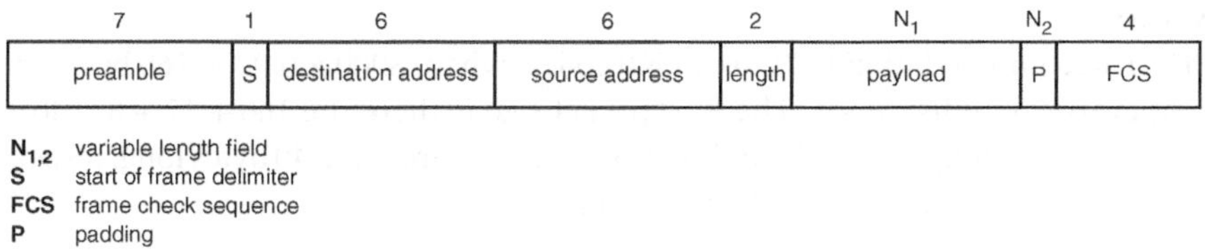

Figure 4.1 Ethernet Header and Payload.

Within the Ethernet frame's structured domain, two key fields dictate the journey of its payload: the 48-bit *Destination Address* (DA) and the matching *Source Address* (SA). The DA shows the intended recipient(s): a single endpoint (unicast), a specific group within the network (multicast), or all the endpoints (broadcast). Meanwhile, the SA identifies the sender and enables response routing. Finally, depending on the frame's adherence to the IEEE 802.3 standard, a dedicated length field quantifies the precious data payload in bytes, ensuring its delivery to the designated destination. This interplay of addresses and size directs the information flow within the Ethernet domain with remarkable precision.

Older Ethernet frames, following the Ethernet II specification, employ a clever trick with the length field. Instead of directly indicating payload size, the field includes an Ethertype value, revealing the type of data it carries. This works because Ethernet's MTU is 1526 bytes. So, lengths equal to or exceeding 1536 (or 0x800 in hexadecimal) signify an Ethertype. For instance, an Ethertype of 1536 reveals an IPv4 datagram tucked inside, while 34525 (0x86DD) indicates IPv6 to the receiver. This allows efficient identification of the encapsulated data without sacrificing space (IEEE 2018).

In IEEE 802.3 frames lacking an Ethertype field, the type of encapsulated data is explicitly identified within a dedicated LLC layer embedded within the payload itself. This distinct layer eliminates reliance on payload length interpretation for data type identification, enabling seamless communication. However, regardless of frame format, all frames must meet a minimum length requirement to ensure efficient collision detection mechanisms. Frames falling short of this threshold are typically padded with additional dummy bits to reach the necessary size. Finally, the last field of every Ethernet frame is the *Frame Checksum* (FCS), a 32-bit block code generated using a *Cyclical Redundancy Checking* (CRC) algorithm. This FCS provides basic error detection during transmission, allowing receivers to identify potential data corruption and initiate retransmission requests if necessary.

Example 4.1 An RTP traffic encapsulates every 20 milliseconds speech using ITU-T Recommendation G.711. Assuming an Ethernet physical and link layer, what is the size of the RTP packets? Assume an 8-byte UDP header length and a 14-byte Ethernet link layer header length. Consider IPv6 as network layer.

Solution:

ITU-T Recommendation G.711 is an NB codec, where $8000 \times 0.02 = 160$ bytes are sent every 20 milliseconds. The RTP packet size is therefore $160 + 12 + 8 + 40 + 14 = 234$ bytes long. The UDP and IPv6 header sizes are 8 and 40 bytes long, respectively. The RTP header size is 12 bytes long.

IEEE 802.2, the LLC protocol layered upon standard IEEE 802.3 frames, offers dual modes for data delivery: connectionless and connection-oriented. These modes cater to different priorities, with connectionless prioritizing speed and simplicity through direct, unacknowledged transmission (acknowledgments can be optionally employed). In contrast, connection-oriented prioritizes reliability by establishing a bidirectional communication channel for error detection and retransmission, ensuring data arrives intact. Figure 4.2 illustrates the key component of the LLC header: the *Service Access Point* (SAP) header. This header contains two identifiers, *Destination Service Access Point* (DSAP) and *Source Service Access Point* (SSAP), which pinpoint the specific upper-layer protocols involved in the communication at both sending and receiving stations. Additionally, the control field within the header specifies the chosen mode (connectionless or connection-oriented) and the messaging mechanism to be used.

Both DSAP and SSAP function as identifiers for the source and destination addresses of the upper layer protocols that are multiplexed under IEEE 802.2. The control field indicates the chosen mode of operation and the messaging mechanism used to enable both connection-oriented and connectionless communication. IEEE 802.2 can be extended through the *Subnetwork Access Protocol* (SNAP), which is also shown in Figure 4.2. SNAP allows for the identification of upper layer datagrams using an organization identifier and an Ethertype field.

Like other wired networking technologies, Ethernet uses full-duplex communication, allowing devices and applications to send and receive data packets at the same time. This characteristic facilitates *Carrier Sense Multiple Access with*

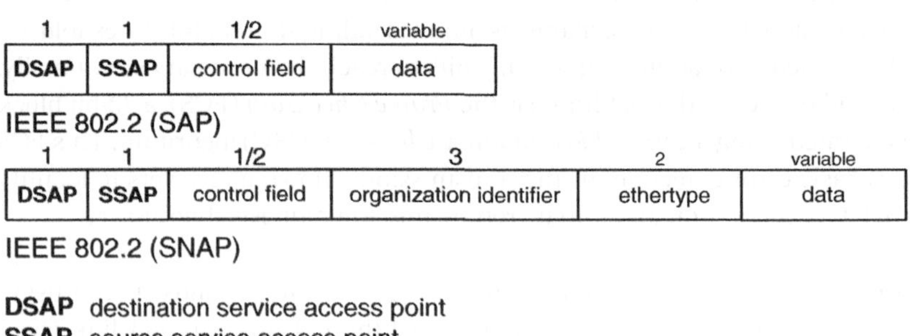

Figure 4.2 IEEE 802.2 SAP and SNAP.

Collision Detection (CSMA/CD), a core mechanism used by the Ethernet MAC layer to manage shared access to the network media. CSMA/CD enables devices to listen for ongoing transmissions before initiating their own, and to detect collisions if multiple devices attempt to transmit simultaneously, reducing data transmission errors.

4.2.2 IEEE 802.11

Wireless Fidelity (Wi-Fi) and IEEE 802.11 refer to wireless networking technologies similar to wired Ethernet. These technologies include various protocols with varying speeds and power usage. For example some Wi-Fi protocols prioritize power efficiency for IoT devices. The IEEE 802.11 standard establishes physical and link-layer communication for both home Wi-Fi networks (WLANs) and smaller, personal area networks (WPANs) relevant to the IoT.

IEEE 802.11 networks operate using three key components: (1) Stations that are devices like laptops or smartphones that connect to the network for internet access, (2) *Access Points* (APs) that act as intermediaries between stations and the wired network and receive as well as transmit data wirelessly to stations within their range, and a (3) Distribution System that refers to the wired infrastructure, such as cables and routers, that connects APs to the broader internet.

Stations are grouped into a *Basic Service Set* (BSS), which essentially defines the coverage area of a single AP. An AP itself uses multiple communication protocols: IEEE 802.11 for wireless communication with stations and another technology like Ethernet for wired communication with the core network. When considering both the wireless and wired components, the entire network setup is called an *Extended Service Set* (ESS) (IEEE 2016).

In IEEE 802.11 networks, a BSS is a group of devices connected to an AP. These devices share a common identifier called a *BSS Identifier* (BSSID) (similar to a network name) to identify the specific wireless network they are connected to. This BSSID, along with other technical details like data rates and signal types, is regularly broadcasted by the AP and participating devices in special messages called beacon frames. These beacons help devices connect, stay synchronized, and share the wireless channel efficiently (Herrero 2021).

The IEEE 802.11 standard defines two ways for devices to connect in a BSS. Ad-hoc mode resembles Figure 4.3, where devices connect directly with each other, similar to how people can talk directly without needing a central point. This type of BSS is called an *Independent BSS* (IBSS) because there is no central AP. Since there is no AP, there is also no connection to the larger wired network (core network) and therefore no ESS. In an IBSS, each device transmits its own beacon frames with randomly generated network names and other technical details. Infrastructure mode works like Figure 4.4, where all devices within the

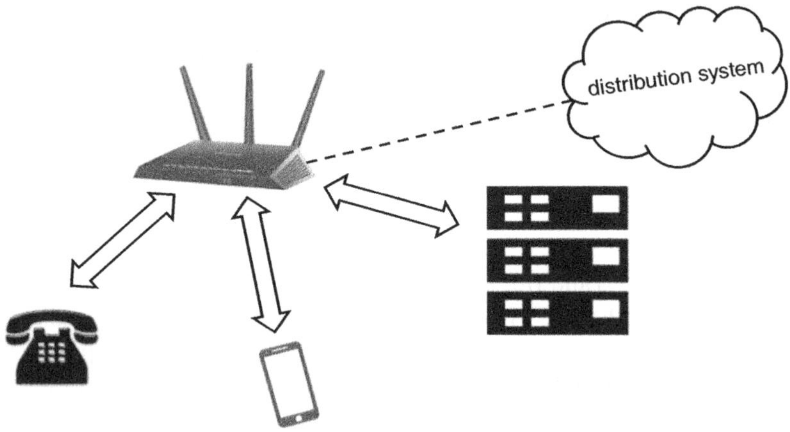

Figure 4.3 Basic Service Set (BSS).

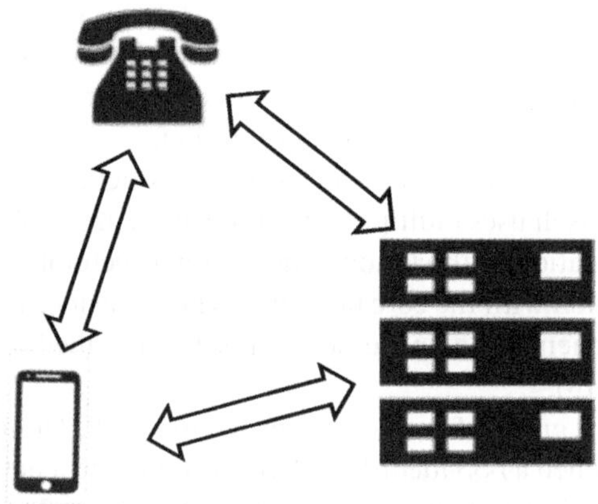

Figure 4.4 Independent Basic Service Set (IBSS).

BSS talk to each other through a central AP, similar to how people can send messages through a designated spokesperson. The AP acts as a gateway, allowing devices to connect to the core network and forming an ESS. A single ESS can even include multiple BSSs with multiple APs, enabling devices across different BSSs to connect with each other through the APs.

The way devices within a BSS communicate depends on whether there is an AP or not. In infrastructure mode, where an AP is present, data packets sent by one device are first received by the AP and then forwarded to the intended recipient. This adds a slightly longer delay to communication compared to ad-hoc mode, where devices speak directly to each other without an intermediary. However, infrastructure mode offers increased reliability because the AP can store data temporarily if the receiver is unavailable due to being out of range, asleep, or malfunctioning. Due to its reliability, infrastructure mode is preferable in RTC.

The very first version of the IEEE 802.11 standard offered two different ways to send data over wireless channels: FHSS and DSSS. Both of these methods operated in the same frequency band, around 2.4 GHz, which is commonly used for other wireless devices like microwave ovens. The standard also included an option for *infrared* (IR) communication, similar to TV remote controls, but this was never widely used. Depending on the settings chosen, the basic version of IEEE 802.11 could send data at either 1 or 2 Mbps, which is relatively slow compared to today's standards. Later updates to the standard changed how data is sent, making it faster, more energy-efficient, and able to reach farther. These improvements were achieved through changes in the power and modulation schemes used for sending and receiving data.

Over the years, several updates have enhanced the IEEE 802.11 standard. In 1999, IEEE 802.11a and 802.11b brought improvements in speed and functionality; in 2003, in 2009 IEEE 802.11n dramatically increased data transfer rates and range; IEEE 802.11g further boosted speed and compatibility; in 2010, IEEE 802.11p facilitated communication in moving vehicles *Wireless Access in Vehicular Environments*; in 2012, IEEE 802.11ad enabled ultra-high speeds (gigabits per second) in the 45–70 GHz range, though availability varies by region; in 2013, IEEE 802.11ac or Wi-Fi 5 brought even faster speeds and better performance in crowded environments; in 2014, IEEE 802.11af or Super Wi-Fi offered a low-rate option for sending data over unused TV signal channels; in 2016, IEEE 802.11ax or Wi-Fi 6 further optimized efficiency and range, particularly for multiple devices connecting simultaneously (rates close to 10 Gbps), also in 2016, IEEE 802.11ah or Wi-Fi HaLow provided broader support for various IoT devices; in 2018, IEEE 802.11aj introduced support for another high-speed band in certain countries; in 2020, IEEE 802.11ba optimized power efficiency for battery-powered IoT devices through *Wake-up Radio* (WUR) technology and narrow-band communication and in 202: IEEE 802.11be known as *Extremely High Throughput* (EHT) or Wi-Fi 7 (rates close to 50 Gbps).

The IEEE 802.11a standard laid the groundwork for several later versions of the wireless networking protocol. It operates in the 5 GHz band, a portion of the radio spectrum dedicated to specific uses like industrial, scientific, and medical applications. To prevent interference between devices, 802.11a divides this band into smaller, non-overlapping channels of 20 MHz each. The number of available channels and the maximum signal strength, measured as *Equivalent Isotropic Radiated Power* (EIRP), can vary depending on regulations in different countries. Often, higher frequency channels are designated for outdoor use, allowing for slightly stronger signals in open spaces.

Table 4.2 IEEE 802.11a.

Modulation	CB/Subc	CB/Symb	CR	Bits/Symbol	Rate (Mbps)
BPSK	1	48	$\frac{1}{2}$	24	6
BPSK	1	48	$\frac{3}{4}$	36	9
QPSK	2	96	$\frac{1}{2}$	48	12
QPSK	2	96	$\frac{3}{4}$	72	18
16-QAM	4	192	$\frac{1}{2}$	96	24
16-QAM	4	192	$\frac{3}{4}$	144	48
64-QAM	6	288	$\frac{2}{3}$	192	48
64-QAM	6	288	$\frac{3}{4}$	216	54

Each 20 MHz channel is further subdivided into 52 even smaller OFDM subchannels, spaced apart at precise intervals. While 48 of these subchannels carry data, the remaining four act as reference points for synchronizing devices on the network. To avoid collisions when multiple devices try to transmit at once, 802.11a allows each device a short delay period (a backoff slot) before attempting to send data. This system ensures efficient use of the shared wireless channel.

Table 4.2 details how different modulation schemes are used in IEEE 802.11a to achieve theoretical transmission speeds between 6 Mbps and 54 Mbps. The table shows the total number of bits encoded in each symbol by combining information from all 48 subchannels used in a symbol. It also specifies the amount of error correction code (FEC) added, which determines how many useful data bits are actually sent in each symbol. This value, along with the symbol transmission time, ultimately determines the actual throughput.

For example, in the last row, the 64-QAM modulation uses 6 bits per subchannel (CB/Subc), resulting in a total of 288 bits encoded in each symbol (CB/Symb). Applying a $\frac{3}{4}$ code rate (or CR) leaves $\frac{288\times3}{4} = 216$ data bits per symbol. Since these symbols are sent every 4 microseconds, the resulting transmission rate is 54 Mbps (calculated as 96 bits divided by 4 microseconds).

Launched alongside IEEE 802.11a, IEEE 802.11b targeted the popular, but congested, 2.4 GHz *Instrumental Scientific and Medical* (ISM) band. This boosted signal penetration through walls and obstacles, but also invited more interference from other devices. Consequently, for similar power consumption, 802.11a typically required more access points to cover the same area. To accommodate existing devices, 802.11b split the 2.4 GHz band into several overlapping 22 MHz channels, with the number ranging from 11 to 14 depending on the region. Additionally, it allowed for up to 31 delay periods of 20 microseconds each for devices to avoid collisions on the shared channel.

Building on the foundation of the original IEEE 802.11 standard, IEEE 802.11b adds additional transmission speeds based on the DSSS technology. Faster speeds are achieved by switching to a different type of modulation called *Differential Quadrature Phase-Shift Keying* (D-QPSK) and *Differential 8-Phase Shift Keying* (D8-PSK). These methods unlock rates of 5.5 Mbps and 11 Mbps, respectively, while using shorter 8-chip codes and a slightly higher symbol rate of 1.375 million per second.

IEEE 802.11a and 802.11b networks do not use a fixed speed for data transfer. Instead, they employ a clever technique called *Dynamic Rate Shifting* (DRS) to automatically adjust the speed based on signal quality. When the connection is rough with lots of interference or obstacles blocking the signal, devices automatically switch to slower speeds. This ensures reliable communication even in challenging environments. However, when the connection is clear and strong, DRS kicks in again, automatically boosting the speed to send data faster. The beauty of DRS is that it happens behind the scenes without any intervention from the user or other software. The physical layer of the network takes care of adjusting the speed based on real-time signal conditions, allowing higher layers like applications to focus on delivering data efficiently.

IEEE 802.11g utilizes the same OFDM technology as 802.11a, but transmits data at the lower frequency of 2.4 GHz. This allows for faster data transfer speeds compared to 802.11b, with rates ranging from 6 to 54 Mbps. Additionally, for compatibility with existing devices, 802.11g radios can also support the slower 802.11b transmission modes. However, this comes at the cost of reduced overall network throughput. To facilitate interoperability with 802.11b devices in the shared 2.4 GHz band, 802.11g radios employ several protection mechanisms. One such mechanism involves a device sending a *Request to Send* (RTS) control frame to the AP before initiating data transmission. The AP then responds with a *Clear to Send* (CTS) frame, granting permission to transmit. This handshake procedure helps minimize collisions with other transmissions, but it introduces additional overhead, impacting network performance.

IEEE 802.11n utilizes MIMO, a technology that employs multiple antennas at both the sender and receiver to increase data rates. Unlike channel bonding, which simply combines existing channels, MIMO leverages spatial diversity where the same signal travels over different paths, boosting strength and overcoming noise. This allows 802.11n to achieve maximum speeds of 600 Mbps. Furthermore, 802.11n doubles its bandwidth compared to previous standards by combining adjacent channels in the 2.4 GHz and 5 GHz bands. To maximize throughput, the technology employs sophisticated algorithms that dynamically adjust channel selection, antenna configuration, and signal encoding based on real-time conditions. This complex interplay of features pushes the boundaries of wireless speed and performance.

Taking IEEE 802.11n's features to the next level, IEEE 802.11ac boasts data rates up to 3.4 Gbps. It achieves this feat by using wider channels (80 MHz and 160 MHz) exclusively in the less congested 5 GHz band. 802.11ac also builds upon MIMO by introducing beamforming, a signal processing technique that focuses wireless signals in specific directions instead of broadcasting them omnidirectionally. This targeted approach minimizes interference and maximizes signal strength for faster, more reliable connections.

Wi-Fi 6, officially called IEEE 802.11ax, builds upon the foundation of 802.11ac to deliver even faster and more efficient wireless connections. Unlike its predecessor, Wi-Fi 6 operates in both the 2.4 GHz and 5 GHz bands, offering wider coverage and flexibility. It also takes MIMO to the next level by enabling two-way communication (uplink and downlink) with multiple antennas, significantly boosting performance and responsiveness. Additional features like trigger based random access and dynamic fragmentation further optimize spectrum usage and reduce overhead, leading to dramatic improvements in overall speed and latency. In essence, Wi-Fi 6 marks a significant leap forward in wireless technology, unlocking a smoother and more reliable experience for users and devices alike.

Designed for cars and roads, IEEE 802.11p builds upon the established technology of 802.11a but focuses on reliable communication between vehicles (V2V) and with infrastructure (V2I). Operating in the same 5 GHz band as its predecessor, 802.11p prioritizes robust connectivity over blazing speeds. To achieve this, it uses wider 10 MHz channels instead of the usual 20 MHz, effectively doubling the transmission time per data packet. This trade-off ensures better signal stability in fast-moving environments at the cost of halving the possible data rate. So, while 802.11a offers speeds ranging from 6 Mbps to 54 Mbps, 802.11p scales this down to a range of 3 Mbps to 27 Mbps, prioritizing clarity and reliability over raw speed for *Vehicle to Everything* (V2X) communication.

Super Wi-Fi, officially known as IEEE 802.11af, takes a different approach to boosting wireless speeds. Instead of relying on dedicated Wi-Fi bands, it taps into unused portions of the TV spectrum (54–790 MHz) with the help of cognitive radio technology. This smart approach dynamically selects the cleanest channels to avoid interfering with existing TV signals. While Super Wi-Fi does not offer the same bandwidth as newer standards like IEEE 802.11ac, it utilizes the same efficient OFDM modulation over 6, 7, and 8 MHz channels, making it a compelling option for extending Wi-Fi coverage and reaching areas with limited spectrum availability.

IEEE 802.11ah (also known as Wi-Fi HaLow) is a specialized Wi-Fi technology designed solely for connecting devices in the IoT. Similar to many traditional Wi-Fi setups, it utilizes star-shaped networks where a central AP acts as a hub for all connected devices. However, unlike conventional Wi-Fi standards, Wi-Fi HaLow operates in a different radio frequency range (915/868 MHz) with narrower channels (1, 2, 4, 8, and 16 MHz). This allows for longer signal range and lower power consumption, making it ideal for battery-powered IoT devices. Note that Wi-Fi HaLow borrows the MIMO OFDM modulation technique from Wi-Fi IEEE 802.11ac but adjusts the timing and size of signal bursts to achieve a variety of data rates while optimizing power usage. This flexibility enables devices to choose the transmission speed that best suits their power limitations and communication needs. Wi-Fi HaLow stands out from past Wi-Fi versions by connecting many devices per network, ideal for large-scale, low-power applications like LPWANs. It boasts longer range, reaching up to 1 kilometer, while consuming minimal power through clever energy-saving techniques.

Unlike wired connections, wireless devices like Wi-Fi cannot listen while talking. This means they cannot sense if another device is transmitting at the same time, leading to collisions where signals overlap and get corrupted. To prevent collisions, wireless devices must take turns using the channel, adding extra time between transmissions. This turn-taking process, compared to instant collision detection in wired connections, adds inherent latency to wireless communication. When several devices all want to talk on the same frequency at once, Wi-Fi relies on a take turns system called *Distributed Coordination Function* (DCF).

Figure 4.5 shows the timing sequence for DCF mode when two devices (B and C) attempt data transmission while another device (A) is already transmitting. After device A's transmission concludes, devices B and C monitor the channel for a fixed waiting period known as the *Distributed Inter Frame Spacing* (DIFS). If the channel remains idle after DIFS, each device independently calculates a random backoff time, denoted as C_w, and initiates transmission only if the channel remains free after this backoff time elapses. The random selection of C_w within a specified interval (C_{wmin}, C_{wmax}) aims to reduce the probability of repeated collisions during retransmissions. It is important to note that C_w is calculated as a multiple of the standard IEEE 802.11 slot time.

While most devices use DIFS for channel access, some prioritize high-importance control frames using *Short Inter Frame Spacing* (SIFS). For predictable

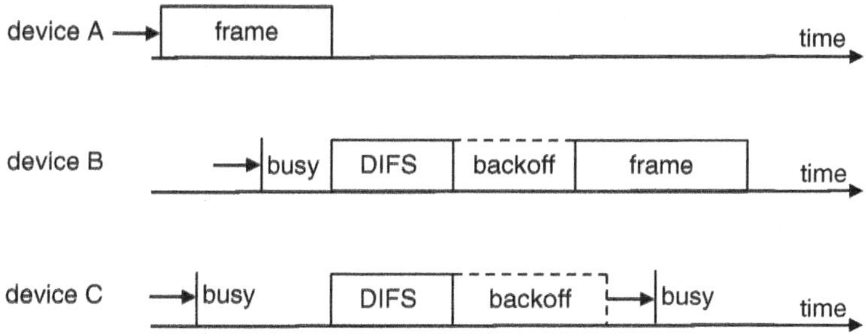

Figure 4.5 IEEE 802.11 Contention under DCF.

access and guaranteed quality of service, devices can turn to PCF. In this controlled approach, the coordinator transmits data with medium *PCF IFS* (PIFS), then instructs other devices on the contention-free period. Stations are sequentially polled to send frames, eliminating collision risk. While part of the initial 802.11 standard, PCF is often used with IEEE 802.11e enhancements. This includes *Enhanced DCF* (EDCF), which assigns three priority levels and corresponding spacings between transmissions *Arbitrary Interframe Spacing* (AIFS), high (high priority, resembling DIFS), medium, and low.

The standard IEEE 802.11 frame (Figure 4.6) carries network data within a MAC header, payload, and checksum. The frame control (FC) defines its type (data, control, or management) and encryption [*Wired Equivalent Privacy* (WEP), *Wi-Fi Protected Access* (WPA) or *Wi-Fi Protected Access II* (WPA2)]. It also manages fragmentation, retransmissions, and airtime reservation for acknowledgments. Addresses 1 and 2 identify receiver and sender, while 3 and 4 support mesh routing and AP connectivity. The sequence control field tracks fragment and frame numbers. Data frames often encapsulate IEEE 802.2 packets (detailed in Section 4.2.1).

Upon activation of its IEEE 802.11 interface, a device initiates a scanning process to discover nearby devices prepared for association. This scanning can occur in either passive or active mode. Passive scanning involves a device sequentially listening for beacon frames transmitted by other devices across IEEE 802.11 channels. These beacon frames convey essential information related to synchronization, modulation, and other physical layer parameters. In contrast, active scanning enables a device to transmit a probe frame specifying the SSID of the desired BSS. Upon receiving this probe, the corresponding AP responds with a probe response. Alternatively, active scanning can include a device broadcasting a probe to all available devices, prompting responses from APs that facilitate the construction of device lists.

Devices compliant with the IEEE 802.11 standard offer two sets of services: station and distribution. Station services handle device-to-device authentication and privacy. Authentication precedes association and can be configured on APs as either open or pre-shared key. Open authentication accepts all devices, offering minimal security and validation. Pre-shared key authentication, on the other hand, requires devices to share a secret password configured beforehand

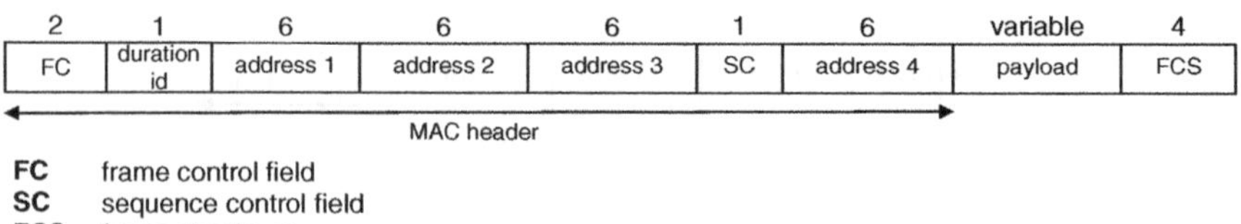

Figure 4.6 IEEE 802.11 Header and Payload.

through an alternate channel like user settings. Deauthentication, mirroring pre-association authentication, allows devices to gracefully disconnect from each other before disassociation. This ensures proper network etiquette and resource management.

Special management frames handle device authentication (before joining a network) and deauthentication (before leaving). Data confidentiality is achieved through encryption using WEP, WPA, and WPA2. Distribution services manage interactions beyond individual BSSs. Association connects devices to APs, enabling APs to deliver traffic. Disassociation allows devices to leave a network, such as when disabling their Wi-Fi. Reassociation lets devices switch APs based on stronger beacon signals. Overall, distribution services govern frame transmission within BSSs and ESSs, even bridging to non-802.11 networks through AP integration.

4.2.3 RTC

The use of SIP and RTP protocols in non-3GPP networks follows the configurations respectively presented in Chapters 2 and 3. Because SIP and RTP were designed to support VoIP scenarios, they can be directly encapsulated over the IP layer.

Figure 4.7 shows how signaling and media stacks integrate SIP and RTP protocols with Ethernet and IEEE 802.11 physical and link layers. In this scenario, the network layer through the IP protocol enables end-to-end connectivity. While the RTP layer relies on UDP transport with SRTP security, the SIP layer can support either UDP or TCP transport and DTLS or TLS security respectively (Schooler et al. 2002; Schulzrinne et al. 2003).

4.2.4 WebRTC

In non-3GPP networks, in the context of WebRTC, signaling is performed over proprietary mechanisms that, relying on JavaScript (JS) APIs, enable endpoints to

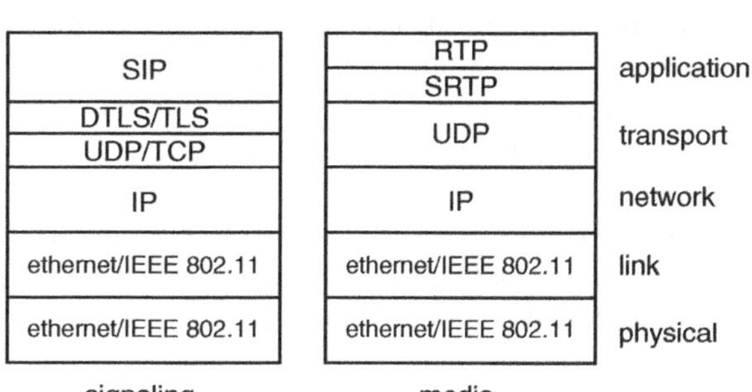

Figure 4.7 SIP and RTP over non-3GPP Networks.

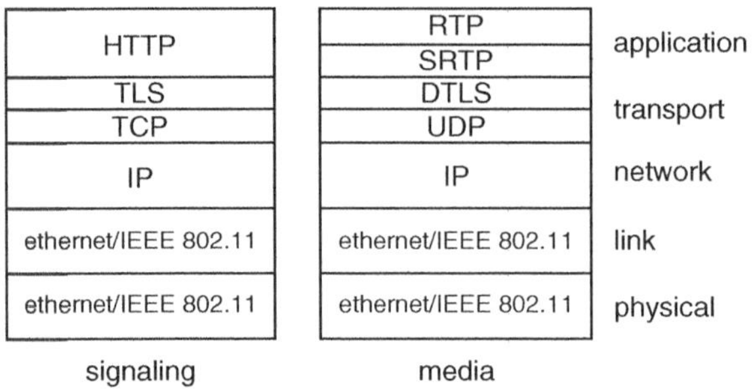

Figure 4.8 WebRTP over non-3GPP Networks.

negotiate media capabilities. This is typically done through the exchange of SDP messages.

Figure 4.8 formalizes the protocol stacks associated with signaling and media in the context of WebRTC. Signaling is carried out by HTTP (introduced in Section 2.5), encrypted by means of TLS, and transported over TCP. On the other hand, media relies on DTLS-SRTP (described in Section 3.4.2), which also relies on UDP transport (Alvestrand 2021).

4.3 3GPP 4G/5G/6G

3GPP networks rely on standards associated with mobile communications. These standards are developed to guarantee that all mobile devices and networks from different manufacturers can interoperate. 4G, also known as *Long Term Evolution* (LTE), corresponds to the fourth generation of mobile communications. Commercial 4G topologies have been deployed since 2009, supporting transmission rates of up to 1 Gbps. 4G introduces *Voice over LTE* (VoLTE) that relies on standard RTC protocols like SIP and RTP for call signaling and media transmission, respectively. Codecs like AMR are common in VoLTE deployments. These mechanisms follow under the umbrella of the *IP Multimedia Subsystem* (IMS) infrastructure. In addition to IMS, 4G provides *Circuit-Switched Fallback* (CSFB) legacy mechanisms, compatible with 2G and 3G architectures, where a circuit switched infrastructure is used instead. As of 2023 and for a few more years, 4G is and will be the dominant mobile technology around the world (Cox 2014).

5G corresponds to the fifth generation of mobile communications. It was first commercially deployed ten years after 4G, in 2019. 5G technologies support transmission rates of up to 20 Gbps. As an alternative to VoLTE, 5G introduces *Voice over New Radio* (VoNR). VoNR supports high-quality codecs, associated with higher sampling rates, like EVS and AMR-WB, that provide a more natural voice experience. VoNR also supports advanced video codecs like ITU-T Recommendation H.265 and ITU-T Recommendation H.264. 5G is slowly gaining market share, and deployments are becoming common (Cox 2020).

6G, as expected, corresponds to the sixth generation of mobile communications. 6G protocols are not fully standardized, and it is still in the early stages of development. 6G technologies are intended to support transmission rates of up to 100 Gbps. It is expected that 6G standards will follow similar mechanisms to those of VoLTE and VoNR to support RTC. 6G technologies are intended to support connected intelligence by supporting a wide range of Artificial Intelligence (AI)-based services. These services are to be deployed end-to-end, covering the access and core networks. The overall goal, when compared to 5G technologies, is not only to dramatically improve transmission rates but also to increase device density as well as reliability by several orders of magnitude. This is accomplished by relying on sub-millimeter bands where signals are modulated in the THz spectrum (Wu et al. 2021).

4.3.1 IMS

IMS was first introduced in 1999 as a component of the initial 3G technology standardization process. Its purpose is to establish SIP as the primary mechanism for enabling multimedia communication. The IMS architecture outlines the roles of SIP servers, proxies, and UACs. It introduces *Call Session Control Functions* (CSCF), which includes (1) Session Border Controllers (SBCs) that are devices that interact with UACs on the network's access side, providing initial call routing and security enforcement through encryption, authentication, and accounting. They also enable lawful intercept capabilities. SBCs comply with a specific CSCF subset known as Proxy CSCF (P-CSCF), (2) *Serving-CSCFs* (S-CSCFs) that define mechanisms for call routing in the context of interaction with user agent servers (UASs). It handles user registration, which is accomplished through the REGISTER SIP method, among other functions, (3) Application Servers (ASs) that support UAS functionality within the CSCF architecture, (4) *Media Resource Functions* (MRFs) that handle media manipulation and interconnects with ASs. The latter consist of two parts: (1) *MRF Controllers* (MRFCs) for signaling support and (2) *MRF Processors* (MRFPs) for media generation (Poikselka et al. 2006).

4.3.2 VoLTE

LTE architectures comply with the layout shown in Figure 4.9. They feature a dual-part structure, comprising a wireless access segment known as *Evolved UMTS Terrestrial Radio Access Network* (EU-TRAN) and a wireline core termed *Enhanced Packet Core* (EPC). It is important to note that *UMTS* stands for *Universal Mobile Telecommunications Framework* (Cox 2014).

Key components within this architecture include the (1) *User Equipment* (UE) that represents devices like smartphones that interact wirelessly with base stations, designated as *EU-TRAN node B* (eNodeB), within the access network, the

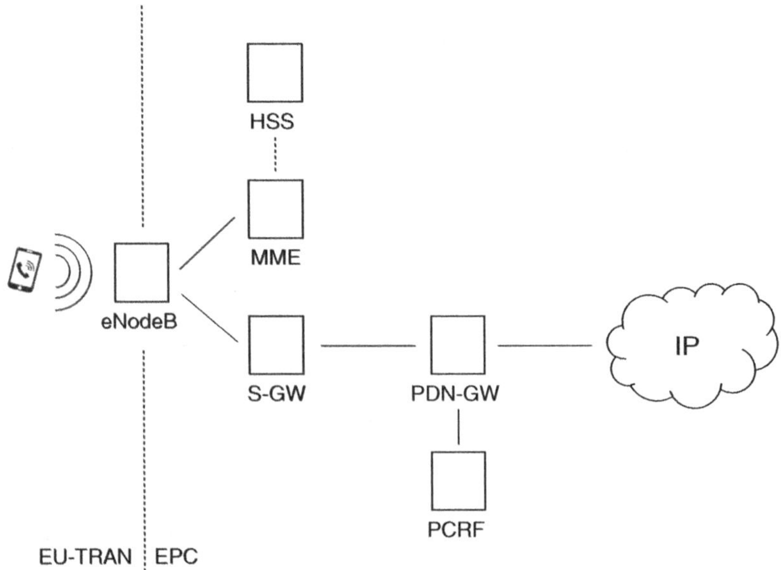

Figure 4.9 LTE Topology.

(2) *Mobility Management Entity* (MME) that resides within the core network and manages signaling and mobility functions, the (3) *Serving Gateway* (S-GW), positioned in the core network, that handles user data traffic (excluding signaling) and interfaces with the Packet Data Network Gateway (PDN-GW), the (4) *Packet Data Network Gateway* (PDN-GW) that provides connectivity to the Internet and is associated with the *Policy and Charging Rules Function* (PCRF), and the (5) *Home Subscriber Server* (HSS) that interacts with the MME.

From the perspective of the physical layer, LTE technologies rely on OFDM for the downlink and *Single Carrier Frequency Division Multiple Access* (SC-FDMA) for the uplink. These signals are transmitted over 44 spectral bands, which are divided into 31 *Frequency Division Duplex* (FDD) bands and 13 *Time Division Duplex* (TDD) bands. LTE networks are split into categories that are associated with the different transmission rates. The higher the category, the higher the transmission rate. These lead to rates between 1 Kbps and 1 Gbps.

Figure 4.10 illustrates the arrangement of SIP and RTP layers when transmitted across the LTE EU-TRAN interface. Note that the link layer comprises three sublayers designed for encapsulating IP traffic. The MAC Sublayer forms the foundation for data transfer. The *Radio Link Control* (RLC) operates in three distinct modes: (1) *Transparent Mode* (TM) that directly forwards frames to the MAC sublayer, (2) *Unacknowledged Mode* (UM) that segments data and adds headers before forwarding, and (3) *Acknowledged Mode* (AM) that resembles UM

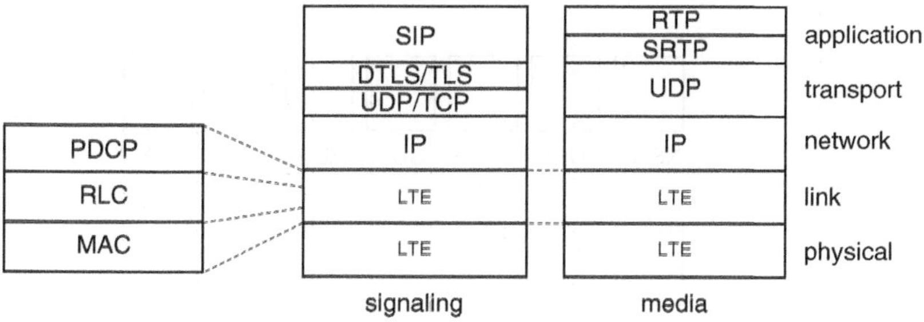

Figure 4.10 SIP and RTP over LTE.

mode but incorporates flow control through acknowledgments. The *Packet Data Convergence Protocol* (PDCP) sublayer compresses IP headers utilizing *Robust Header Compression* (ROHC), standardized in IETF RFC 3095 and fortifies security through ciphering, deciphering, integrity protection, and verification. Note that the acronym RLC is also used to describe Run Length Coding in Chapter 1.

4.3.3 VoNR

5G architectures follow the topology shown in Figure 4.11. The wireless access network is called *5G Radio Access Network* (5G RAN) and the wireline core network is called *5G Core*. UEs interact with the base station *Next Generation node B* (gNB). The gNB connects to the *Access and Mobility Management Function* (AMF) that provides signaling and to the *User Plane Function* (UPF) that enables connectivity to the Internet. The UPF supports the *Session Management Function* (SMF) that guarantees that data packets are routed correctly. The AMF supports multiple functions: (1) *Network Slice Selection Function* (NSSF) to enable network selection, (2) *Network Exposure Function* (NEF) to enable security, (3) *Network Repository Function* (NRF) to support network function registration, (4) *Unified Data Management* (UDM) to manage user data in a similar fashion to the HSS in LTE, (5) *Authentication Server Function* (AUSF) to support user authentication, (6) *Policy Control Function* (PF) to specify control policies, and (7) *Application Function* (AF) to enable applications to interact with other network functions and resources. 5G differentiates in user and control planes based on functionality. While the user plane deals with the transmission of data, the control plane has to do with the signaling traffic (Nath et al. 2023).

5G architectures follow the topology outlined in Figure 4.11. They feature a two-part structure. The *5G Radio Access Network* (5G RAN) is a wireless access network that facilitates communication between UEs and base stations, known as gNB. On the other hand, the 5G Core is the wireline core network comprising several key components including the *Access and Mobility Management Function*

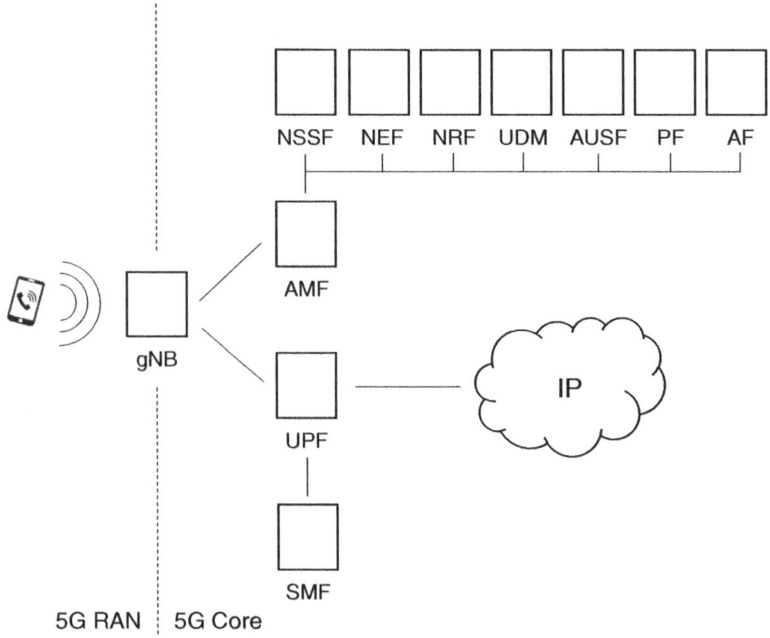

Figure 4.11 5G Topology.

(AMF) that manages signaling and interacts with the *User Plane Function* (UPF) to enable Internet connectivity. UPF, in turn, routes data packets correctly and collaborates with the *Session Management Function* (SMF). The AMF supports multiple functions: (1) *Network Slice Selection Function* (NSSF) that enables network selection, (2) *Network Exposure Function* (NEF) that enforces security, (3) *Network Repository Function* (NRF) that registers network functions, (4) *Unified Data Management* (UDM) that manages user data, similar to the HSS in LTE, (5) *Authentication Server Function* (AUSF) that authenticates users, (6) *Policy Control Function* (PF) that specifies control policies, and (7) *Application Function* (AF) that allows applications to interact with other network functions and resources.

5G networks differentiate between user and control planes based on functionality. The User Plane handles data transmission and the Control Plane manages signaling traffic. While 4G utilizes MIMO technology to boost transmission speeds, 5G takes it a step further with Massive MIMO. This advanced technique employs a larger number of antennas to achieve even faster data rates and improved spectral efficiency. Unlike 4G, 5G combines OFDMA with *Non-orthogonal Multiple Access*

(NOMA) to further enhance efficiency. 5G offers three distinct operational modes tailored for specific needs:

- *Enhanced Mobile Broadband* (eMBB): This mode prioritizes high speeds, exceeding 10 Gbps even in challenging environments, making it ideal for data-intensive applications.
- *Ultra-reliable Low-Latency Communication* (URLLC): This mode focuses on reliability and minimal delays, aiming for transmission rates at or above 100 Mbps with an end-to-end latency below 50 milliseconds, important for real time communication and critical control systems.
- *Massive Machine-Type Communication* (mMTC): This mode caters to a large number of low-power devices, typical of the Internet of Things (IoT), with low transmission rates primarily in uplink scenarios.

5G defines three frequency ranges for its operation, partially overlapping with existing 4G bands. In total, approximately 80 bands are deployed across three ranges: (1) Low bands (below 1 GHz) that offer extensive coverage at the expense of lower speeds, (2) Mid bands (between 1 GHz and 6 GHz) that provide a balance between coverage and speed, and (3) High bands (24 GHz to 40 GHz), also known as *millimeter waves* (mmWaves) that deliver the highest speeds but suffer from limited coverage due to signal attenuation.

Example 4.2 An RTP traffic encapsulates every 20 milliseconds speech using ITU-T Recommendation G.711. Assuming a VoNR scenario, what is the size of the RTP packets? Assume 5-byte RLC, 5-byte PDCP, 4-byte SDAP, and 4-byte MAC header sizes. Consider an IPv4-based network layer.

Solution:
ITU-T Recommendation G.711 is an NB codec, where $8000 \times 0.02 = 160$ bytes are sent every 20 milliseconds. The RTP packet size is therefore $160 + 12 + 8 + 20 + 5 + 5 + 4 + 4 = 218$ bytes long. Note that 12 bytes correspond to the RTP header length and 8 bytes correspond to the UDP layer header size. The IPv4 header length is 20 bytes long.

Figure 4.12 shows the RTC protocol stack within a 5G environment. It bears a notable resemblance to the stack illustrated in Figure 4.10, with the distinguishing feature of an added upper sublayer within the link layer, designated as the *Service Data Adaption Protocol* (SDAP). This protocol bolsters the attainment of QoS objectives. It is important to note that the 3GPP protocol stacks exhibit a degree of similarity, particularly in regard to their link and upper layers. This characteristic fosters full software compatibility when deploying applications across both 4G

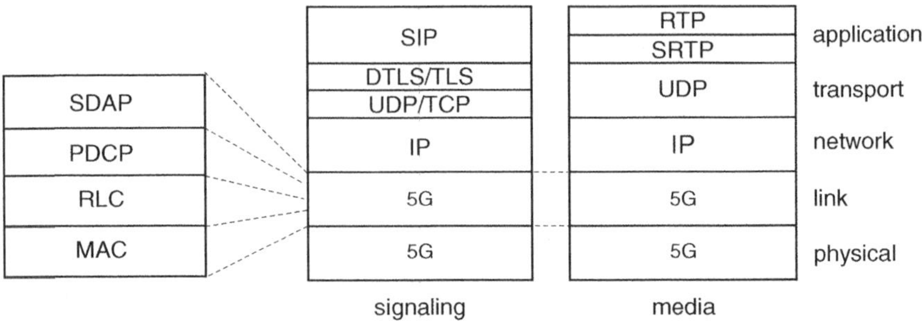

Figure 4.12 SIP and RTP over 5G.

and 5G technologies. A comparable level of compatibility is anticipated for future 6G technologies as well.

4.4 IoT

In real-world IoT scenarios, various limitations like transmission power, channel bandwidth, and deployment costs influence the choice of communication solutions. Note that the transmitted signals are affected by factors like channel noise, interference, and multipath fading, impacting their strength upon reaching the receiver. This attenuation is reflected in the SNR, which plays an important role in data transmission. According to the Shannon-Hartley channel capacity theorem, higher SNR values generally allow for faster data rates (Cover and Thomas 2006; Popovski 2020). It is important to note that achieving error-free transmission at higher rates often requires appropriate channel encoding techniques. However, in low-power IoT networks, prioritizing battery life often leads to lower signal strength and SNR, consequently limiting transmission rates.

As shown in Figure 4.13, IoT networks link applications and devices that interact with physical assets. These assets are accessed through sensing or actuation by the IoT system, which comprises devices, applications, and communication networks. Typically, IoT networks utilize both wireless and wired links employing various physical and link-layer technologies, but they all rely on IPv6 for connectivity.

The communication between devices and applications in these networks can follow two approaches. In the one-hop scenario, devices transmit packets directly to the application. In the multi-hop scenario, devices send packets through intermediate devices before they reach the application. This multi-hop approach is characteristic of capillary networks, originally introduced in the context of *Wireless Sensor Networks* (WSNs) and now widely adopted in various IoT technologies (Holler et al. 2014).

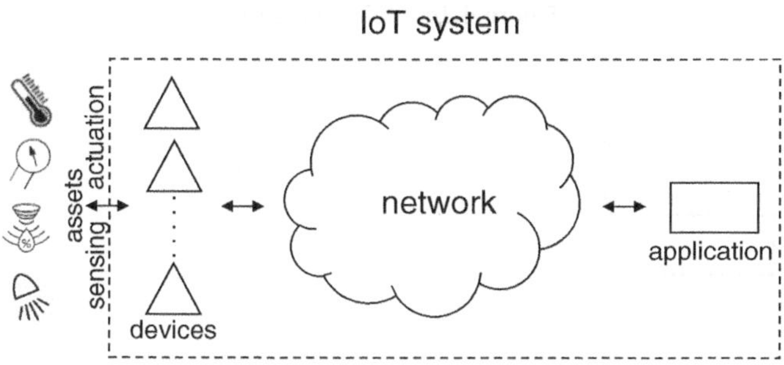

Figure 4.13 IoT Networks.

Traditionally, networks are categorized based on their geographical coverage. The three main types are Local Area Networks (LANs), *Metropolitan Area Networks* (MANs), and Wide Area Networks (WANs). A LAN typically covers a small area like an office, lab, or building, relying on its own infrastructure for high-speed data transfer. Conversely, a WAN spans a large geographical region and relies on external infrastructure, resulting in lower throughput compared to a LAN. A MAN occupies an intermediate size between LANs and WANs, also relying on external infrastructure but operating at higher speeds, often connecting LANs and WANs. Finally, an *Internet Area Network* (IAN), even larger than a WAN, connects endpoints within a cloud environment.

Advancements in virtualization have enabled the fusion of WANs and *Software Defined Networks* (SDNs) into *Software Defined WAN* (SD-WANs). These networks offer reliable long-distance WAN access at lower costs by utilizing IP tunnels established over the public Internet. Each of these network types can utilize either wireless or wired connections. For instance, WWAN and WLAN are the wireless counterparts of WAN and LAN, respectively. The presence of a "W" prefix generally signifies a wireless network type, while its absence indicates a standard wired configuration.

As previously mentioned in this chapter, networking in the context of IoT primarily refers to the access network where devices reside and perform sensing or actuation. Figure 4.14 shows a typical IoT access network. Generally, the devices communicate with an intermediary device like an IoT gateway, border router or clusterhead. This intermediary acts as the interface between the access network and the core network, also known as the Internet edge. The core network typically leverages a standard IP architecture, providing global connectivity to the applications. These applications, in turn, may be hosted within cloud environments like *Amazon Web Services* (AWS).

Access-side networks in the IoT primarily deploy constrained and embedded devices, belonging to a category of low-rate, short-range LAN subcategories. One

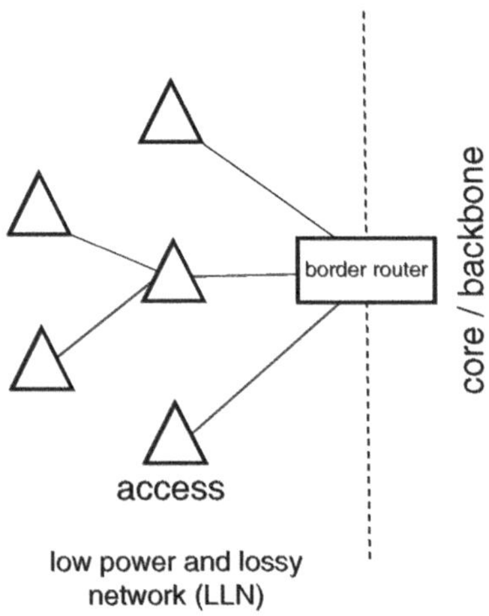

Figure 4.14 IoT Access Network.

such subcategory is the *Home Area Network* (HAN), specifically designed for IoT communication within residences and buildings. Another widely used and important subcategory in the IoT context is the *Personal Area Network* (PAN). PANs prioritize battery life by employing low transmission rates and offering reduced signal coverage. Note that a significant portion of IoT solutions leverage WPANs due to their efficient power consumption. WPANs are among the most prevalent network types in IoT scenarios. The other major category of IoT networks falls under the WAN classification due to their extended signal range. These technologies, known as LPWAN, prioritize low power consumption to extend battery life. While inherently wireless, LPWANs are not referred to as LPWWAN, as one might expect, as indicated in Chew (2019).

Since both WPANs and LPWANs operate wirelessly, battery power serves as their primary energy source. Consequently, minimizing power consumption becomes important for maximizing battery life. This low-power design, however, comes at a cost: weaker signal strength and consequently, a lower SNR. Low SNR translates to an increased likelihood of packet loss and reduced transmission rates. As illustrated in Figure 4.14, the majority of IoT technologies fall under the category of *Low Power and Lossy Networks* (LLNs) as defined in Ersue et al. (2014).

Various PAN physical and link-layer mechanisms, ranging from wireless options like IEEE 802.15.4 and *Bluetooth Low Energy* (BLE) to wired solutions like ITU-T G.9903 PLC and *Master-Slave/Token-Passing* (MS/TP), can coexist in an IoT environment. Given that IoT PANs and WPANs involve devices communicating with each other or gateways, they can be deployed in various configurations, including buses, rings and master-slave setups. In most cases, these technologies reside within proprietary network stacks independent of the Internet for communication.

The emergence of the IoT has driven the development of network layer protocols compatible with these diverse mechanisms to facilitate extensive IPv6 communication. Specifically, 6LoWPAN, alongside other adaptation mechanisms known as 6Lo technologies, enables IPv6 connectivity atop physical and link-layer protocols like IEEE 802.15.4. This fosters uniform support for IoT applications regardless of the underlying mechanism. Additionally, several LPWAN stacks coexist within various full and hybrid IoT environments. These are primarily wireless technologies, including LoRa and NB-IoT.

4.4.1 WPANs

WPANs consists of a diverse array of technologies operating within a limited range, typically up to a few kilometers. These technologies provide transmission rates starting at low Kbps, sufficient for handling IPv6 communication. Examples include several IEEE 802.11 variants discussed in Section 4.2.2, along with IEEE 802.15.3, IEEE 802.15.4, BLE, ITU-T G.9959, and DECT ULE. While certain IEEE 802.11 subtypes natively support IPv6 by directly carrying IPv6 datagrams, other technologies require an additional adaptation layer to enable IPv6 transmission.

IEEE 802.15.3 stands out for its ability to achieve high data rates, reaching up to 55 Mbps. This speed is achieved through a modulation technique called 64-QAM and operates in the unlicensed 2.4 GHz ISM band. In practical deployments, IEEE 802.15.3 networks are implemented as piconets. Each piconet is controlled by a central device called a *Piconet Coordinator* (PNC). This coordinator synchronizes all devices within the piconet by transmitting periodic beacon signals.

In contrast to IEEE 802.15.3, ITU-T G.9959 operates at the physical layer and is optimized for very low transmission rates. Its foundation lies in the established consumer technology Z-Wave of ITU (2015). Unlike piconets, ITU-T G.9959 networks can be arranged in mesh topologies. This allows data to hop through intermediate nodes as repeaters, extending network coverage but at the cost of increased overall latency due to multi-hop transmissions.

Digital Enhanced Cordless Telecommunications Ultra Low Energy (DECT ULE), managed by the ULE Alliance, aims to achieve a unique combination of features: long-range and mid-rate transmission rates with minimal power consumption and low latency. This combination enables extended battery life exceeding five years due to low power usage and infrequent data transmissions (low duty cycle). DECT ULE offers typical coverage of up to 60 meters indoors and 500 meters outdoors. In terms of network topology, DECT ULE utilizes a star architecture with direct, single-hop communication between devices and a central base station functioning as an IoT gateway. A single DECT ULE network can support up to 400 connected devices.

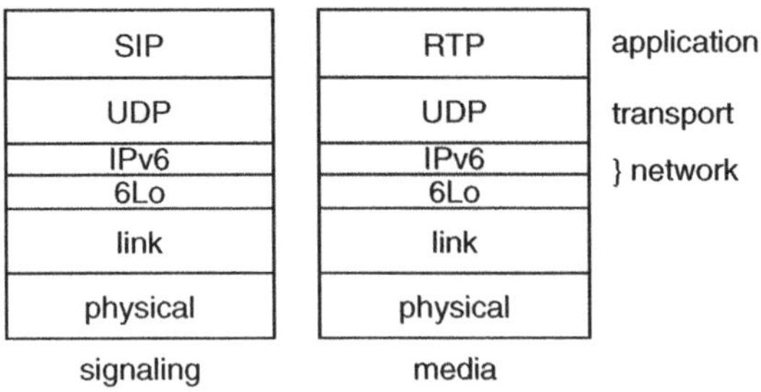

Figure 4.15 RTC WPAN Protocol Stack.

Within the WPAN category, *Near Field Communications* (NFC) stands out as a unique technology for short-range, contactless communication. NFC enables devices to exchange information through direct physical proximity, facilitating convenient and secure transactions as well as access to digital content. Its reliance on established contactless card mechanisms ensures backward compatibility with existing infrastructure.

Figure 4.15 shows the two protocol stacks employed for media and signaling within WPAN scenarios. Irrespective of the specific physical and link-layer technology in use, such as IEEE 802.15.4, BLE or ITU G.9903 a 6Lo layer is incorporated within the network layer to facilitate IPv6 adaptation. The transport layer relies on UDP, and the application layer aligns with SIP and RTP, as previously introduced in Chapters 2 and 3.

It is key to note that UDP is the favored transport protocol within IoT contexts due to its superior efficiency compared to TCP. The fundamental issue with TCP transport lies in its assumption that packet loss is primarily caused by network congestion, resulting in intermediate devices like routers discarding packets when their queues become saturated. However, packet loss within IoT environments is predominantly attributable to low SNR scenarios, where signals are significantly degraded by noise. In low-volume traffic scenarios, TCP's approach involves further rate reduction, intentionally slowing down packet transmission to mitigate congestion. However, in many of these IoT scenarios, it is often more advantageous to implement very rapid retransmissions, providing an FEC-like mechanism capable of circumventing channel-induced loss.

The subsequent subsections provide in-depth explanations of IEEE 802.15.4 and 6LoWPAN technologies, with a specific focus on their roles in enabling the integration of SIP and RTP protocols.

4.4.1.1 IEEE 802.15.4

The IEEE 802.15.4 specification defines a set of physical and link-layer technologies designed for use in WPANs (IEEE 2020). It is a preferred choice for enabling ultra-low power consumption, leading to extended battery life in LLNs. IEEE 802.15.4 also serves as the foundation for standalone standards like ZigBee

(Zig 2015), ISA 100.11a (ISA 2011), and WirelessHART (IEC 2016). These technologies are well-established protocol stacks, integral to many cyber-physical system solutions. ZigBee focuses on home automation and smart energy scenarios, while WirelessHART and ISA 100.11a target industrial automation and control. These standards utilize IEEE 802.15.4 in conjunction with proprietary upper layers that typically do not provide native IP connectivity.

Modern IoT scenarios, as illustrated in Figure 4.16, often leverage IEEE 802.15.4 in conjunction with IETF protocols to achieve efficient end-to-end IPv6 connectivity. This approach prioritizes ultra-low power consumption by restricting both signal reach and data transmission volume, consequently reducing transmission speeds. Furthermore, power consumption is optimized through duty cycle management, which involves controlling device transitions to power-down and sleep states.

The initial IEEE 802.15.4 standard employed DSSS modulated with OQPSK in the 2.4 GHz band, offering 250 Kbps across 16 channels. Earlier versions using BPSK provided slower rates: 40 Kbps in the US across ten 915 MHz channels and 20 Kbps in Europe on a single 868 MHz channel. Modern revisions have expanded 868 MHz and 915 MHz band support to 100 Kbps and 250 Kbps, respectively, offering increased flexibility for low-power wireless communication.

The IEEE 802.15.4 link layer handles essential wireless networking tasks like managing connections, verifying data integrity, and ensuring security. It also defines two device types: *Full Function Devices* (FFDs) act as central hubs or regular participants in any network layout, while *Reduced Function Devices* (RFDs) simply relay data within simpler setups like star and P2P configurations. This distinction offers flexibility, with FFDs managing complex networks while RFDs simplify communication in basic arrangements.

Figure 4.17 demonstrates how different FFD and RFD configurations can be combined. A straightforward P2P connection can even evolve into a complex mesh network accommodating up to 64000 devices. In such cases, efficient routing often relies on responsive methods which can outperform traditional table-based routing approaches.

IEEE 802.15.4 employs a structured approach to media access, combining both contention-based and contention-free periods within a timeslot-based system.

Figure 4.16 IEEE 802.15.4 and IETF Protocols.

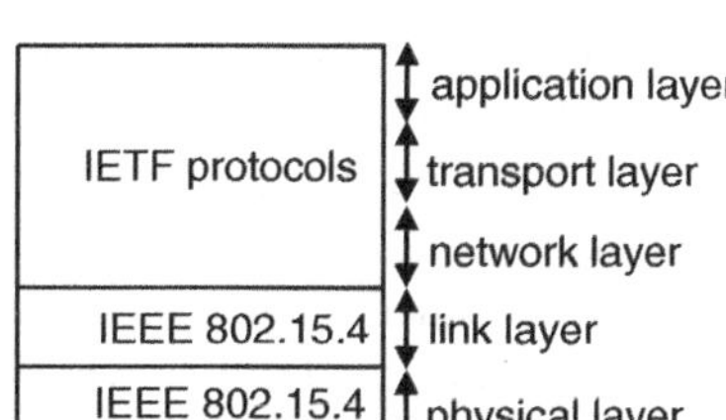

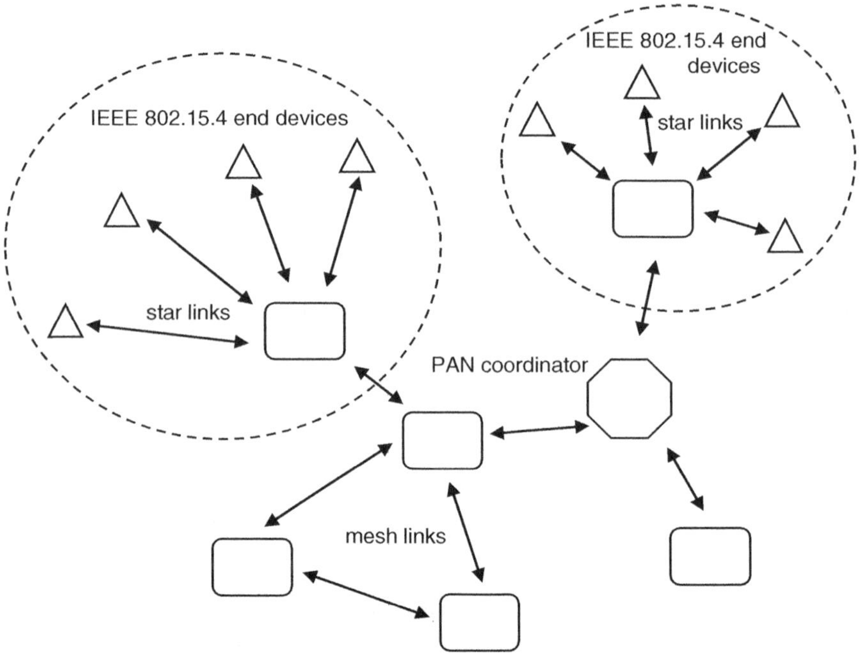

Figure 4.17 IEEE 802.15.4 Topologies.

This hybrid strategy supports both *Carrier Sense Multiple Access with Collision Avoidance* (CSMA/CA) and *Time Division Multiple Access* (TDMA) mechanisms for efficient communication. Each cycle (superframe) starts with a beacon broadcast from the central coordinator. This beacon synchronizes devices, shares configuration details, and tells them if they can join the network. Then, the available time is split into two sections. In the free-for-all section, devices compete for access using a CSMA/CA approach to minimize collisions. Similarly, in the scheduled section, the coordinator pre-assigns specific timeslots to certain devices, ensuring reliable transmission for applications with strict bandwidth needs.

This structure balances flexibility for general communication with guaranteed access for critical tasks. Unlike some other systems, joining this network typically involves manual action on the device itself, like pressing a button. The original IEEE 802.15.4 standard relied on a single communication channel, which made reliability unreliable, especially in multi-hop networks where messages have to hop between devices. This was a problem for some industrial applications that depend on timely communication. The IEEE 802.15.4e update introduced a solution called *Time Synchronized Mesh Protocol* (TSMP) with *Time Slotted Channel Hopping* (TSCH). Synchronization is important for this, so TSMP uses two methods. In the acknowledgment-based method, the receiver tells the sender how far off their clocks are, allowing the sender to adjust and stay in sync. In the frame-based method the receiver adjusts its own clock based on the message arrival time, keeping everyone on the same beat. In simpler terms, TSMP gives devices a predictable schedule and avoids channel jams, making multi-hop communication more reliable for time-sensitive applications.

IEEE 802.15.4 networks support four types of messages: (1) data packets that are the main movers and shakers, carrying the actual information like sensor readouts, media, or chat messages, (2) acknowledgments that are used to ensure messages arrive safely, (3) control messages that are used to join the network, leave it, or take turns transmitting, and (4) beacon frames that set the network parameters and tell devices when to start sending.

Key Components of an IEEE 802.15.4 Frame are illustrated in Figure 4.18. The 7-bit Frame Length specifies the total size of the frame, including all fields and checksum bytes. The maximum allowed frame size is 127 bytes. A reserved Bit is set aside for potential future extensions to the protocol. The 16-bit *Frame Control Field* (FCF) contains multiple flags and settings that control frame handling. The FCF includes: (1) PAN ID Compression that indicates whether the 16-bit PAN Identifier is included within the MAC addresses or not, (2) ACK Request that signals that the sender expects an acknowledgment frame from the receiver to confirm successful delivery, (3) Frame Pending that alerts the receiver that more

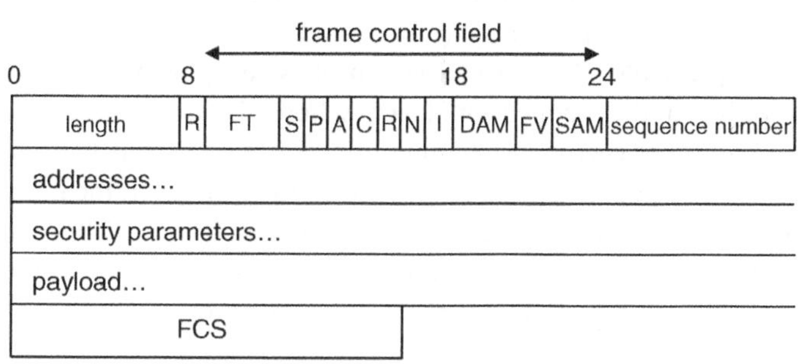

C PAN ID compression
N sequence number suppression
I information element present
A ACK request
P frame pending
S security enabled
FT frame type
SAM source address mode
FV frame version
DAM destination address mode
FCS frame checksum
R reserved

Figure 4.18 IEEE 802.15.4 Header and Payload.

frames are coming from the sender, promoting efficient channel usage, (4) Security Enable that indicates that the frame carries security parameters, (5) Frame Type that specifies the nature of the frame, (6) Frame Version that identifies the specific IEEE 802.15.4 revision under which the frame was constructed, (7) *Source Address Mode* (SAM) and *Destination Address Mode* (DAM) indicate how the source and destination MAC addresses are encoded, and (8) Sequence Number Suppression and IE Present Bits that control the inclusion of optional frame elements. Note that Source Address and Destination Address specify the sending and receiving devices, respectively. In addition, the optional *Information Elements* (IEs) are used to convey additional information between devices, such as security-related data. IEs are authenticated but not encrypted.

Note that the frame structure prioritizes efficiency with compact fields and versatile control flags. The FCF offers diverse functions for acknowledgments, security, addressing, and frame type identification. IEs provide a flexible mechanism for device-specific information exchange. IEEE 802.15.4 addresses can be long addresses where every device has a unique 64-bit code built-in, like a permanent digital ID, for global reach anywhere or they can be short addresses where the network coordinator) can assign shorter, 16-bit names for quicker communication within smaller areas.

If both addresses are in the same network, the long network label (PAN ID) can be skipped, making the address field smaller (4 to 20 bytes). While IEEE 802.15.4 can offer link-layer security (when enabled), it does not fully follow the end-to-end principle. This principle emphasizes placing security functions like encryption at the application layer, where they can cover the entire communication path between devices. In contrast, 802.15.4 security often relies on hardware/firmware implementations limited to the link layer, meaning connections must be re-secured every time data hops between devices. This can impact efficiency by adding latency and reducing throughput. Therefore, while 802.15.4 provides security features, it is important to be aware of its limitations compared to the end-to-end approach.

When security is activated within the IEEE 802.15.4 *Frame Control Field* (FCF), an auxiliary security header is appended to the frame following the address field (as shown in Figure 4.19). The frame concludes with a 16-bit CRC block code serving as the FCS.

This security header incorporates several key elements. The 8-bit Security Control Field incorporates a 3-bit Security Level field that determines the extent of protection applied to the frame's payload, including encryption and/or message authentication. The 32-bit Frame Counter field prevents unauthorized message replays and maintains precise sequencing for chain encryption techniques. The *Key Identifier Mode* (KIM) field specifies the method for devices to ascertain the necessary keys for security processing. The optional Key Control Fields includes

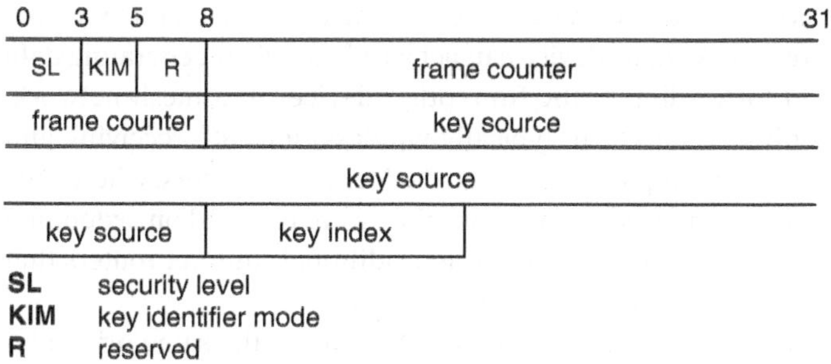

Figure 4.19 IEEE 802.15.4 Security Header.

the 64-bit *Key Source* (KS) field that typically designates the originator of a group key and the 8-bit *Key Index* (KI) field that isolates a specific key emanating from a given source.

4.4.1.2 6LoWPAN

Direct utilization of IPv6 with most popular IoT physical and link-layer technologies is unfortunately not possible. Protocols like 6LoWPAN and its derivatives operate between network and link layers, adapting and translating IPv6 datagrams into a format suitable for transmission over LLNs. Essentially, 6LoWPAN grants native IPv6 support to resource-constrained and low-power devices with minimal overhead, as documented in IETF RFC 4944 Montenegro et al. (2007). This native IPv6 support, as a consequence, grants connectivity to other Internet-based hosts and routers. While 6LoWPAN was initially designed for full IPv6 integration with IEEE 802.15.4, many of its functionalities have been adopted by other physical and link-layer technologies. Hence, the term 6LoWPAN network typically refers to any LLN leveraging 6LoWPAN principles.

The 6LoWPAN architecture consists of several independent WPANs. These WPANs operate as access stub networks, meaning their traffic never directly reaches other networks. All devices within a WPAN possess IPv6 addresses that share a common prefix, typically obtained through RA ND messages. Three types of WPANs exist. Simple WPANs connect to the IP core network through dedicated edge routers. Ad hoc WPANs directly connect devices without access to the IP core, forming closed-loop networks. Extended WPANs connect to the IP core through multiple edge routers linked by a backbone network.

An edge router acts as the gateway between the access network (the WPAN) and the core network. It manages data flow by: (1) Performing 6LoWPAN adaptation to translate data formats for transmission over the WPAN, (2) Facilitating ND for device communication within the WPAN, and (3) Handling other tasks like IPv4-to-IPv6 translation for communication with the core network. Devices in a WPAN can act as either hosts or routers, depending on the datagram's source and destination addresses. A smart device can act as a host when generating data and as a router when forwarding traffic from other devices in a mesh network. Most devices have only one interface (typically wireless) for both receiving and transmitting datagrams. This applies even when devices act as routers, where they simply forward datagrams on the same interface they were received on, extending network coverage. This differs significantly from traditional Internet routers that receive traffic on one interface and forward it to another.

Communication with the core network in WPANs follows the same principles as any other IP communication. Each WPAN device has a unique IPv6 address and can send and receive datagrams. However, due to limited processing power in constrained devices, the types of transport and application protocols are restricted. While protocols like UDP can operate, TCP and HTTP (Belshe et al. 2015; Fielding et al. 1999) are generally discouraged due to their resource demands. TCP's reliance on complex state machines and HTTP's large message size make them computationally expensive and unsuitable for most WPAN devices. Nevertheless, their usage is not entirely impossible, but less common.

Figure 4.20 compares a standard IPv6 protocol stack and a 6LoWPAN-based stack, showing key differences. Each stack utilizes distinct protocols at the physical and link layers, catering to their respective network characteristics. Direct use of IPv6 on top of IEEE 802.15.4 is not possible. The 6LoWPAN adaptation layer acts as a bridge, enabling IPv6 to function over this specific link layer. Due to performance limitations, TCP is not recommended for use in 6LoWPAN-based networks. In contrast to the standard IPv6 stack commonly employing HTTP over TCP, the 6LoWPAN stack typically utilizes *Constrained Application Protocol* (CoAP) over UDP. This difference is driven by the resource constraints and performance considerations within 6LoWPAN environments.

The 6LoWPAN protocol stack differs from the standard IPv6 stack in two key ways. First, it includes a dedicated 6LoWPAN adaptation layer positioned between the IPv6 and IEEE 802.15.4 link layers. This layer links these protocols, enabling

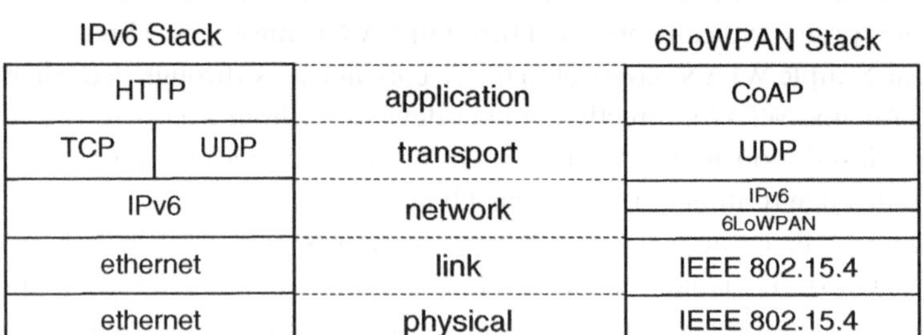

Figure 4.20 6LoWPAN Protocol Stack.

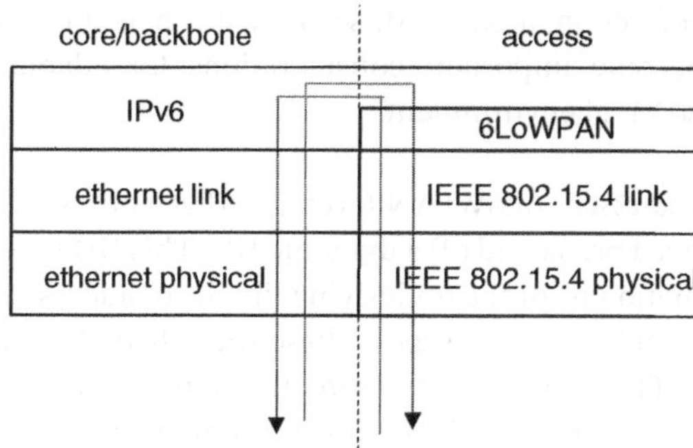

Figure 4.21 6LoWPAN Protocol Translation.

efficient IPv6 communication over low-power networks. Second, many 6LoWPAN implementations tightly integrate the UDP and IPv6 layers, providing APIs for convenient access to network and transport parameters. This integration sometimes leads to the 6LoWPAN adaptation layer being represented as part of the IPv6 layer, although it remains a distinct functional component.

Figure 4.21 shows a fundamental conversion process occurring at a WPAN edge router. This process translates data between the wired Ethernet core network and the wireless access network provided by the IoT-oriented IEEE 802.15.4 standard. Incoming IPv6 data from the core network over the Ethernet interface is first encapsulated into smaller 6LoWPAN datagrams before being transmitted through the IEEE 802.15.4 interface. Conversely, incoming 6LoWPAN data received on the IEEE 802.15.4 interface is unpacked back into standard IPv6 datagrams before being forwarded to the Ethernet side. This bidirectional translation is highly efficient and can operate in either a stateless or stateful manner depending on the chosen 6LoWPAN compression scheme. Further details regarding these compression mechanisms will be discussed later in this chapter.

While designed for the IEEE 802.15.4 standard, 6LoWPAN can be customized to operate with other link-layer technologies through minor adjustments. Specifically, it requires functionalities like data framing (explained in Section 2.2.1), one-to-one data delivery (unicast transmission), and unique device identifiers. These identifiers, along with the SAA) process, are used to establish unique IPv6 addresses for each device. Additionally, due to minimum IPv6 fragment sizes, 6LoWPAN performs its own fragmentation to ensure efficient data transmission on link layers with smaller MTUs.

Larger link layer frames with at least 60-byte payloads are desirable for two reasons: they minimize the number of fragments 6LoWPAN needs to manage, and they reduce the likelihood of datagram loss due to network issues, especially when fragments are numerous. 6LoWPAN compression further contributes by shrinking IPv6 and UDP headers, maximizing payload size within each frame. While not essential for all link layers, reliable transmission via error detection and correction, along with security mechanisms like encryption and authentication, become important considerations for adaptation beyond the standard IEEE 802.15.4 environment.

Addresses 6LoWPAN leverages IPv6 addresses built from two components: (1) a fixed prefix and (2) a dynamic IID. This IID is generated through SAA using information from the underlying link layer address and parameters received in 6LoWPAN ND RA messages. These specialized ND messages, optimized for WPANs by IETF RFC 6775, differ from traditional ND by accommodating power-saving cycles and sleeping devices while minimizing multicast usage.

More importantly, this mapping enables efficient IPv6 header compression in 6LoWPAN. Since most of the 128-bit source and destination addresses can be inferred from known values, they do not need to be transmitted in every datagram. This is particularly relevant for common IoT link-layer technologies like IEEE 802.15.4, which typically use 64-bit MAC addresses and configurable short addresses (8-bit or 16-bit), often assigned by the PAN coordinator. Both types of addresses can be used to construct the 64-bit IID within 6LoWPAN.

To minimize manual setup and enable deployment in remote locations, IoT devices rely on automated initialization through the link layer. This process goes beyond assigning basic local addresses; it attempts to secure a unique, globally routable address for each device using SAA. The 6LoWPAN edge router plays an important role by exchanging specialized *Neighbor Discovery* (ND) messages. These messages not only facilitate SAA but also allow the router to maintain a comprehensive device registry. By assigning simpler IPv6 addresses based on this registry, the router ultimately simplifies network operation and reduces reliance on human intervention for managing large-scale IoT deployments.

Figure 4.22 illustrates a complete IP network combining a traditional IPv6 core with a 6LoWPAN access network. The core network uses the prefix *2001:15 ::/64*, while the 6LoWPAN access network has its own unique prefix of *2001:20::/64*. This access network consists of five devices connected to an edge router. These devices can act as either routers or hosts based on their location and processing capabilities. Note that end devices within the access network do not directly connect to the edge router; they rely on mesh routing to establish communication. The edge router serves as the bridge between the access network and the core network, allowing it to access a server running on the address 2001:15::2110.

The bootstrapping process kicks off when the edge router broadcasts the *2001:20::/64* prefix to the devices using 6LoWPAN ND RA messages. Devices combine this prefix with their unique IID, typically derived from their 64-bit IEEE 802.15.4 MAC addresses, to create their initial IPv6 addresses. To further streamline this process, devices register themselves with the edge router, again using 6LoWPAN ND. In response, the router assigns them 16-bit IIDs, which

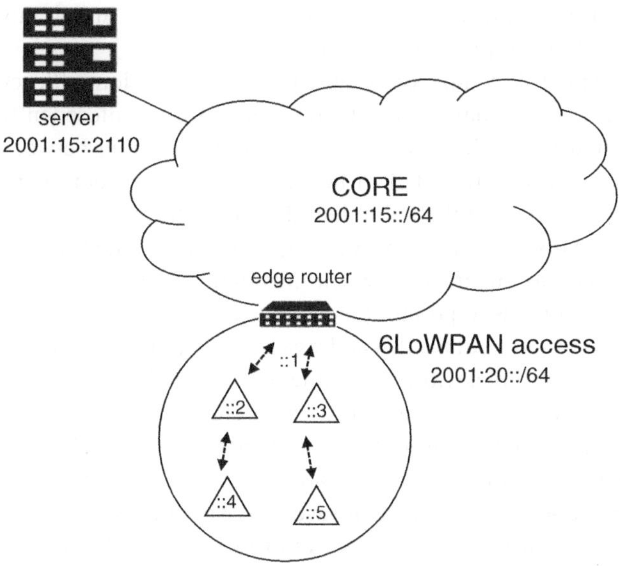

Figure 4.22 6LoWPAN Example.

are combined with the prefix to generate more concise IPv6 addresses. The edge router itself designates 2001:20::1 as its own access interface address and randomly assigns addresses *2001:20::2* through *2001:20::5* to the devices on the network.

Thanks to the use of shorter, 16-bit addresses within the WPAN, 6LoWPAN can efficiently compress network layers. Within the WPAN, devices only need to specify their own unique address and their neighbor's to communicate. For sending data outside the WPAN, full addresses are needed. The edge router acts as a translator, using its internal registry to convert the shorter address back into the full address needed for external communication, similar to looking up a house number in a directory. Ultimately, the edge router handles all the address conversions and compression, allowing WPAN devices to focus on their local tasks more efficiently.

Header Format The 6LoWPAN layer executes a field-level mapping of IPv6 and upper layer headers to generate a compressed 6LoWPAN header. This compression operates in either a stateless or stateful mode, contingent upon whether the compressed datagram itself encapsulates sufficient information for the reconstruction of the original IPv6 and transport layer headers.

Stateless compression functions by compressing each datagram header autonomously, leveraging solely intra-datagram redundancies. This approach is characterized by its relative simplicity, as it obviates the need for mechanisms to maintain inter-datagram context. However, this simplicity is inherently accompanied by a trade-off in compression efficiency.

In contradistinction, stateful compression augments compression efficacy by capitalizing on redundancies that span both the current and antecedently transmitted datagrams. It accomplishes this by establishing associations between redundant information in uncompressed IPv6 headers and context identifiers, which are then transmitted in lieu of the redundant data, thereby engendering a reduction in header size. This modality, however, requires coordination among the communicating devices to ensure the synchronization of their respective contexts, as this synchronization is important for accurate decompression.

6LoWPAN headers usually skip bulky IPv6 addresses because they can be figured out in other ways. For example, within the same network, a device's 128-bit address can be shrunk to a shorter 16-bit one. This stateless compression uses hints inside each datagram to figure out the full address. But there is another trick called stateful compression, where a common address shared by many datagrams gets a special code instead of being sent every time. This saves space but needs the sender and receiver to agree on this code upfront. Usually, a built-in protocol called 6LoWPAN ND helps devices share codebooks for these shortcuts.

Limited power in some link layers, like IEEE 802.15.4, creates two major hurdles for 6LoWPAN. First, devices cannot always talk directly to their neighbors; they need relay points (mesh routing) to pass messages. This requires 6LoWPAN to add extra information to packets, like who sent it initially and who it is ultimately for. Second, these link layers were built for closed systems and lack a clear indicator of what type of data they carry (like IPv6 or something else). Unlike Ethernet, which has a dedicated field for this. To overcome these challenges, 6LoWPAN employs a special scheme alongside packet fragmentation and header compression. These are the core functionalities that make 6LoWPAN tick.

To compensate for the absence of a type field at the link layer, 6LoWPAN introduces a unique 8-bit field at the beginning of every datagram, called the dispatch value. This dispatch value functions as a datagram type indicator, fulfilling a role similar to the Ethertype field in Ethernet frames (IETF RFC 4944) (Montenegro et al. 2007). Although the dispatch value is 8 bits long, it often leverages only a few of the most significant bits to signify the datagram type. The remaining bits within the dispatch value are repurposed to convey additional header information.

Table 4.3 outlines the specific dispatch values assigned to various 6LoWPAN header types. For instance, a dispatch value starting with *10* denotes a mesh header, whereas a value starting with *01000001* designates an uncompressed IPv6 header. It is important to note that the dispatch value *01000000* holds a special purpose: it signals that the dispatch value itself extends beyond a single byte. Any dispatch values not explicitly listed in the table are considered reserved for future use.

Table 4.3 6LoWPAN Dispatch Values.

Pattern	Header Type		Order
00		Not a LoWPAN frame (NALP)	
01	000001	Uncompressed IPv6	4th
	000010	HC1 stateless compression	
	010000	BC0 broadcast	2nd
	1	IPHC stateful compression	4th
	111111	Extension Escape (ESC)	
10		MESH header	1st
11	000	FRAG1 initial fragment	3rd
	100	FRAGN subsequent fragments	
	1010	RFRAG recovery fragments	

Figure 4.23 Uncompressed IPv6 Header through 6LoWPAN.

Figure 4.23 shows an IPv6 header that is transmitted uncompressed through 6LoWPAN. This is achieved by prepending the header with a dispatch value of *01000001*. It is important to note that the majority of IPv6 header fields are aligned within 32-bit boundaries. This alignment serves an important purpose: it enables 32-bit and 64-bit processors to access these fields exceptionally efficiently, using a single clock transition. However, the addition of the 8-bit dispatch value introduces a slight disruption to this alignment. For certain fields, such as the flow label, this misalignment requires an extra clock transition, which can lead to a marginal decrease in processing speed. Fortunately, this displacement has a less pronounced impact on many embedded IoT devices that employ 8-bit microcontrollers where memory is, of course, aligned on an 8-bit boundary.

In regular IPv6, there is a next header field that does the same job. After this label, 6LoWPAN squishes some of the message details (compressed fields) and puts them first. Then, it adds any leftover information that could not be shrunk (uncompressed fields) directly after, like names of devices outside the network (IPv6 addresses). These leftover details are called in-line because they follow right along after the compressed ones.

To grasp how 6LoWPAN compresses headers, let us examine a simple example. Figure 4.24 illustrates the headers employed when application traffic is encapsulated using UDP over IPv6. These headers, IPv6 and UDP, have fixed lengths of 40 and 8 bytes, respectively, resulting in a total of 48 bytes for each datagram, irrespective of the specific field values. In contrast, Figure 4.25 demonstrates the

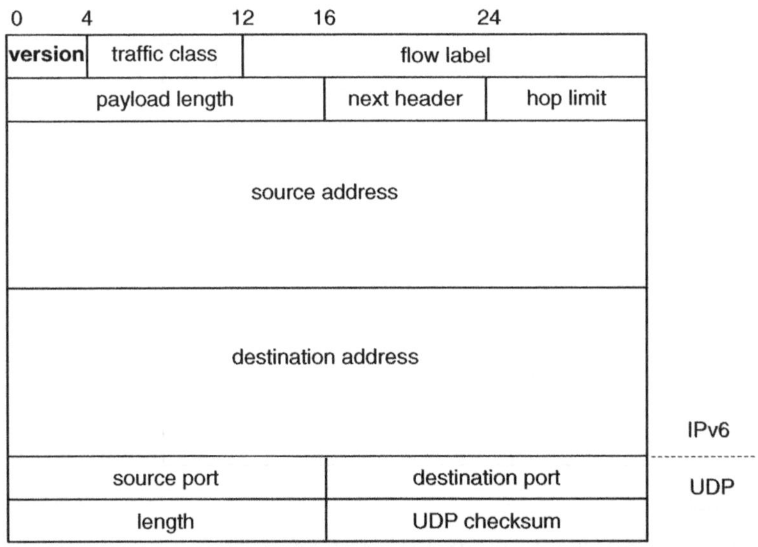

Figure 4.24 UDP over IPv6.

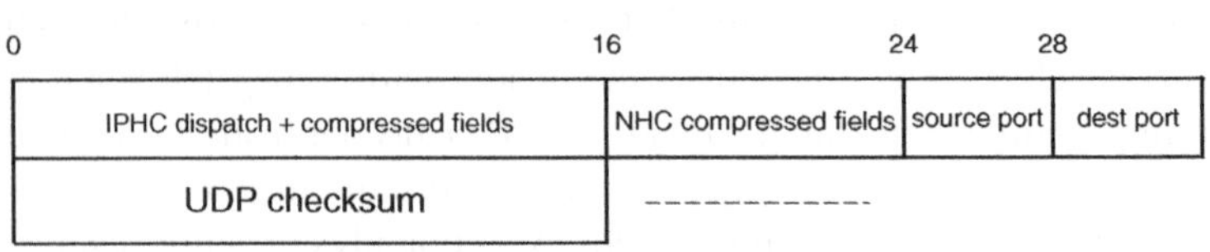

Figure 4.25 UDP over IPv6 over 6LoWPAN.

maximum compression that 6LoWPAN can achieve with these same headers. The compressed 6LoWPAN header spans a mere 6 bytes, starting with a 2-byte stateful *IP Header Compression* (IPHC) dispatch value accompanied by compressed fields. This is followed by a 1-byte *Next Header Compression* (NHC) UDP compressed header, which incorporates a 1-byte field to represent both the source and destination UDP ports. It is important to note that uncompressed UDP ports invariably occupy 2 bytes. The 6LoWPAN header concludes with the 2-byte UDP checksum field, transmitted in-line without compression. This remarkable compression technique yields a compression ratio of 1:6.

6LoWPAN offers two header compression options: stateless and context-based. Stateless compression is simpler, removing unnecessary fields and shrinking addresses based on known information like link-layer MACs. This works well for general messages. Context-based compression, however, leverages the context of ongoing communication between devices, allowing it to completely omit global unicast IPv6 addresses from headers. This saves even more space but requires keeping track of the context, making it more complex and suitable for specific data flows. In essence, 6LoWPAN chooses the right compression method based on the message and its context for optimal efficiency.

6LoWPAN uses two optional stateless header compression layers for efficiency: *Header Compression 1* (HC1) and *Header Compression 2* (HC2). HC1 focuses on shrinking the IPv6 header by removing redundant fields and compressing addresses based on known information. It is included using *01000010* as a dispatch value in the main header. Additionally, an HC2 flag within HC1 can tell the network that there is an even more specialized HC2 header following, which helps compress certain transport layer headers like UDP. This layered approach lets 6LoWPAN customize the compression based on the information at hand, maximizing space savings without adding unnecessary complexity.

6LoWPAN's HC2 header specifically targets the UDP header for further compression. Two initial bits, when set, indicate compressed source and destination ports. Otherwise, those ports are sent directly (in-line). Compressed ports utilize the four least significant bits of a specific range (61616–61631). Next, a field controls how the UDP payload length is handled: either inferred from the frame or transmitted in-line. Note that the UDP checksum remains an in-line field.

While stateless compression, defined in IETF RFC 4944, was the initial approach, it faced limitations. Enter stateful compression, covered by IETF RFC 6282, which tackles these head-on. The key difference? Stateful compression leverages contextual information shared by related data streams, something stateless compression lacked. For instance, smart device data sent to an application often shares the same destination address. Stateful compression allows both devices to agree on a shorter context identifier for this address, saving precious spacc in header compression. Instead of transmitting the full address every time, a compact identifier suffices, making communication more efficient for repetitive data flows. Because stateful compression is a superset of stateless compression, stateful compression can be used as a replacement to the latter too.

6LoWPAN stateful compression relies on shared context information between devices for efficient header shrinking. However, RFC 6282 only defines the compression format, leaving context negotiation to other mechanisms like ND. To ensure successful compression, both sender and receiver's context information must be synchronized. Due to potential connectivity and power issues, this context is divided into independent slots for flexibility. Higher-layer protocols like UDP or TCP, which use checksums based on pseudo headers including IPv6 addresses, can detect and prevent context synchronization errors. Incorrect checksums lead to datagram drops and retransmissions (for TCP).

Similar to stateless compression, stateful compression also utilizes the IPHC header and can optionally employ NHC for further header shrinking. Figures 4.26 and 4.27 illustrate the header formats: one with IPHC alone and another combining IPHC with NHC. Note that stateful compression utilizes only 3 out of 8 dispatch value bits to signal the presence of IPHC. This frees up the remaining 5 bits for encoding other essential header fields, maximizing space savings within the compressed header.

6LoWPAN's stateful compression header, IPHC, begins with a 2-bit TF field. When set to *11*, traffic class and flow label are assumed zero. Otherwise, TF encodes combinations of the 2-bit *Explicit Congestion Notification* (ECN), the 6-bit *Differentiated Services Code Point* (DSCP), and 20-bit flow label. Note that padding ensures 8-bit alignment for efficient encoding. A final next header flag indicates

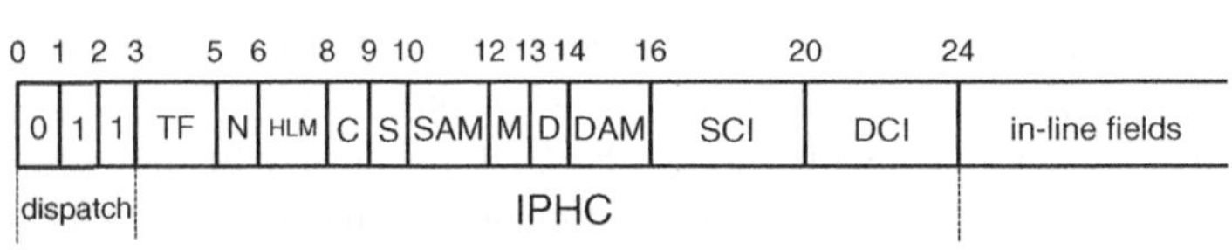

TF	traffic class and flow label encoding
N = 0	NHC does not follow
HLM	hop limit encoding
C	include context information fields
S	context-based source address encoding
SAM	source address mode
M	multicast destination address
D	context-based destination address encoding
DAM	destination address mode
SCI	source context identifier (if C = 1)
DCI	destination context identifier (if C = 1)

Figure 4.26 Network Layer IPHC.

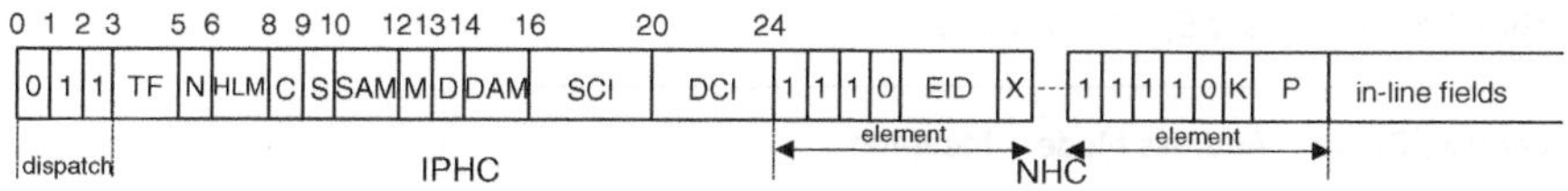

TF traffic class and flow label encoding
N = 1 NHC follows
HLM hop limit encoding
C include context information fields
S context-based source address encoding
SAM source address mode
M multicast destination address
D context-based destination address encoding
DAM destination address mode
SCI source context identifier (if C = 1)
DCI destination context identifier (if C = 1)
EID IPv6 extension header identifier
X more NHC elements
K UDP checksum removal
P UDP source and destination port encoding

Figure 4.27 Network IPHC with Transport NHC.

whether an optional NHC header follows. If NHC is absent, the standard 8-bit IPv6 next header field appears directly after IPHC. This concisely explains IPHC's structure and encoding options.

The 1-bit source context flag tells if the address is local or needs an SAM code from Table 4.4. Next, a multicast flag checks if it is sending to multiple destinations. Finally, a destination context flag and a DAM code (again, from Table 4.4) pack the destination address efficiently. If it is a multicast message, Table 4.5 shows how address bits are cleverly combined to save space. When context encoding is active, 4-bit context identifiers for both source and destination are used to specify the respective address prefixes, efficiently representing them within the compressed header. If an NHC header is present, it consists of multiple 8-bit elements. One element type, starting with the sequence *1110*, incorporates a 3-bit *Extension*

Table 4.4 Unicast S/SAM and D/DAM Values.

Context-based	Address Mode	Meaning
0	00	full unicast address
0	01	FE80:0:0:0:inline
0	10	FE80:0:0:0:0:0:0:inline
0	11	FE80:0:0:0:link-layer
1	01	context[0...63]:inline
1	10	context[0...111]:inline
1	11	context[0...127]

Table 4.5 Multicast D/DAM Values.

Context-Based	Address Mode	Meaning
0	00	full unicast address
0	01	FFxx::xx..xx
0	10	FExx::xx..xx
0	11	FE02::00xx
1	00	FFxx:context[0...31]:xx..xx

Identifier (EID), which maps to specific header fields. This EID is followed by a flag indicating whether another element is present in the NHC header.

Another NHC element type, identified by *11110*, targets UDP headers. This element includes a flag: if true, the UDP checksum is sent directly (in-line); otherwise, it is recalculated at the receiving end when converting back to an IPv6 header. Next, a 2-bit field controls source and destination port encoding (see Figure 4.28). Ports can be: sent fully (in-line), encoded as an 8-bit offset within a specific range (61440–61695), or, similar to HC2, encoded as a 4-bit offset within a narrower range (61616–61631). This flexible approach optimizes space based on the port values being transmitted.

Figure 4.29 illustrates how combining IPHC and NHC headers can efficiently compress UDP over IPv6 using link-local addresses. This approach squeezes the header down to roughly 48 bits, saving a byte compared to using HC1 and HC2. Two additional bytes can be shaved off by ditching the NHC checksum and

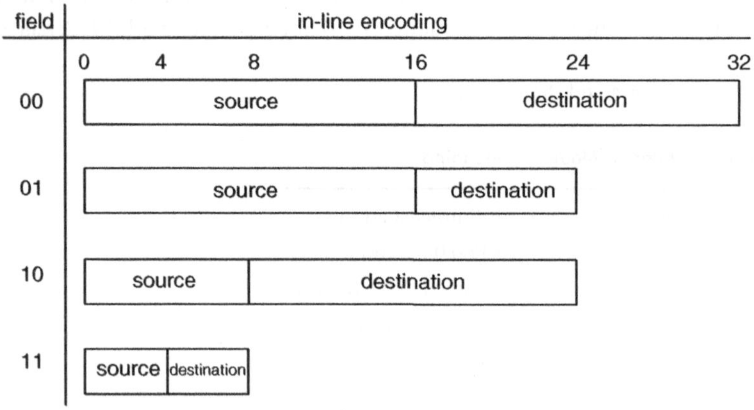

Figure 4.28 NHC Port Number Encoding.

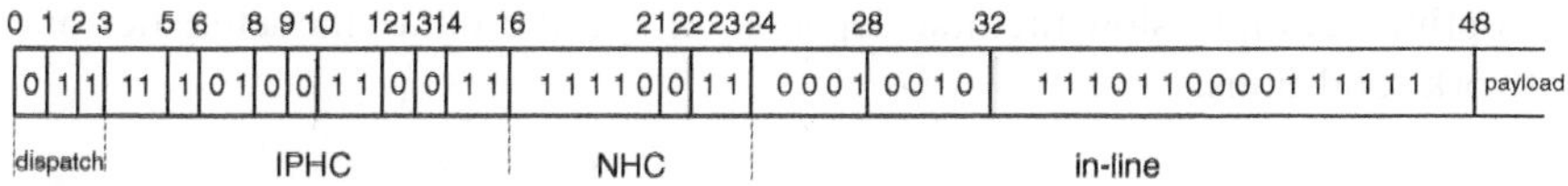

TF	11	neither traffic class not flow label encoding
N = 1	1	NHC follows
HLM	01	hop limit is 64
C	0	no context information
S	0	no context-based source address encoding
SAM	11	link layer source address
M	0	unicast destination address
D	0	no context-based destination address encoding
DAM	11	link layer destination address
UDP	11110	UDP transport
K	0	transmit checksum in-line
P	11	4-bit source and destination port encoding
source	0001	UDP source port (61441)
destination	0010	UDP destination port (61442)
checksum	11 ..11	checksum

Figure 4.29 6LoWPAN Datagram with Link-Local Addresses.

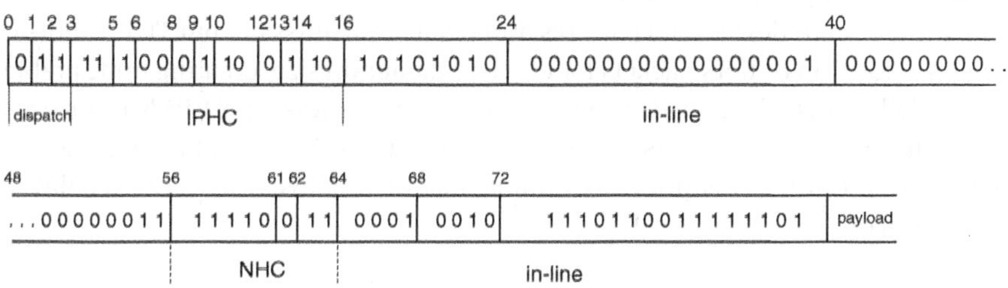

TF	11	neither traffic class not flow label encoding
N = 1	1	NHC follows
HLM	00	hop limit is 1
C	0	use default context
S	1	context-based source address encoding
SAM	10	context + 16-bit in-line source address
M	0	unicast destination address
D	1	context-based destination address encoding
DAM	10	context + 16-bit in-line destination address
HLM	00000010	hop limit (2)
source address	0000000000000001	16-bit short source address
destination address	0000000000000011	16-bit short destination address
UDP	11110	UDP header follows
source port	0001	source port (61441)
destination port	0010	destination port (61442)
checksum	11 .. 01	checksum

Figure 4.30 6LoWPAN Datagram with Unicast Addresses.

trusting lower layer protection. However, Figure 4.30 showcases IPHC and NHC with global unicast addresses. Here, a default context and in-line encoding of 16-bit short addresses are employed. This offers flexibility for different address types, though the resulting header size might be slightly larger. Ultimately, the optimal compression method depends on specific data characteristics and network conditions.

TCP and 6LoWPAN Unlike UDP, which uses stateless and stateful header compression (HC2 and NHC) in 6LoWPAN, there is no official TCP compression scheme yet. While TCP offers reliable data transfer with *Automatic Request* (ARQ), its complexity and retransmissions can be detrimental in resource constrained devices and loss-prone IoT environments. Frequent retransmissions in limited-capacity networks can lead to collisions and exacerbate packet loss, making received data useless and causing further application-layer losses. So, network packet loss gets amplified by retransmissions that ultimately harm application reliability.

TCP tackles packet loss by slowing down transmission, assuming congestion on intermediate devices. However, in IoT environments with low SNR and minimal traffic, this approach hurts performance. Frequent, rapid retransmissions act like an FEC mechanism, directly counteracting channel-induced loss and mitigating the need for congested-flow adaptations.

Real-time smart device data requires immediate transmission, but TCP's built-in Nagle's algorithm buffers packets for efficient datagram size, creating undesirable latency. While disabling Nagle can improve responsiveness, 6LoWPAN prioritizes speed and does not offer session management with TCP, only allowing uncompressed TCP traffic. This prioritizes immediate data delivery over the reliability guarantees of TCP.

6Lo and 6LoBTLE 6LoWPAN's success in adapting IPv6 to IEEE 802.15.4 paved the way for similar adaptations on other low-power network technologies like BLE, DECT ULE, NFC, and more. These adaptations, collectively known as *IPv6 over Networks of Resource Constrained Nodes* (6Lo) technologies, are the core focus of the IETF's 6Lo working group, dedicated to standardizing IPv6 connectivity for resource-constrained devices (Gomez et al. 2017).

These adaptations fine-tune 6LoWPAN's features to suit their specific physical and communication characteristics. Figure 4.31 shows the different technology families and their corresponding adaptation mechanisms. For example, BLE uses 6LoBTLE, a modified version of 6LoWPAN, while DECT ULE sticks to plain 6LoWPAN. This variety reflects the need to tailor IPv6 support to the unique properties of each low-power network.

While technologies like ZigBee, BLE, NFC, and DECT ULE offer complete communication stacks with custom network, transport, and application layers, 6Lo overlays these with standardized IETF IP protocols, essentially adding IPv6 connectivity to existing networks. However, each technology's unique physical and link layer presents challenges from slow data rates and limited message sizes to

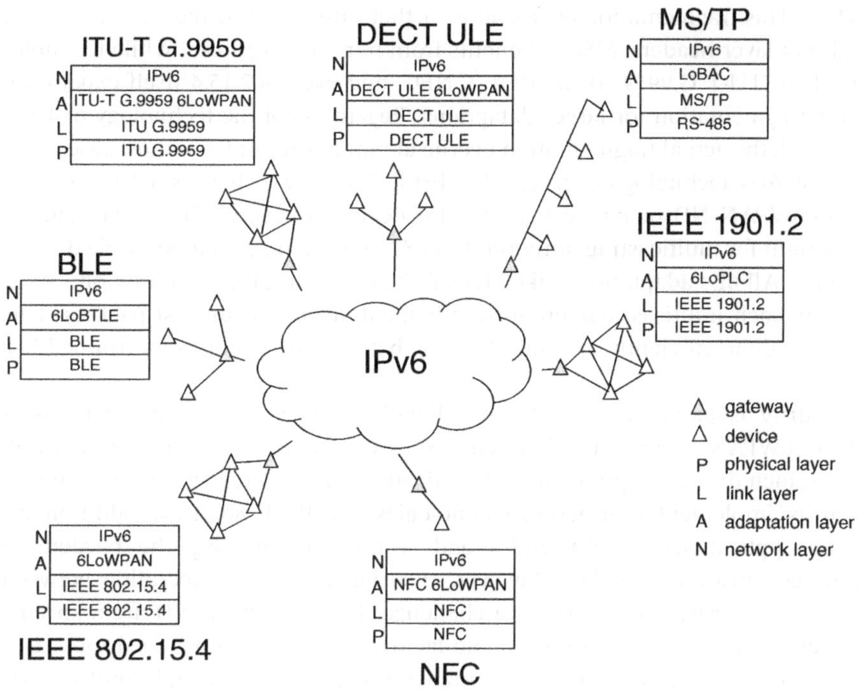

Figure 4.31 Summary of 6Lo Families.

lack of fragmentation and uneven uplink/downlink capabilities, posing hurdles for seamless IPv6 integration.

Both 6LoWPAN and other 6Lo technologies employ header compression to shrink packet size. This can be stateless (simple) or stateful (more complex but leveraging flow information for better compression). All 6Lo options apart from 6LoWPLC (built on IEEE 802.15.4) adapt 6LoWPAN's header compression to their specific link layer needs. ITU-G.9959, MS/TP LoBAC, and NFC make minor tweaks related to their link layer address size. For single-hop star networks like BLE and DECT ULE, header compression is simpler because the gateway always knows the source (originator) and destination (itself) of packets. This allows suppressing the IPv6 source and destination addresses in some cases.

6LoWPAN relies on fragmentation because its underlying link layer (IEEE 802.15.4) lacks this capability. However, most other 6Lo technologies have native link-layer fragmentation, making it unnecessary in their IPv6 adaptation layers. Moreover, the high MTUs of technologies like MS/TP (2032 bytes) eliminate the need for fragmentation altogether. Conversely, BLE and DECT ULE, with considerably smaller payloads, face significant fragmentation due to their limited

MTUs. This fragmentation overhead is further influenced by the size of physical and link layer headers. MS/TP benefits from minimal overhead, while technologies like ITU-T G.9959, IEEE 1901.2, NFC, and even 802.15.4 itself experience some fragmentation for larger datagrams. Regardless of the technology or layer involved, the actual fragmentation overhead remains minimal in most cases.

Most 6Lo technologies, except for IEEE 1901.2 which uses DHCPv6, rely on 6LoWPAN ND or native IPv6 ND for device discovery. This is particularly important for multicasting, important in IoT for sending data to multiple devices at once. All 6Lo adaptations, like 6LoWPAN, rely on multicast emulation, transmitting each multicast datagram to individual devices as unicast packets. This must be done efficiently to avoid draining batteries in resource-constrained IoT devices.

Security is paramount for 6Lo, and luckily, most adaptation mechanisms like 6LoWPAN leverage the inherent security features of their underlying link layers, including encryption and authentication. This layered approach simplifies security implementation across technologies like IEEE 802.15.4. Additionally, many applications opt for end-to-end security, minimizing the burden on resource-constrained nodes. However, ensuring efficient co-operation between different security layers remains a challenge. DoS attacks on ND are a specific concern, where attackers impersonate legitimate devices with forged messages. Mitigation strategies include prioritizing router messages and limiting ND transmission rates. Fragmentation further complicates security. Attackers can send incomplete datagrams, wasting memory on pending fragments, or inject fake or duplicate fragments to corrupt incoming data.

4.4.2 LPWANs

WPAN technologies operate at low power; they offer sufficient transmission rates for native IPv6 support (Al-Sarawi et al. 2017). However, WPANs suffer from limited range, typically reaching only a few hundred meters due to their direct hop-to-hop communication. LPWAN technologies aim to extend device coverage but sacrifice SNR, leading to lower transmission rates and smaller MTU (Farrell 2018). These limitations impede full IPv6 support and adoption of derived protocols like CoAP and *Routing for Low Power* (RPL) on these devices. Consequently, most LPWAN deployments use hybrid approaches with proprietary access network stacks that rely on IoT gateways to bridge the gap to IP networks.

Traditional mobile cellular technologies like 5G offer the highest performance, but their high power consumption makes them unsuitable for most IoT applications. Similarly, while IEEE 802.11-derived technologies achieve comparable throughput, their power drain is often excessive for IoT environments. Conversely, IoT WPAN technologies like IEEE 802.15.4 and BLE prioritize power efficiency at the expense of both throughput and range, typically offering low data rates and very limited coverage.

LPWAN bridges this gap by extending the range of WPANs while maintaining low power consumption, enabling multi-year battery life for devices that transmit small data packets infrequently over long distances. Different LPWAN solutions vary in their network topology, device coverage, battery life, interference resilience, node density, security features, and communication direction (unidirectional or bidirectional). These factors are tailored to specific application needs. The following section introduces several prominent LPWAN technologies relevant to the IoT.

Several LPWAN technologies are available, including: *Long Range* (LoRa), SigFox, the *DASH7 Alliance Protocol* (D7AP), Weightless, Narrowband IoT (NB-IoT), *Narrowband Fidelity* (NB-Fi), IQRF, *Random Phase Multiple Access* (RPMA), Telensa, *Sensor Network Over White Spaces* (SNOW, Nwave, Qowisio), *IEEE 802.15.4k Task Group* (TG4k), *IEEE 802.15.4g Task Group* (TG4g), LTE-M and *Extended Coverage GSM IoT* (EC-GSM-IoT).

The following subsections provide a detailed exploration of how LoRa and 6Lo technologies can be combined to facilitate the integration of SIP and RTP protocols.

4.4.2.1 LoRa

LoRa is a comprehensive protocol stack designed for LPWAN applications, enabling devices to operate for over a decade on a single battery charge. As illustrated in Figure 4.32, LoRa's architecture includes a physical layer, link layer, and application layer. Note that it omits network and transport layers due to its focus on LoRa-to-LoRa device communication and the absence of native IP support. Further, each layer comprises multiple sublayers, each serving distinct functions.

LoRa doubles down on security with both link and application layer protection. At the link layer, AES-128 encryption paired with a unique 64-bit IEEE EUI-64

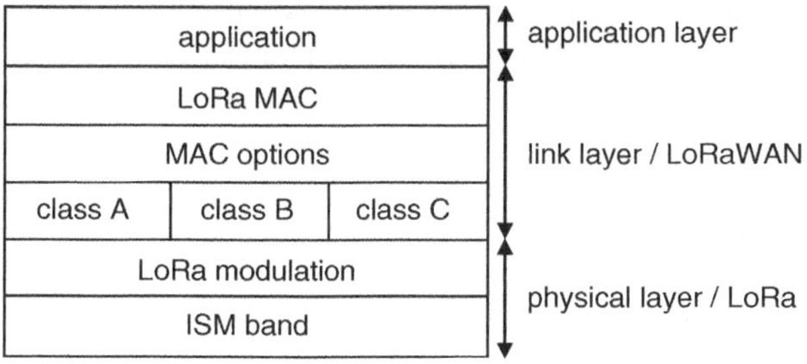

Figure 4.32 LoRa Stack.

key ensures secure network communication through device authentication and data encryption. The application layer adds another layer of defense by separately encrypting device data with AES-128.

LoRa leverages the power of *Chirp Spread Spectrum* (CSS) across 433 MHz and 915/868 MHz ISM bands. This modulation technique optimizes power efficiency and facilitates exceptional range, exceeding 10 kilometers at low data rates (though specific coverage varies depending on physical obstacles). A single gateway can potentially cover hundreds of square kilometers, with the maximum supported devices and speed inversely proportional to range. For example, an 8-kilometer range with 250 bps supports 50 devices, while a 2.5-kilometer range with 5 Kbps supports 700. Note that data rates also vary based on ISM band: 250 bps to 50 Kbps for Europe's 868 MHz band and 980 bps to 21.9 Kbps for the US 915 MHz band. This difference stems from the number of available subchannels (10 for 868 MHz, over 64 for 915 MHz) and bandwidth per subchannel (125–500 kHz for both uplink and downlink in the 868 MHz band, 125 kHz uplink and 500 kHz downlink for the 915 MHz band).

LoRaWAN Alliance (2017), the heart of LoRa networking, is not just about connecting devices, it defines the entire network architecture and access protocols. Unlike traditional WPANs that rely on intricate mesh networking for long-range communication, LoRaWAN adopts a simpler approach: direct device-to-gateway communication. This eliminates the need for data aggregation through intermediate nodes, preserving battery life and network capacity for resource-constrained devices.

To compensate for the lack of meshing, LoRaWAN sacrifices speed for distance. Devices transmit at lower rates, but their signals can reach farther and penetrate interference more effectively. Redundancy is built in by associating each device with multiple gateways. This ensures that even if a packet gets lost during transmission, another copy likely reaches the server. The server plays an important role in this simplified architecture. It verifies packet integrity, eliminates redundant copies, and takes care of complex decisions, offloading the burden from resource-limited gateways. This approach also simplifies mobile communication, eliminating the need for complex handover mechanisms when devices switch between gateways. In essence, LoRaWAN prioritizes simplicity and efficiency, making it a powerful tool for long-range networking in resource-constrained environments.

LoRaWAN's MAC layer adopts a straightforward yet effective approach rooted in a predecessor of CSMA/CA called ALOHA. This protocol liberates devices from the need to wait for channel availability before transmission. Should a collision occur, the device simply pauses for a random interval before retrying. This simplicity aligns perfectly with LoRaWAN's ethos of minimizing power consumption, as it eliminates the need for network synchronization. Additionally, transmission rates are influenced by factors such as the spreading factor of CSS modulation. Devices closer to gateways, typically enjoying stronger links, transmit at higher speeds, while those farther away conserve power by transmitting slower. LoRaWAN excels in its scalability. Networks can start with a single gateway and seamlessly expand with additional gateways as capacity demands grow.

LoRaWAN devices operate within three distinct classes (A, B, and C) that impact their power consumption, interaction with gateways, and the trade-off between battery life, latency, and throughput. Class A is the default and most power-efficient option for all devices. Gateways can only send downlink packets in response to uplink traffic from the device. This minimizes power consumption and extends battery life. However, it also leads to higher latency for downlink communication. Class B offers a balance between power consumption and latency. Gateways can send downlink packets on a scheduled basis, allowing for proactive communication with devices. While it consumes more power than Class A, it offers lower latency for downlink messages. Class C provides the lowest latency and highest throughput but also consumes the most power. Gateways can send downlink packets continuously unless they are receiving uplink traffic from the device. This constant listening mode offers the fastest response times but significantly reduces battery life.

Choosing the appropriate class depends on the specific application requirements. For situations where battery life is paramount and latency is less critical, Class A might be the best option. Conversely, applications that require RTC might benefit from Class C, despite its higher power consumption. Class B offers a middle ground, suitable for scenarios where both power efficiency and timely downlink communication are important.

IPv6 adaptation for LoRa uses the protocol stacks shown in Figure 4.33. Observe the similarities with Figure 4.15. Fundamentally, WPAN and LPWAN technologies exhibit a common layered structure, with the primary distinction, as anticipated, residing in the physical and link layers. Within the LoRa context, 6LoBTLE serves as the adaptation mechanism, furnishing functionalities akin to 6LoWPAN, albeit without fragmentation support.

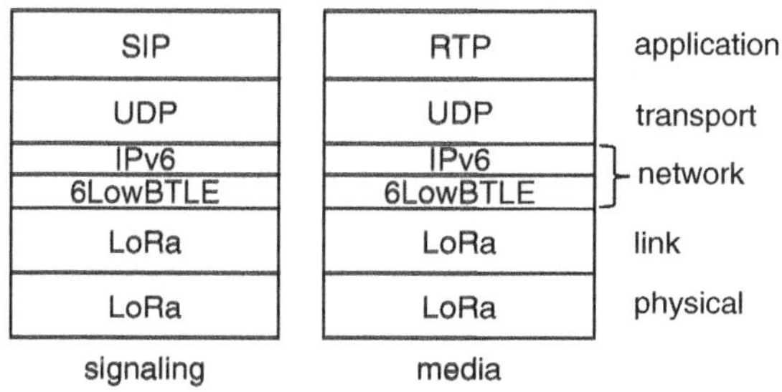

Figure 4.33 Media and Signaling over LoRa.

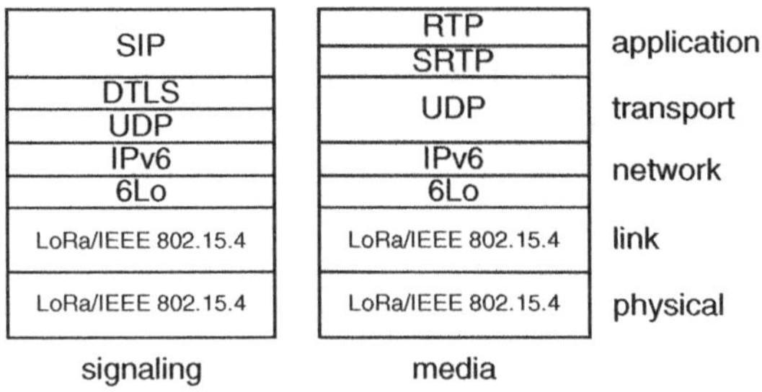

Figure 4.34 SIP and RTP over IoT Networks.

RTC IPv6 adaptation facilitates the utilization of IPv6 at the network layer and UDP at the transport layer, thereby enabling the transmission of the non-3GPP RTC protocols expounded upon in Section 4.2.3.

Figure 4.34 illustrates the integration of SIP and RTP protocols with the physical and link layers of LPWAN LoRa and WPAN IEEE 802.15.4 within signaling and media stacks. In this context, 6Lo adaptation facilitates end-to-end connectivity through the IPv6 protocol. While the RTP layer relies on UDP transport with SRTP, the SIP layer exclusively supports UDP transport with DTLS security.

Example 4.3 An RTP traffic encapsulates every 20 milliseconds speech using ITU-T Recommendation G.711. Assuming an IEEE 802.15.4 physical and link layer, what is the size of the RTP packets? Assume a 6-byte 6LoWPAN header length and a 10-byte IEEE 802.15.4 link layer header length.

Solution:
ITU-T Recommendation G.711 is an NB codec, where $8000 \times 0.02 = 160$ bytes are sent every 20 milliseconds. The RTP packet size is therefore $160 + 12 + 6 + 10 = 188$ bytes long. Note that 12 bytes correspond to the RTP header length. The 6LoWPAN header respectively holds through IPHC and NHC all the information typically included in the IPv6 and UDP headers.

4.5 Putting All Together

Figure 4.35 shows a network topology including various technologies, including 3GPP (LTE, 5G, IoT) and non-3GPP (Wi-Fi and Ethernet). Note that irrespective of the underlying physical and link layers, all protocol stacks converge at the SIP and RTP application layers, enabling unified communications. This functionality is facilitated by the presence of Internet connectivity.

The 3GPP protocols are capable of supporting either IPv4 or IPv6 and interact with the base stations eNodeB (for LTE) and gNB (for 5G). The laptop utilizes IEEE 802.11 for media transmission and reception via a Wi-Fi router. In an IoT scenario,

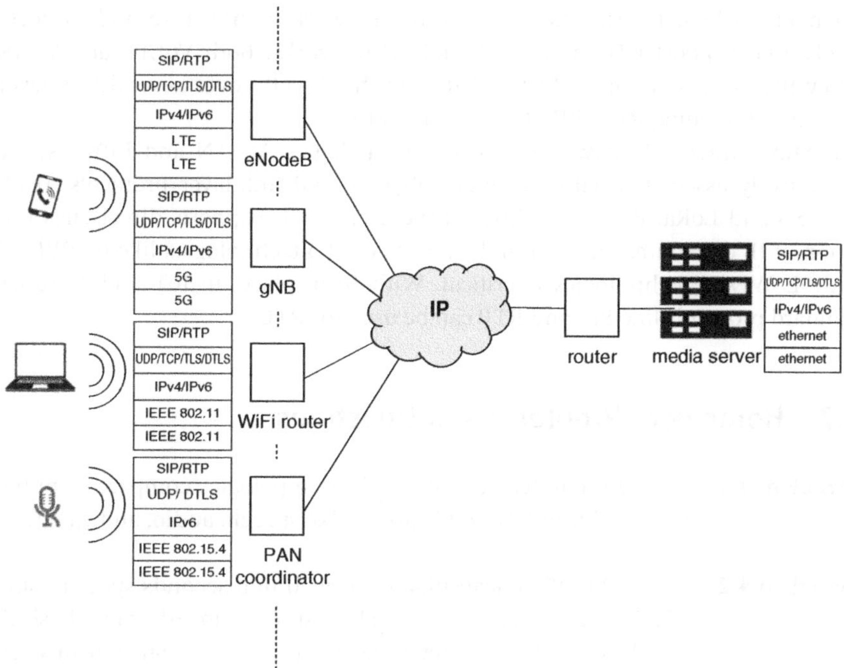

Figure 4.35 SIP and RTP over Telecommunication Networks.

the microphone leverages IEEE 802.15.4 transport to transmit LBR audio through a PAN coordinator. The interaction is directed toward a media server situated on the core side of the network, which, through Ethernet connectivity, facilitates media processing using *Automatic Speech Recognition* (ASR) and *Text-To-Speech* (TTS) functionalities. It is important to note that the media server can operate either in a cloud-based or on-premises environment.

4.6 Summary

The IP protocol and other upper layer protocols like UDP, SIP, RTP, HTTP, DLTS, and RTCP require link layers that enable the encapsulation of the corresponding datagrams. Under non-3GPP networks, two main technologies are key: wired Ethernet and wireless Wi-Fi. Both Ethernet and Wi-Fi (standardized as IEEE 802.11) support compatibly high speeds in the context of Local Area Network (LAN) topologies. IP traffic can be directly encapsulated over both Ethernet and Wi-Fi. In this context, SIP and RTP enable the establishment and termination of RTC sessions, while WebRTC provides the same functionality for web-based applications.

Under 3GPP networks, IMS defines the network architecture and protocol stacks that support RTC in 4G, 5G, and 6G networks. Both VoLTE and VoNR allow the encapsulation of SIP and RTP traffic directly over several link-layer protocols, including RLC, PDCP, SDAP, and MAC.

Finally, under IoT networks, two main topologies, WPAN and LPWAN, are respectively associated with two main physical and link-layer protocols: IEEE 802.15.4 and LoRa. Because neither of these protocols supports direct encapsulation of IP datagrams, adaptation by means of 6Lo technologies like 6LoWPAN and 6LoWPAN technologies is critical. With IP in place, in IoT architectures, standard protocols like SIP and RTP can be used for RTC.

4.7 Homework Problems and Questions

Problem 4.1 How do non-3GPP, 3GPP, and IoT topologies compare from the perspective of RTC applications like speech, audio, and video?

Problem 4.2 An RTP traffic encapsulates every 20 milliseconds speech using ITU-T Recommendation G.711. Assuming an Ethernet physical, link layer and IPv4 datagrams, what is the actual transmission rate? Assume a 6-byte 6LoWPAN header length and a 10-byte IEEE 802.15.4 link layer header length.

Problem 4.3 Neither Ethernet nor the IEEE 802.11 families require any type of adaptation to encapsulate IPv4 traffic; why is this so?

Problem 4.4 Why do IEEE 802.11 and many other wireless technologies include four address fields in their frames? What is the point of this?

Problem 4.5 How large is an Ethernet frame that carries an RTP packet that encapsulates a single 6.7 Kbps AMR speech frame in bandwidth-efficient mode?

Problem 4.6 What is the actual transmission of a session with RTP packets carrying a single 6.7 Kbps AMR speech frame in bandwidth-efficient mode over VoLTE?

Problem 4.7 What is the actual transmission of a session with RTP packets carrying two 16.4 Kbps Primary Rate EVS frames over VoNR?

Problem 4.8 While the transmissions rates obtained from transmitting a single EVS session over VoLTE and VoNR are comparable, where is the real difference between VoLTE and VoNR?

Problem 4.9 In the context of transmitting an uncompressed IPv6 header through 6LoWPAN, what purpose does the dispatch value of 01000001 serve, and how does it impact the alignment of header fields?

Problem 4.10 Why does 6LoWPAN prioritize speed and only allow uncompressed TCP traffic, even though it sacrifices the reliability guarantees provided by TCP?

Problem 4.11 In IEEE 802.3 frames lacking an Ethertype field, how is the type of encapsulated data explicitly identified within the payload?

Problem 4.12 An ITU-T Recommendation G.711 speech stream gets encapsulated over multiple frames in 364-byte RTP packets. If the network frame loss is 2%, what is the datagram loss for the following scenarios?
(a) IEEE 802.15.4
(b) Etherne

Problem 4.13 An RTP traffic encapsulates speech using ITU-T Recommendation G.711. Assuming an IEEE 802.15.4 physical and link layer, what is the actual transmission rate?

Bibliography

Al-Sarawi, S., Anbar, M., Alieyan, K., and Alzubaidi, M. (2017). Internet of things (IoT) communication protocols: review. In 2017 8th International Conference on Information Technology (ICIT), pp. 685–690.

Alvestrand, H.T. (2021). Overview: real-time protocols for browser-based applications. RFC 8825, January. Available from: https://www.rfc-editor.org/info/rfc8825. (Accessed March 15th, 2024)

Belshe, M., Peon, R., and Thomson, M. (2015). Hypertext transfer protocol version 2 (HTTP/2). RFC 7540, May. Available from: https://rfc-editor.org/rfc/rfc7540.txt. (Accessed March 15th, 2024)

Chew, D. (2019). "Protocols of the Wireless Internet of Things," in *The Wireless Internet of Things: A Guide to the Lower Layers* , IEEE, 2019, pp. 21–45, doi: 10.1002/9781119260608.ch2.

Cover, T.M. and Thomas, J.A. (2006). Elements of information theory (*Wiley Series in Telecommunications and Signal Processing*). USA: Wiley-Interscience. ISBN 0-471-24195-4.

Cox, C. (2014). *An introduction to LTE: LTE, LTE-advanced, SAE, VoLTE and 4G mobile communications*. 2nd ed. Wiley Publishing. ISBN 1-118-81803-2.

Cox, C. (2020). *An introduction to 5G: the new radio, 5G network and beyond*. Wiley. ISBN 978-1-119-60269-9. Available from: https://books.google.com/books?id=aJsMEAAAQBAJ. (Accessed March 15th, 2024)

Ersue, M., Bormann, C., and Keranen, A. (2014). Terminology for constrained-node networks. RFC 7228, May. Available from: https://rfc-editor.org/rfc/rfc7228.txt. (Accessed March 15th, 2024)

Farrell, S. (2018). Low-power wide area network (LPWAN) overview. RFC 8376, May. Available from: https://rfc-editor.org/rfc/rfc8376.txt.313 (Accessed March 15th, 2024)

Fielding, R., Gettys, J., Mogul, J., Frystyk, H., Masinter, L., Leach, P., and Berners-Lee, T. (1999). RFC 2616, hypertext transfer protocol – http/1.1. Available from: http://www.rfc.net/rfc2616.html. (Accessed March 15th, 2024)

Gomez, C., Paradells, J., Bormann, C., and Crowcroft, J. (2017). From 6LoWPAN to 6Lo: expanding the universe of IPV6-supported technologies for the internet of things. *IEEE Communications Magazine*, 55, 12. DOI: 10.1109/MCOM.2017.1600534.

Haykin, S. (2009). *Communication systems*. 5th ed. Wiley Publishing. ISBN 0-471-69790-7.

Herrero, R. (2021). Fundamentals of IoT communication technologies. *Textbooks in telecommunication engineering*. Springer International Publishing. ISBN 978-3-030-70079-9. Available from: https://books.google.com/books?id=k70rzgEACAAJ. (Accessed March 15th, 2024)

Holler, J., Tsiatsis, V., Mulligan, C., Avesand, S., Karnouskos, S., and Boyle, D. (2014). *From machine-to-machine to the Internet of Things: introduction to a new age of intelligence*. 1st ed. USA: Academic Press, Inc. ISBN 0-124-07684-X.

IEC. (2016). IEC 62591:2016 Industrial networks – wireless communication network and communication profiles - WirelessHART. Standard, International Electrotechnical Commission, Switzerland, 3.

IEEE. (2016). IEEE standard for information technology—telecommunications and information exchange between systems local and metropolitan area networks specific requirements – part 11: Wireless LAN medium access control (MAC) and physical layer (PHY) specifications. IEEE Std 802.11-2016 (Revision of IEEE Std 802.11-2012), pp. 1–3534.

IEEE. (2018). IEEE standard for Ethernet. IEEE Std 802.3-2018 (Revision of IEEE Std 802.3-2015), pp. 1–5600.

IEEE. (2020). IEEE standard for low-rate wireless networks. IEEE Std 802.15.4-2020 (Revision of IEEE Std 802.15.4-2015), pp. 1–800, 2020.

International Telecommunication Union (ITU). (2015). Telecommunication Standardization Sector of ITU. ITU-T G.9959: Short range narrow-band digital radiocommunication transceivers – PHY, MAC, SAR and LLC layer specifications. Technical report, 1.

ISA. (2011). ANSI/ISA-100.11a-2011. Wireless systems for industrial automation: process control and related applications. USA: Standard, International Society of Automation, 8.

LoRa Alliance. (2017). LoRaWAN 1.1 specification, April. Available from: https://lora-alliance.org/sites/default/files/2018-04/lorawantm specification -v1.1.pdf. (Accessed March 15th, 2024)

Montenegro, G., Hui, J., Culler, D., and Kushalnagar, N. (2007). Transmission of IPv6 packets over IEEE 802.15.4 networks. RFC 4944, September. Available from: https://rfc-editor.org/rfc/rfc4944.txt. (Accessed March 15th, 2024)

Nath, S.V., Kara, O., and Simlai, A. (2023). Mastering 5g network design, implementation, and operation: the comprehensive guide to understanding, designing, deploying, and managing 5G networks. Packt Publishing, Limited. ISBN 978-1-838-98010-8. Available from: https://books.google.com/books?id=38ZQzwEACAAJ. (Accessed March 15th, 2024)

Poikselka, M., Niemi, A., Khartabil, H., and Mayer, G. (2006). The IMS second edition: IP multimedia concepts and services. Hoboken, NJ, USA: John Wiley & Sons, Inc. ISBN 0-470-01906-9.

Popovski, P. (2020). *Information-Theoretic view on wireless channel capacity*, pp. 201–234.

Schooler, E., Rosenberg, J., Schulzrinne, H., Johnston, A., Camarillo, G., Peterson, J., Sparks, R., and Handley, M.J. (2002). SIP: Session initiation protocol. RFC 3261, July. Available from: https://rfc-editor.org/rfc/rfc3261.txt. (Accessed March 15th, 2024)

Schulzrinne, H., Casner, S.L., Frederick, R., and Jacobson, V. (2003). RTP: a transport protocol for real-time applications. RFC 3550, July. Available from: https://rfc-editor.org/rfc/rfc3550.txt. (Accessed March 15th, 2024)

Wu, Y., Singh, S., Taleb, T., Roy, A., Dhillon, H.S., Kanagarathinam, M.R., and De, A. (2021). 6G mobile wireless networks. *Computer communications and networks.* Springer International Publishing. ISBN 978-3-030-72777-2. Available from: https://books.google.com/books?id=4E4 EAAAQBAJ. (Accessed March 15th, 2024)

Zig. (2015). ZigBee specification. Standard. USA: The ZigBee Alliance, 8.

Part II

Building Topologies

5

Non-3GPP Networks

5.1 Introduction

This section is about the practical implementation of non-3GPP networks using readily available tools. Following the theoretical foundations covered in Section 4.2, now the focus is on building and analyzing real-world network stacks. For stack creation and deployment, the capabilities of Netualizer, a powerful protocol stack virtualization framework, are leveraged. Netualizer enables users to design and implement diverse networking scenarios and solutions. In the context of this book, Netualizer serves as the perfect platform to prototype and experiment with non-3GPP network configurations. Complementing Netualizer for traffic analysis is Wireshark, the industry-standard network protocol analyzer. Wireshark can be used to capture and dissect network traffic, providing valuable insights into the inner workings of implemented stacks.

5.1.1 Netualizer

Netualizer, an acronym for Network Virtualizer, serves as a powerful tool for virtualizing and building protocol stacks (L7TR). Its functionality is readily accessible through an installer downloadable from https://www.l7tr.com. This installer deploys two key components. One component is the controller that acts as a centralized entity that manages and configures Netualizer agents. The other component is an agent that is a local software service that resides on a device and executes various protocol stacks as instructed by the controller. Note that Netualizer offers free usage for personal and educational purposes, making it ideal for exploring the concepts presented in this book.

Figure 5.1 illustrates an example topology facilitated by Netualizer. The controller oversees five agents, configuring each with tailored protocol stacks to handle their specific communication needs, including communication

Hands-On Real Time Communications: A Practical Guide to RTC Protocols in Non-3GPP, 3GPP, 4G/5G/6G and IoT Networks, First Edition. Rolando Herrero.

Companion website: www.wiley.com/go/herrero/RTCProtocols

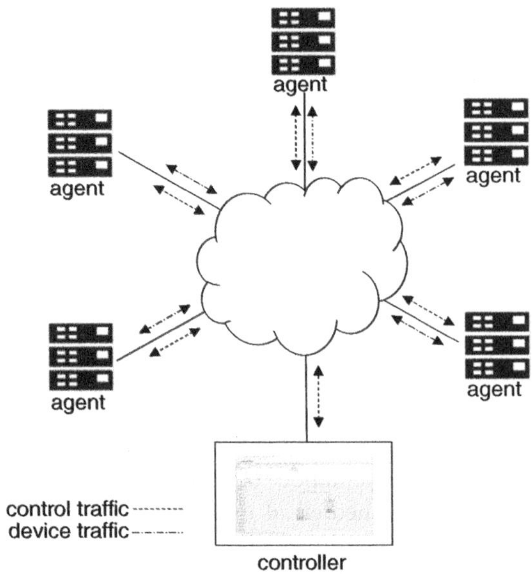

Figure 5.1 Example Netualizer Deployment.

with the controller itself. This communication, including agent provisioning, configuration, and programming, utilizes dedicated management interfaces. Conversely, regular data exchanged between agents and other endpoints typically flows through specialized networking interfaces. This separation minimizes potential interference between control and data traffic.

Once configured with their protocol stacks, the agents establish a bidirectional communication channel with the controller. This involves the controller sending commands and receiving event notifications from the agents. Additionally, the controller can leverage Lua scripts for more complex interactions (Ierusalimschy 2006). These Lua scripts can be directly executed on the controller or pushed down to individual agents for local execution. This choice depends on the desired level of flexibility and control. Lua's suitability for this role comes from its main features: lightweight, high-level, and supporting multiple programming paradigms. Furthermore, its low memory and complexity requirements make it particularly appealing for resource-constrained environments, such as low-end IoT devices. For instance, a standard 8-bit device equipped with Netualizer can readily execute Lua scripts, enabling it to handle diverse IoT scenarios. Note that this script compatibility extends to more powerful 32-bit and 64-bit devices, offering additional versatility. It is worth noting that other scripting languages, like Python, are often too demanding for low-end devices due to their heavier resource requirements.

The Netualizer controller features an *Integrated Development Environment* (IDE) dedicated to building, deploying, and debugging communication and networking projects. Each project serves as a container for interacting with multiple agents. These agents, in turn, host two types of components: configurations and Lua scripts. Configurations define sets of protocol stacks compliant with the layered architecture advocated by the IETF. Each stack includes layers like physical, link, network, transport, and application (Sethi and Sarangi 2017). While the physical layer relies on hardware and modulation mechanisms specific to the agent, all other layers are implemented as software modules. Lua scripts comprise sequences of statements designed to automate tasks and enhance traffic generation and processing through custom logic. This script execution can occur directly on the controller or can be pushed to individual agents for localized control.

As previously mentioned, the basic Netualizer installation deploys a local agent alongside the controller on the same computer. This local agent configuration is perfectly suited for prototyping protocol stacks, enabling the demonstration and understanding of proof-of-concept networking scenarios. Consequently, all network topologies presented in this book utilize a standalone setup with a co-located local agent and controller.

Figure 5.2 shows the initial screen of the Netualizer controller. This IDE presents a comprehensive interface with menus, edit boxes, and various panes and windows designed to show real-time information regarding traffic, packets, commands, and events as they are transmitted, processed, and received. By default, the controller automatically launches a local agent, as previously mentioned. It also establishes a virtual Ethernet network interface, existing solely within the context of the computer running Netualizer. This interface, named NT, possesses IPv4 and IPv6 addresses of *192.168.21.5* and *2001::21:5*, respectively.

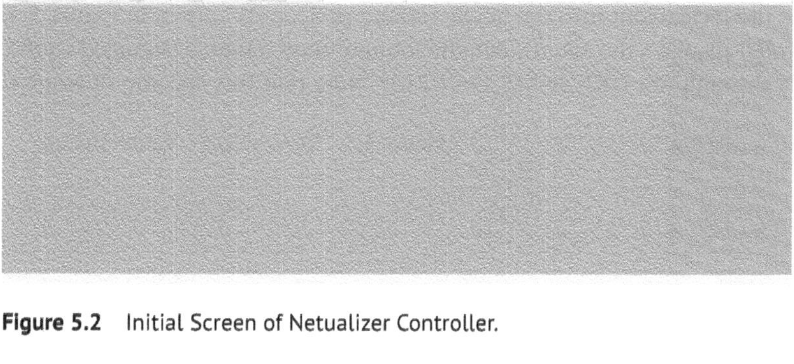

Figure 5.2 Initial Screen of Netualizer Controller.

The combination of the local agent and this interface facilitates the deployment of protocol stacks within the internal networks *192.168.21.0* and *2001::/64*. Also shown in Figure 5.2, the controller's *User Interface* (UI) incorporates a series of menus, including File and Agents. Additionally, it features several docked windows, each serving a distinct purpose. The agent window displays a list of agents currently connected to the controller. The watcher window enables the monitoring of variables and counters. The events window presents events in chronological order as they occur. The commands window displays commands in the sequence of their execution. The results window provides a platform for applications to output pertinent results. The Project Explorer window organizes and shows all configurations and scripts within the project. Finally, the output window exhibits the standard output generated by scripts.

To initiate the creation of a new project within Netualizer, start by navigating to the *File* menu. Then select the *New* option and choose the *Project* option from the available choices. Netualizer will then prompt to provide a name for the project using a dialog window, as illustrated in Figure 5.3. While *newProject* serves as the default name, any other preferred name can be assigned instead. Project files are identified by the *.prj* extension and are stored within individual directories housed within the Netualizer/projects directory. This directory, in turn, resides within the user's directory. In the event that the specified project directory already exists during the creation process, Netualizer will explicitly request permission to overwrite its contents. Project files serve as containers for important information pertaining to the topology being constructed. This includes (1) Details regarding the various agents involved in the topology and (2) File paths indicating the locations of corresponding configuration files (bearing the .cfg extension) and scripts (bearing the .lua extension). As seen in Figure 5.3, the project creation dialog window incorporates a checkbox labeled *Attach Local Agent*. It is imperative to enable this checkbox for projects aligned with the concepts presented in this book. By activating this flag, Netualizer is instructed to create the project and utilize the integrated local agent for the generation of protocol stacks.

Completing the project creation step, Netualizer automatically searches for and integrates the local agent into the newly formed project. As shown in Figure 5.4, the controller displays the agent's default configuration panel, designated *config*, accessible through the URL *172.26.176.1:2105*. Note that this specific IP address

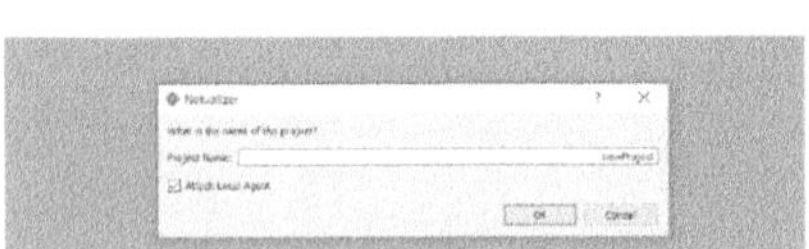

Figure 5.3 New Project Creation Dialog.

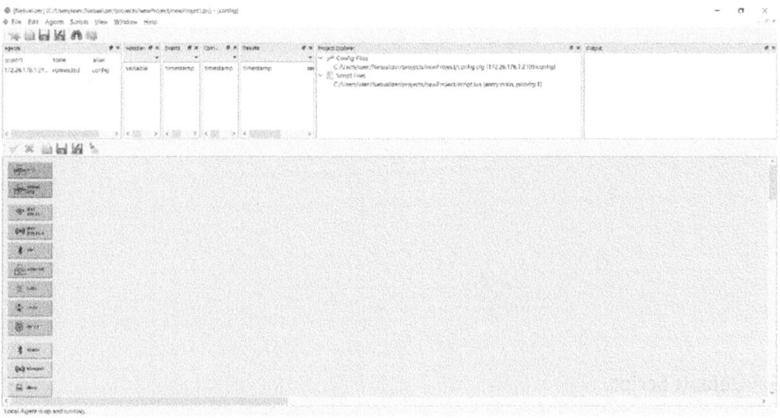

Figure 5.4 Default Agent Configuration.

and port combination (2105 for the management TCP port) solely applies to the local agent configuration and will vary for other agents and deployment scenarios, usually depending on the underlying platform hosting the agent. Netualizer's *Project Explorer* simplifies managing project components. It grants immediate access to both the main configuration file (*config*) and the default script (*script.lua*), while conveniently displaying key parameters alongside. A highlight is the *main* function, automatically called when the script starts. Additionally, script priority is displayed, enabling the execution of multiple scripts concurrently, prioritized based on their assigned settings. This streamlined interface helps effortlessly configure the project's behavior.

In Figure 5.4, the configuration panel shows a comprehensive set of layers supported by the agent. These layers, which can be arranged to construct protocol stacks, are categorized based on their functionality and distinguished by color. Physical layers shown in green facilitate direct interaction with physical communication mediums. Link layers shown in orange manage the exchange of data frames between adjacent network nodes. Network layers shown in gold are responsible for routing and delivering data packets across networks. Although not shown in the Figure, transport layers shown in light blue ensure reliable and ordered transmission of data between applications Application layers shown in beige provide interfaces for user-facing applications and services. Utility layers shown in thistle offer supplementary functions to simplify the creation of specialized protocol stacks or enhance application layer capabilities. Examples include integration with digital sensors, actuators, and cloud services like AWS

Figure 5.5 Default Script.

or Azure Amazon Corporation, Microsoft Corporation. While the configuration panel initially appears empty upon attaching an agent, users can construct protocol stacks by intuitively clicking and dragging desired layers from the extensive list located on the left side of the panel.

Initiating a double-click action on the default script item within the Project Explorer shows the script itself within a dedicated editor window. This window, as illustrated in Figure 5.5, incorporates various buttons designed to facilitate text manipulation, including search and replace functionality. The default script, presented below, serves as the starting point for customization:

```
--Netualizer Script Template

function main()
    clearOutput();
end
```

Upon executing the project, the designated entry point, the *main* function, is triggered. This function then calls upon a function named *clearOutput*, which diligently clears the output window within the Netualizer controller. It is essential to note that the Lua programming language, employed within Netualizer scripts, employs a double dash symbol (–) to initiate line comments. While a comprehensive exploration of Lua programming ventures beyond the scope of this book, a wealth of informative Lua documentation is readily accessible on the Internet, serving as a valuable resource for further learning. For seamless navigation between configuration panels and script editors, simply click the corresponding items within the Project Explorer. To preserve any modifications made to the project, such as the addition, removal, or editing of configurations and scripts, choose the *Save* option located within the *File* menu. Importantly, if a project remains unsaved upon exiting the controller, a prompt will appear to offer a final opportunity for saving the work.

5.1.2 Wireshark

Wireshark is a widely used protocol analyzer and network traffic monitoring tool that supports a broad range of standardized IoT and non-IoT networking technologies. The installation package, available for download at https://www.wireshark.org, installs the necessary drivers and executables to enable packet capture functionality. Wireshark is open-source software and can be used free of charge (Wireshark).

Figure 5.6 displays the main window of Wireshark. Within this window, under the *Capture* heading, a list of available network interfaces is presented. These interfaces can be utilized for network traffic monitoring. To initiate traffic capture on a specific interface, double-click on its name. For instance, double-clicking on the *NT* interface will start traffic capture on that interface. To capture traffic on multiple interfaces simultaneously, hold down the *Shift* key while selecting the desired interfaces. Alternatively, access the *Capture* menu and select the *Options* option to choose interfaces for capture.

Figure 5.7 illustrates that initiating network traffic capture on the *NT* interface can also be accomplished by selecting the *Start* option within the *Capture* menu. To halt the ongoing capture process, choose the *Stop* option, also located within the *Capture* menu.

Wireshark offers a feature called protocol filtering, which allows users to focus on specific protocols that are pertinent to their analysis. This feature functions by applying a Boolean expression to each captured packet. If the expression evaluates to *false* for a particular packet, that packet is excluded from the displayed results. Figure 5.8 shows the Wireshark filter edit box, where users can input the

Figure 5.6 Wireshark Network Interface Selection.

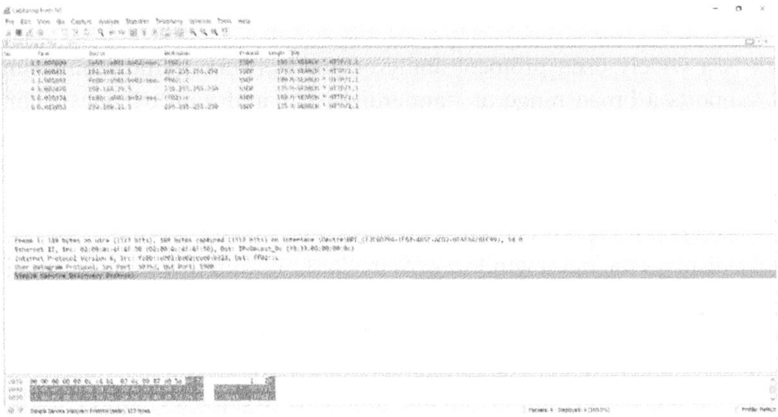

Figure 5.7 Traffic Capture on the NT Interface.

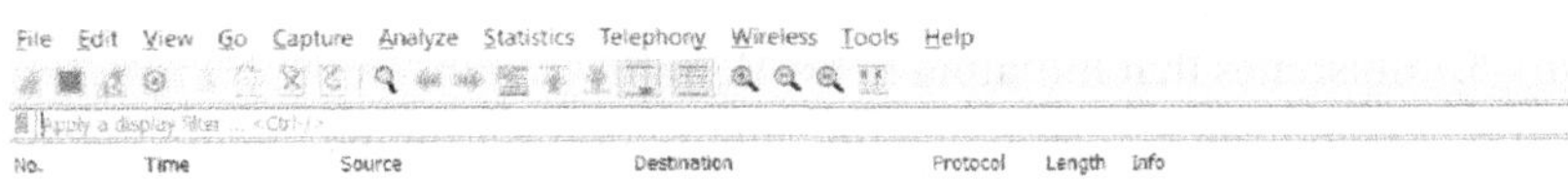

Figure 5.8 Wireshark Filter Option.

desired filter expression. This box is situated above the packet list pane, as shown in Figure 5.7.

5.2 Lower Layers

This section focuses on techniques for simulating the behavior of physical, link, and network layers within non-3GPP network protocols.

5.2.1 Physical and Link Layers

The objective of this section is to create two IP stacks using Netualizer and then conduct a ping test between them. The ping process, as outlined in Section 2.2.2.1, involves the exchange of two specific ICMP messages: (1) echo requests and (2) echo replies (RFC 1981b). One endpoint initiates the process by sending an echo request message to the other endpoint. The recipient then responds with an echo reply message, which serves to estimate the RTT and, consequently, help evaluate the overall network reliability. The Netualizer configuration is designed to accommodate both IPv4 and IPv6 network stacks operating on a single local agent.

Begin by creating a new project. Follow the instructions outlined in Section 5.1.1. Choose a name that accurately reflects the project's purpose, such as *network-Layer*. Upon successful project creation and local agent attachment, proceed to the configuration panel window. Click the green *phy* button situated on the left side to initiate the creation of a physical layer. Netualizer will then display a comprehensive list of all network interfaces currently available at the agent. This includes not only conventional physical layers like Ethernet and WiFi, but also any additional hardware-based radios that may be connected to the agent. For this specific project, select the *NT* interface that holds the IPv4 address *192.168.21.5*, as shown in Figure 5.9.

Position the newly created physical layer within the configuration panel. With the physical layer established, the next step involves the addition of a link layer. Since the chosen *NT* interface corresponds to an Ethernet physical layer, an Ethernet link layer is also required. To initiate its creation, click the orange *ethernet* button and designate it with the name *ether*, as demonstrated in Figure 5.10 (IEEE 2018). It is essential to note that most layers, excluding physical layers, require a unique name for reference purposes when accessed from Lua scripts. While *ether* is a recommended name for this particular layer, there is the flexibility to choose any other suitable name.

Proceed to position the Ethernet link layer directly above the previously created physical layer. This action constructs a fundamental 2-layer stack. It is very important to note that layers form a cohesive connection when placed in a valid hierarchical order. For instance, an Ethernet physical layer and an Ethernet

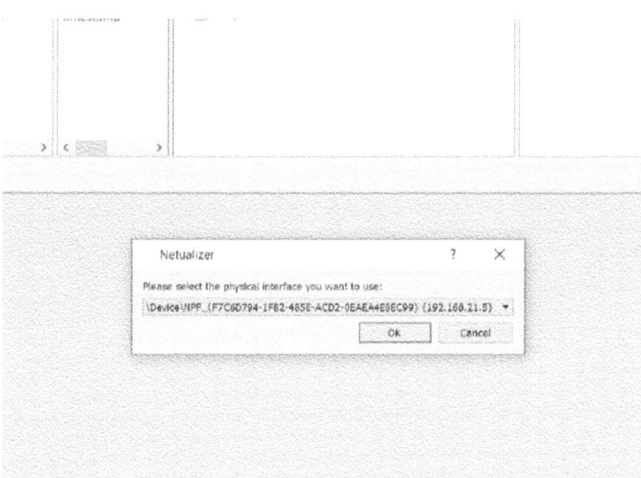

Figure 5.9 Selecting the Physical Layer.

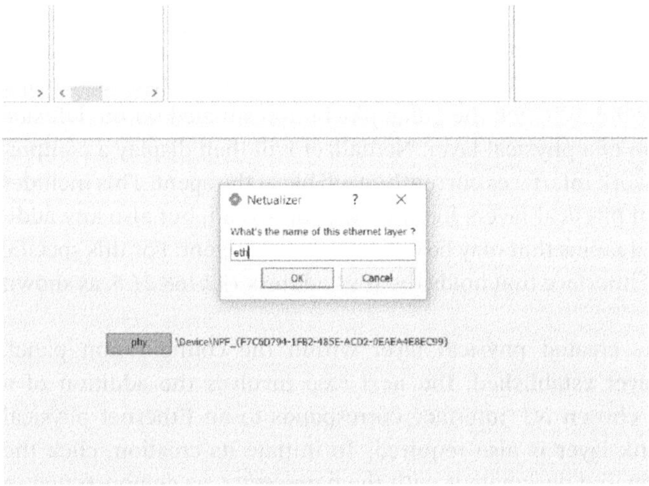

Figure 5.10 Creating the Ethernet Layer.

link layer would automatically connect, whereas an IEEE 802.15.4 physical layer would not be compatible with an Ethernet link layer. Within IETF protocol stacks, the only supported network layer is IP. To incorporate an IP layer, click the gold *IP* button and assign it the name *ip*, as demonstrated in Figure 5.11. While there is flexibility to choose any other name, using a relevant name like *ip* is generally recommended for clarity and simplicity (RFC 1981a).

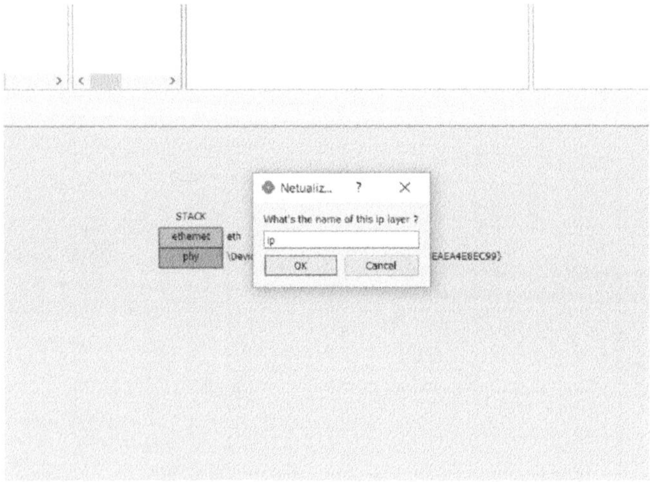

Figure 5.11 Creating the IPv4 Layer.

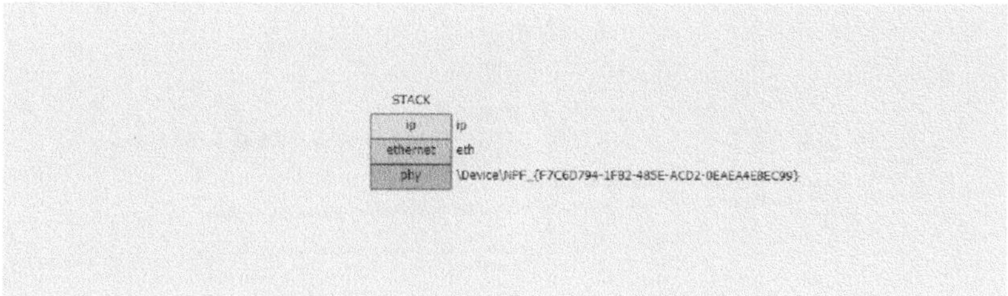

Figure 5.12 The 3-Layer IPv4 Stack.

The arrangement of these three layers–the physical, link, and network layers–establishes the straightforward 3-layer stack shown in Figure 5.12. It is important to note that this configuration possesses the full capability necessary to support ICMP echo requests and replies. While ICMP packets are encapsulated within the payload of IP datagrams, their generation and processing are specifically handled by the IP layer.

To establish a second 3-layer stack, there are two primary approaches. One approach consists of creating a new stack from the beginning. An alternative is to duplicate an existing stack through the copy and paste functions. Since a stack has already been constructed, proceed with the second option. To initiate the duplication process, first select the existing stack. Right-click on it to reveal a menu of available actions. Choose the *copy* option, as illustrated in Figure 5.13. This action will place a copy of the stack within the clipboard. Note that it is possible to generate an unlimited number of stack copies by repeatedly pasting from the clipboard

To position the copied stack, right-click within a desired location near the original stack and select the *paste* option from the displayed menu. As it can be observed in Figure 5.14, the two 3-layer stacks will now be shown in close proximity within the configuration panel. However, it is essential to note that the spatial arrangement of stacks holds no functional significance and merely serves the purpose of organization during the configuration process. To ensure the uniqueness of layer

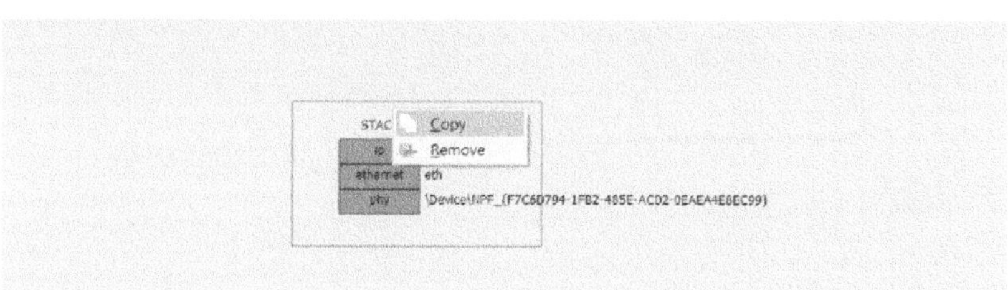

Figure 5.13 Stack Copy.

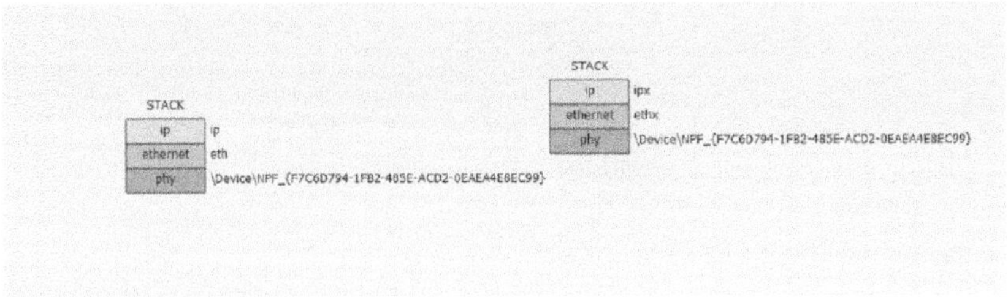

Figure 5.14 Stack Pasting.

names, which is a mandatory requirement, pasted stacks undergo an automatic renaming process. The names of their layers will differ from those of the original stack. Specifically, an *x* is appended to the end of each layer name to prevent any potential conflicts with the existing layer names.

While stacks are initially assigned the default name *STACK*, there is flexibility to modify these labels to better reflect their intended purpose or characteristics. To initiate a name change, double-click on the existing stack label. Upon double-clicking, Netualizer's configuration panel will prompt to input the desired new name for the stack. As demonstrated in Figure 5.15, proceed to change the default name of the original stack from *STACK* to *FIRST*. It is important to remember that, as with layers and projects, there is the freedom to choose any name that identifies the stack, as long as it maintains uniqueness within the configuration. Apply the same renaming process to the copied stack. Double-click on its stack label and replace the default name *STACK* with *SECOND*. It is important to note that, unlike layer names, stack names do not play a role within Lua scripts. Consequently, there is no strict requirement to maintain the uniqueness of stack names for functional purposes. However, using distinct names for each stack aids in clarity and organization within the configuration panel.

To ensure seamless communication between the *FIRST* and *SECOND* stacks, it is essential to assign distinct MAC addresses to their respective Ethernet layers. If both stacks shared an identical MAC address, it would create conflicts and

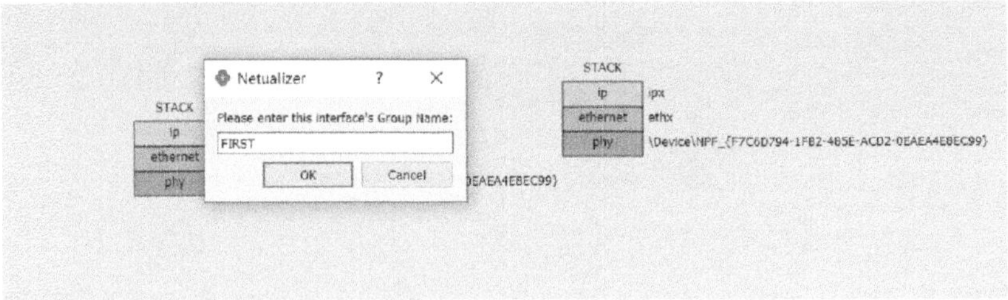

Figure 5.15 Stack Label Change.

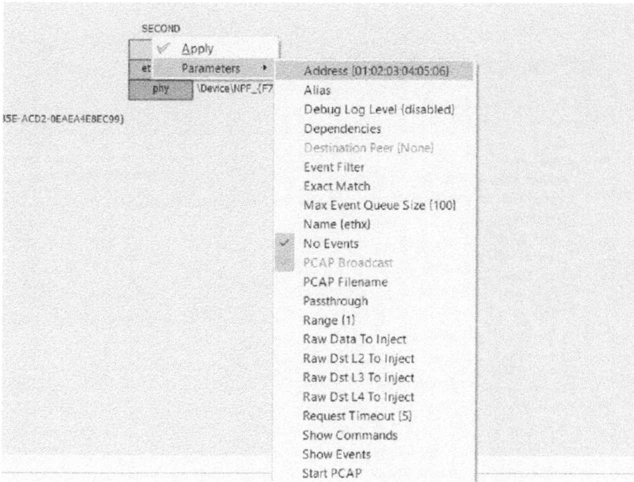

Figure 5.16 MAC Address Change on the *SECOND* Stack.

prevent successful communication (IEEE 2018). As illustrated in Figure 5.16, proceed to modify the default MAC address of the *SECOND* stack's Ethernet layer. Change it from *01:02:03:04:05:06* to a different value, such as *01:02:03:04:05:07*. The only stipulation is that the new address must follow two criteria. The address must be 48 bits long and its format must follow the standard Ethernet MAC address specifications. Note that Netualizer automatically assigns a default MAC address of *01:02:03:04:05:06* to the first Ethernet link layer that is created within a project. Any subsequent Ethernet link layers will be assigned incrementally increasing MAC addresses. For example, the second Ethernet link layer is *01:02:03:04:05:07*, the third Ethernet link layer address is *01:02:03:04:05:08* and so on. While these default addresses can be manually adjusted to any other valid MAC address, it is important to remember that copying and pasting a link layer preserves the original layer's MAC address. Therefore, it is important to manually modify the MAC address of the copied layer to ensure uniqueness and prevent communication conflicts.

5.2.2 Network Layers

While the IP layers in both stacks are initially configured as IPv4 (RFC 1981a), also configure them for IPv6 (Deering and Hinden 2017; Graziani 2012) due to its importance in many RTC environments. However, let us first focus on IPv4 configuration by setting IPv4 addresses and examining other parameters. Similar to the link layer, the first time an IP layer is created on top of an *NT* network interface physical layer, it is assigned the address *192.168.21.10*. Subsequent network

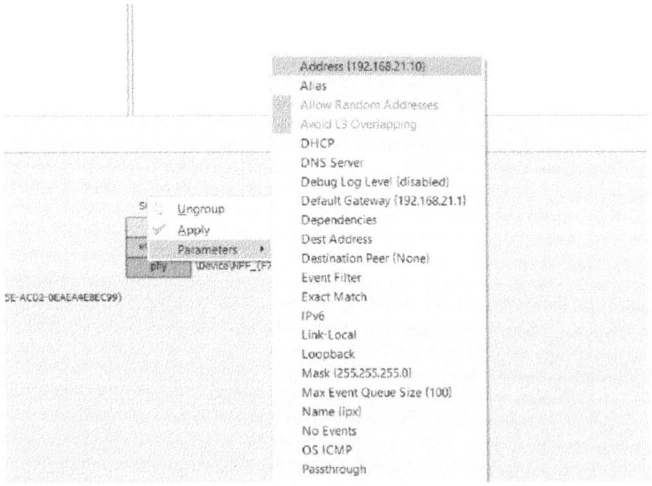

Figure 5.17 Parameters of the IP Layer.

layers receive increasing IPv4 addresses. For instance, a second IP layer gets *192.168.21.11*, a third gets *192.168.21.12*, and so on. It is very important to note that when copying stacks, as done in this chapter, the pasted network IP layer retains the address of the copied layer. To prevent address collisions, the address of the pasted layer must be manually changed.

Figure 5.17 illustrates all the protocol options available when right-clicking on the IP layer *ipx* within the *SECOND* stack. The layer displays several critical parameters. The IPv4 address is set *192.168.21.10* matching the IP layer of the *FIRST* stack. The *DHCP* Flag is not set indicating a static IP address. The DNS server address is blank and does not specify DNS server configured. The *Default Gateway* is *192.168.21.1* serving as the default route for network traffic. The *Destination Address* is left blank to open it for specific destination settings. The *IPv6* flag is not set indicating IPv4 operation. The Network *Mask* is set to *255.255.255.0* specifying a network prefix of *24* under IPv4.

Addresses assigned to IP layers on top of physical layers associated with *NT* interfaces are considered static, hence the unset DHCP flag. The IP mask and addresses create the *192.168.21.0* subnet, corresponding to a private Internet network as described in Section 2.2.2.1. To test connectivity between stacks, change the IPv4 address of the SECOND stack from *192.168.21.10* to a different address within the subnet, such as *192.168.21.11*. Before sending any network traffic between these two stacks, the configuration in the configuration panel must be applied to the agent. To do this, click the *Run Suite* option within the *Scripts* menu. This action also executes the default script on the controller, which clears the output screen and terminates.

Figure 5.18 Suite Execution.

Once the configuration is deployed to the agent, the two stacks transition into an active state. This means the configuration panel window indicates this change by graying out the layer buttons and stacks, as shown in Figure 5.18. Users are prevented from modifying or moving any components of the active stacks within the configuration panel.

To initiate network traffic testing, begin traffic capture by referring to Section 5.1.2 for instructions on how to start capturing traffic on the *NT* interface using Wireshark. Initiate pinging by right-clicking on the IP network layer named *ip* within the *FIRST* stack and selecting the *Ping* option. Then enter 192.168.21.11 as the destination IP address, targeting the *SECOND* stack. This process is illustrated in Figure 5.19. Upon sending an ICMP request, the controller displays a progress dialog measuring the time it takes for the corresponding ICMP reply to arrive. This real-time value reflects the instantaneous RTT between the first and second stacks. It is important to note that the standard ICMP request interval of 4 seconds is configurable and can be adjusted based on the specific characteristics of the underlying physical layer.

Figure 5.20 shows a sequence of ICMP requests and replies as captured on Wireshark. The first pair of ICMP requests/responses are carried in frame numbers 0 and 1 while the second pair are carried in frame numbers 3 and 4. Note that Wireshark also indicates whether there is a matching response or request for respective ICMP request or responses. Wireshark also shows source

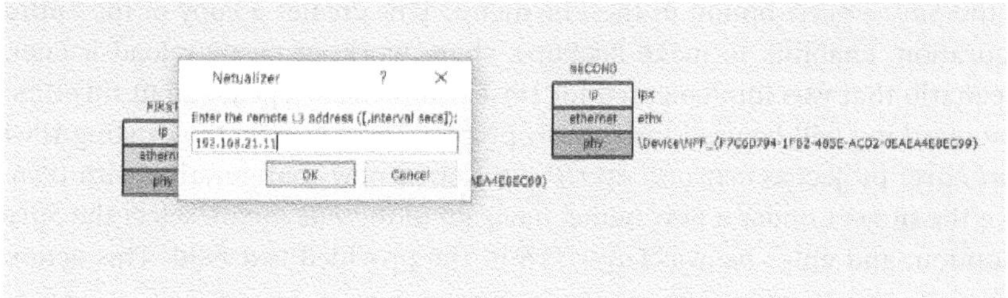

Figure 5.19 Configuring the Destination Address for Pinging.

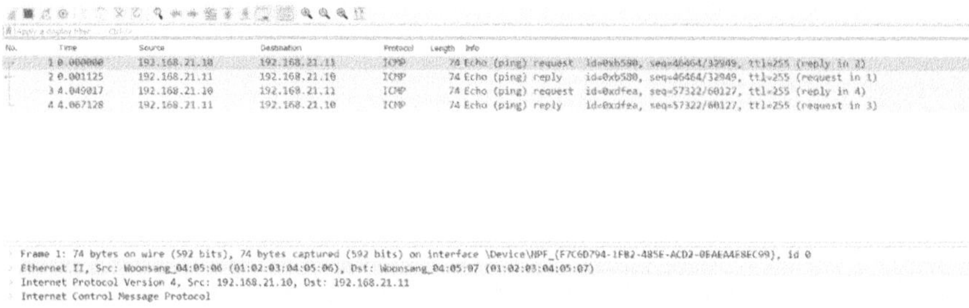

Figure 5.20 ICMP Echo Packets.

and destination IP addresses, ICMP packet identifiers and sequence numbers. Figure 5.20 presents a sequence of captured ICMP requests and replies as observed in Wireshark.

Example 5.1 Consider Figure 5.20, what is the RTT?

Solution:

The figure shows the exchange of two echo requests/responses. The first RTT is 0.112 ms while the second RTT is 1.8811 ms. The average RTT is $0.9965 \approx 1$ ms.

A few key observations are that the first ICMP request and its corresponding reply are captured in frames 0 and 1. Similarly, the second ICMP request and reply pair appear in frames 3 and 4. Note the request-response matching where Wireshark indicates whether each request has a matching response and vice versa. When looking at the packet details, Wireshark displays the source and destination IP addresses, ICMP packet identifiers, and sequence numbers for each captured frame. Upon completion of the configuration on the agent, immediate deactivation is very important. Clicking the red cross symbol (Figure 5.18) instantly triggers deactivation of all layers and stacks within the configuration panel, ensuring efficient resource utilization and preventing unnecessary processing burden.

To preserve the work and enable future access, remember to save the project. Click the *Save Project* option in the *File* menu. This creates a copy of the entire configuration, enabling to make backups, share work, or easily reload it later. The scenario that was implemented for IPv4 will be now repeated, but this time IPv6 networking will be used instead. To prepare for this, save the existing IPv4 *networkLayer* project as *networkLayerIPv6* to start a new configuration with IPv6. To save the project under a new name, navigate to the *File* menu, select the *Save As…* option, and enter *networkLayerIPv6* in the provided text field. This action will initiate a prompt within the configuration panel, inquiring *What name do you want to save this project as?.*

When looking at IPv6 addressing, the main difference compared to IPv4 is the type of address that is used. To set up static addresses on the two main stacks (*FIRST* and *SECOND*), follow the following steps. Find the layer named *ip* and right-click on it. In the options, tick the *IPv6* checkbox to tell Netualizer that this is the address type. Look at Figure 5.21 and match the address there: set the *FIRST* stack to *2001::21:10*. Ignore all other settings, just leave them as they were for now. Turn on IPv6 for the whole *SECOND* stack. Finally, give the *SECOND* stack its own address: *2001::21:11*. That's it! Now two stacks are up and running with different IPv6 addresses. Remember that these addresses can be customized to fit specific needs.

With the essential configuration steps for IPv6 support completed, perform the following actions. Save the configuration to ensure that the changes are preserved. Initiate the suite's execution as described earlier in this section. Reactivate traffic capture within Wireshark by navigating to the *Capture* menu and selecting the *Restart* option.

Let us see if those IPv6 addresses are working. From the *FIRST* stack's *ip* layer, choose the *Ping* command and enter the IPv6 address for the *SECOND* stack (*2001::21:11*). Just like with IPv4, the handy Netualizer tool will show the RTT for each ping based on the echo requests and responses sent using ICMPv6 (as defined in RFC 4443). If everything is set up correctly, a value like the

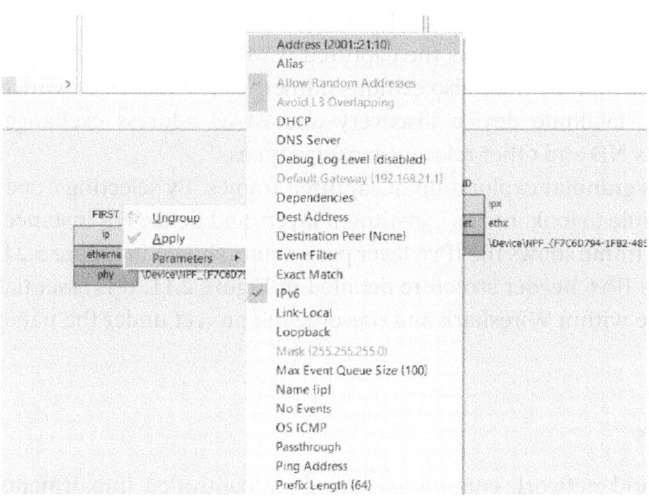

Figure 5.21 IPv6 Address Configuration.

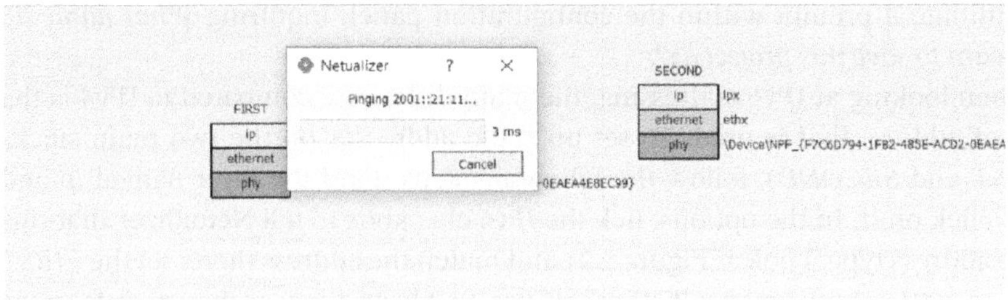

Figure 5.22 SECOND Stack Ping.

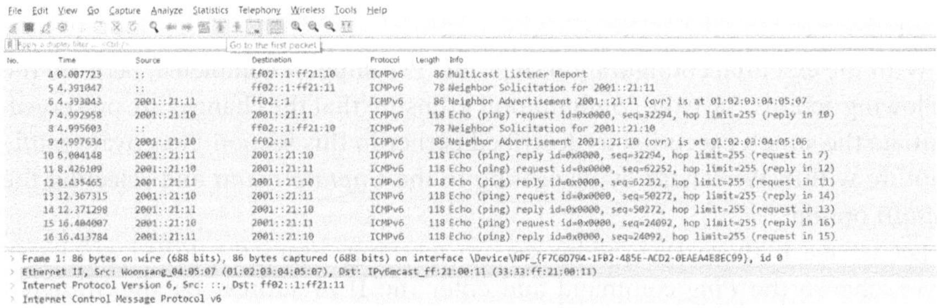

Figure 5.23 ICMPv6 Echo Packets.

3-millisecond RTT shown in Figure 5.22 should be seen. This confirms that data can successfully travel between the two stacks using IPv6.

Using Wireshark, Figure 5.23 shows the captured ICMPv6 packets exchanged between devices. Note that the trace also exhibits numerous ND messages, which leverage ICMPv6 to facilitate device discovery and MAC address exchange. Section 2.2.2.2 details ND and other relevant message types.

Wireshark enables granular exploration of captured frames. By selecting a specific frame, it is possible to look into its constituent layers and fields. For instance, clicking on the third frame shows the IPv6 layer parameters shown in Figure 5.24, which align with the IPv6 header structure detailed in Figure 2.11. It is essential to stop traffic capture within Wireshark and preserve the project under the name *networkLayerIPv6*.

5.2.3 Impairments

To replicate real-world network conditions, introduce controlled impairments like packet loss and latency. Leverage the IPv4 project from Section 5.2.2 as a foundation. Begin by reloading the *networkLayer.prj* file, accessible via the

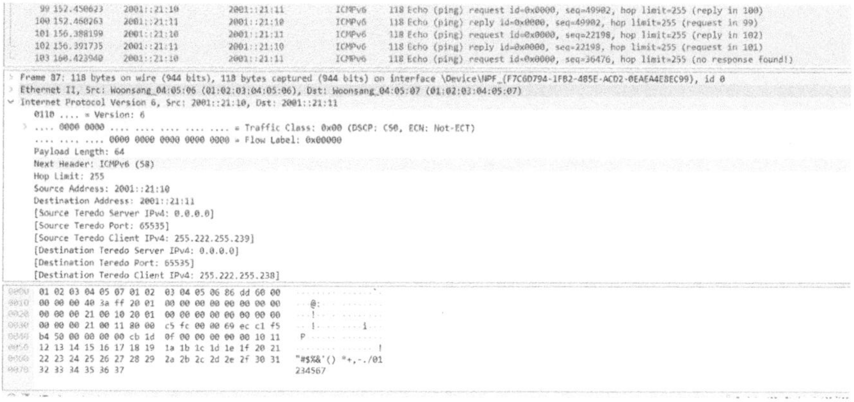

Figure 5.24 Third Frame Details.

Recent Projects option within the *File* menu. To preserve the integrity of the original project, save the loaded configuration and scripts as a new project titled *networkLayerImpair*. The objective is to augment the existing IPv4 network layer project with controllable network impairments, such as latency and loss, to experimentally observe their impact on ICMP packet exchange and RTT estimations.

Netualizer offers a unique utility layer called *impair*, specifically designed to introduce controlled packet loss and latency in both incoming and outgoing traffic. To utilize this feature, insert an *impair* layer between the IP network layer (*ip*) and the Ethernet layer (*ether*). It is important to note that impairment layers can be flexibly positioned within a stack to selectively affect packets traversing it in either direction. Begin the process by double-clicking the *ip* layer to detach it.

To proceed, generate an impairment layer by clicking the *impair* button situated on the configuration panel's left side. Assign it the name *impair*, as demonstrated in Figure 5.25. Position the newly created *impair* layer directly above the *ether* layer. Finally, reposition the detached *ip* layer back to its original position atop the stack, ensuring it rests above the *impair* layer within the *FIRST* stack.

Figure 5.26 shows both stacks, incorporating arrows that clearly indicate the traffic's direction relative to the *FIRST* stack. The transmitting interface (*TX*) of the *impair* layer is responsible for generating packets that follow the outgoing flow. Conversely, the receiving interface (*RX*) of the *impair* layer is tasked with receiving packets that constitute the incoming flow.

ICMP echo request packets, originating at the *ip* layer within the *FIRST* stack, undergo encapsulation. This process involves passing through the transmitting interface (*TX*) of the *impair* layer, followed by the *ether* and *phy* layers. Upon

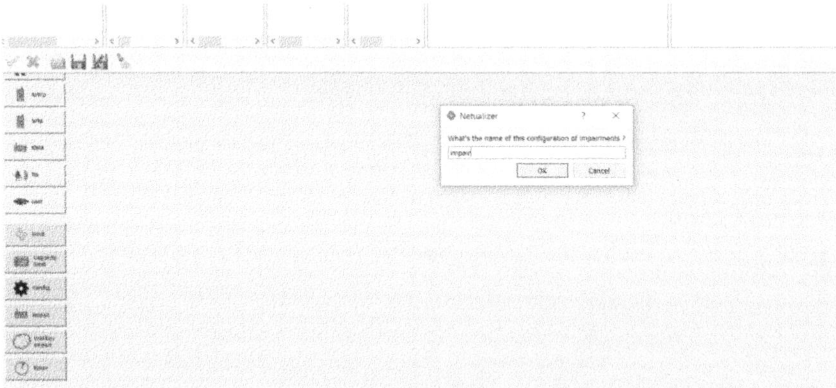

Figure 5.25 Impairment Layer Creation.

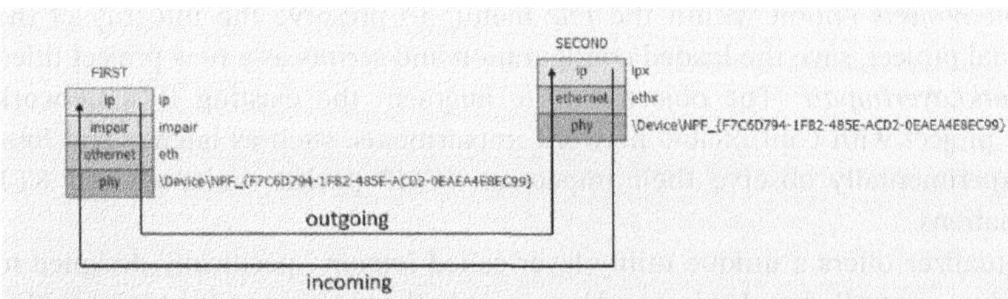

Figure 5.26 Traffic Direction on Stacks.

reaching the *phy* layer of the *SECOND* stack, these frames are sequentially decapsulated by the *etherx* and *ipx* layers. The IP network layer (*ipx*) processes the ICMP request and generates an ICMP echo response, which is then encapsulated by the *etherx* and *phy* layers. Once the frame arrives back at the *FIRST* stack, it undergoes decapsulation by the *ether* layer and passes through the receiving interface (*RX*) of the *impair* layer. The resulting packet is subsequently forwarded to the *ip* network layer for processing. Initiate traffic capture on Wireshark, utilizing the *NT* interface. Concurrently, launch the suite to facilitate configuration transfer to the agent. Right-click on the impair layer, followed by selecting *Configure Parameters*. Within the *Configure Parameters* option, select the *TX Delay* parameter. As demonstrated in Figure 5.27, enter the value *1000* to designate a 1000-millisecond delay for packets in transit between the IP and ether layers.

Figure 5.28 shows the outcome of pinging the 192.168.21.10 address of the *FIRST* stack from the *ipx* network layer of the *SECOND* stack. As anticipated, the measured RTT is now approximately 1 second (or 1025 milliseconds to be precise).

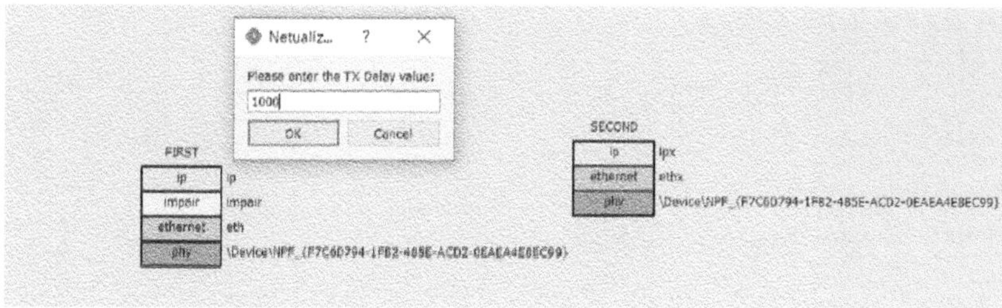

Figure 5.27 Configuration Transmission Latency.

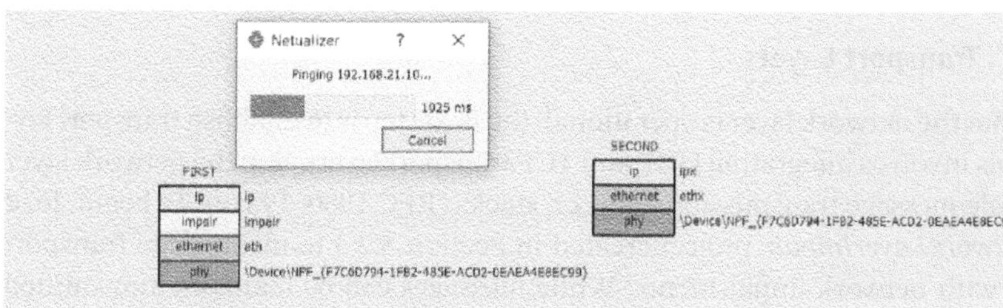

Figure 5.28 1-second Latency SECOND Stack Ping.

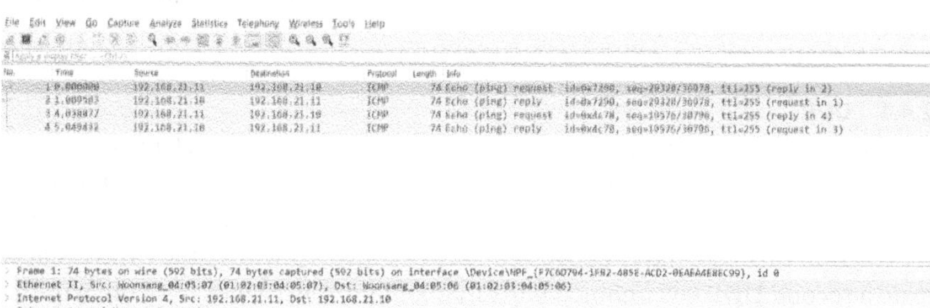

Figure 5.29 1-second Latency ICMP Echo Packets.

Figure 5.29 displays the captured Wireshark network trace containing four ICMP packets. Frame number 1, captured at time 0.000000, corresponds to the ICMP echo request with ID 0x7290. Its corresponding response is captured in frame number 2 at time 1.009583. The second ICMP echo request, with ID 0x4c78 and transmitted 4 seconds after the first, is captured in frame number 3 at time 4.038877. Its response is captured in frame number 4 at time 5.049432. For both transactions, with IDs 0x7290 and 0x4c78, the calculated RTT values are 1.009583 and 1.010555 seconds, respectively. Terminate the traffic capture process within Wireshark and proceed to save the project.

Figure 5.30 UDP Layer Selection.

5.2.4 Transport Layers

Now that the network layer is operational, the next step is to establish transport layers. This involves integrating UDP and TCP transport layers atop the network layer to enable message transmission between stacks (RFC 1980; 1981c). To begin, load the *networkLayerImpair* project created in Section 5.2.3 to incorporate transport layers with network impairments. While messages can be manually transmitted using the agent configuration panel's user interface in Netualizer, automation is possible through Lua scripting. This allows for the creation of scripts that can send a high volume of messages, facilitating the verification of loss and latency effects.

5.2.4.1 UDP

Open the *networkLayerImpair* project and save it as *transportLayerImpair* to reflect the transport layer modifications. Refer to Figure 5.30. Click the UDP button on the left side of the configuration panel and create a new layer named *udp*. Position this UDP layer above the network layer *ip* within the *FIRST* stack. Create a second UDP layer named *udpx* to maintain consistency with the naming convention of the *SECOND* stack (which was originally generated by copying and pasting the *FIRST* stack). By default, the Netualizer controller automatically assigns port numbers 4000 and 4001 to the transport layers *udp* and *udpx*, respectively.

It is important to note that any additional UDP transport layer created will be configured with sequentially increasing port numbers. Since UDP is inherently connectionless and provides a minimal layer on top of IP, the *udp* layer must be configured to direct outgoing segments to port 4001 of the *udpx* layer. To achieve this, right-click the *udp* layer, select the *Dest Port* parameter, and set it to 4001.

To ensure datagrams are forwarded to the *SECOND* stack, configure the destination IP address of the *ip* layer. Right-click on the *ip* layer. Select the *Dest*

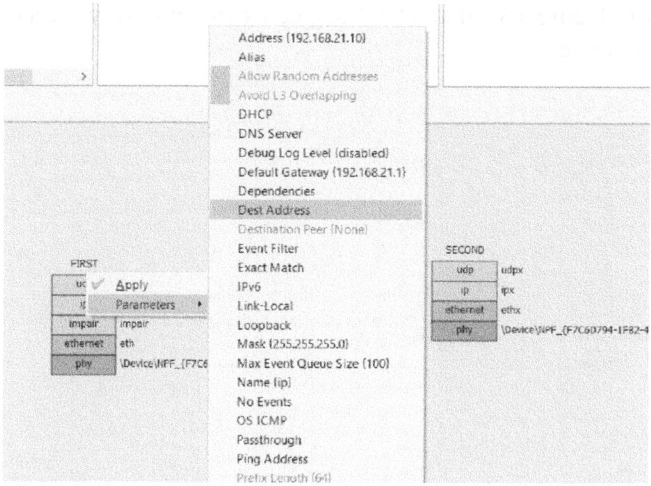

Figure 5.31 IP Destination Address Configuration.

Address parameter, as shown in Figure 5.31. Set the destination IP address to *192.168.21.11*.

To inject a UDP segment and capture traffic, launch the *networkLayerImpair* project suite within Netualizer. Utilize Wireshark to initiate traffic capture on the *NT* interface. Right-click on the *udp* layer and choose the *Inject Raw Data* option, as shown in Figure 5.32. When prompted for the raw data to be transmitted, enter *Hello World* (or another relevant string). Netualizer's user interface will request Layer 3 and Layer 4 addresses, offering the option to override the default IPv4 address (192.168.21.11) and UDP port (4001) previously configured. These prompts will appear as two distinct questions: *What's the destination L3 address?* and *What's the destination L4 port?*. Maintain the default values by clicking OK in response to both questions.

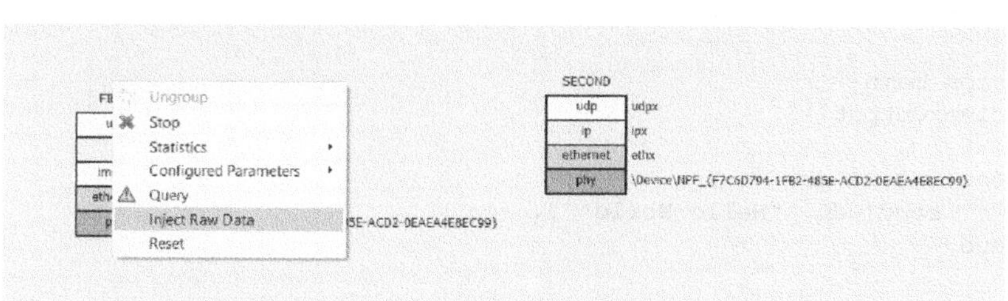

Figure 5.32 Raw Data Injection.

Example 5.2 Refer to Figure 5.33, if 100 UDP segments are sent every second, what is the transmission rate?

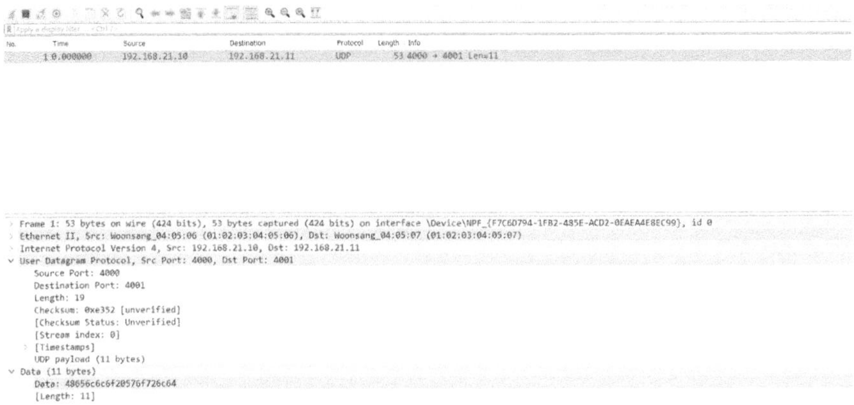

Figure 5.33 Captured UDP Packet.

Solution:
The frame size is 53 bytes or $53 \times 8 = 424$ bits per segment. The transmission rate is therefore $R = 424 \times 100 = 42.4$ Kbps.

Figure 5.33 displays the captured UDP segment in Wireshark. To isolate UDP traffic, use the string *udp* in the display filter. It shows an UDP segment transmitted from IPv4 address *192.168.21.10* on port 4000 to IPv4 address *192.168.21.11* on port 4001. Note that the UDP header structure aligns with the description in Figure 2.12, detailing source and destination port numbers, segment length, and checksum. Wireshark also presents the 11-byte *Hello World* segment payload (message).

To automate the transmission of UDP messages, double-click on the *script. lua* entry within the Project Explorer. Then, insert the following script into the editor:

```
--Send ten UDP messages

function main()
    clearOutput();

    for i=1,10 do
        send(udp, "Hello World" .. i);
    end
end
```

The script complies with standard Lua syntax and features a loop that transmits ten distinct messages, *Hello World 1* through *Hello World 10,* down the *udp* layer.

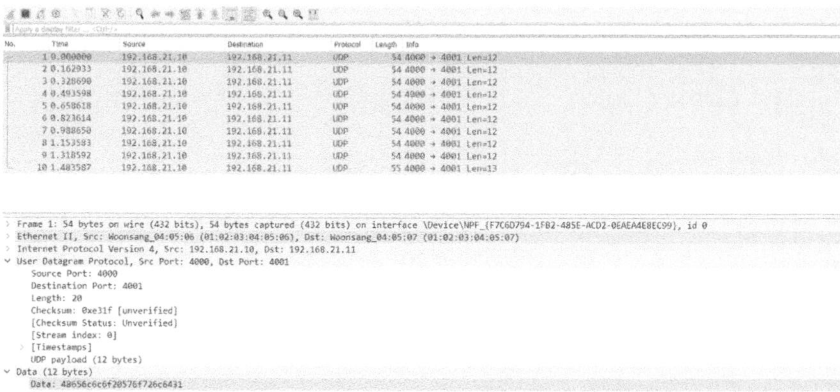

Figure 5.34 UDP Packets.

It executes the *main* function, which in turn calls the send function within a loop that iterates ten times. The *send* procedure accepts two parameters: (1) the name of the UDP layer responsible for generating the message and (2) the message itself, represented as a string. It is very important to note that the messages are directed to the destination port 4001 and IPv4 address *192.168.21.11*, as previously configured in the *udp* and *ip* layers, respectively.

To proceed, save the script, resume Wireshark traffic capture, and execute the suite. The trace captured after the transmission of all ten packets from the *FIRST* to the *SECOND* stack is shown in Figure 5.34. While the configuration is operating, adjust the transmission packet loss within the *impair* layer to 50%. To accomplish this, locate the *TX Loss* option within the configured parameters of the impair layer and enter the value 50. Subsequently, verify that Wireshark continues to capture traffic and initiate a rerun of the suite. This action will prompt the *FIRST* stack to transmit ten datagrams. It is important to note that due to the agent configuration already being deployed and active, a reload will not occur.

Figure 5.35 presents the traffic captured by Wireshark. In alignment with expectations, only five messages were successfully captured, as the remaining five were discarded by the network through the *impair* layer. This observed datagram loss ratio directly corresponds to the 50% packet loss probability that was configured within the *impair* layer.

5.2.5 TCP

Compared to UDP, TCP has significantly more intricate functionality, including packet retransmission and congestion control mechanisms. Now, use the *transportLayerImpair* UDP project from Section 5.2.4.1 as the foundation for

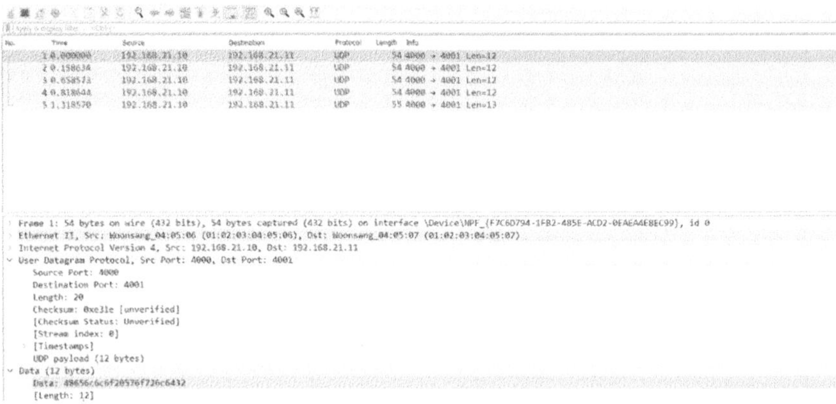

Figure 5.35 UDP Packets with 50% Loss.

building a new TCP-based project, here named *transportLayerImpair*. Duplicate the *transportLayerImpair* project and rename it to *transportLayerImpair* (or another appropriate name). Detach both UDP layers (*udp* and *udpx*). Replace the detached UDP layers with two new TCP layers. To add a TCP layer, click the light-blue TCP button positioned above the UDP button (see Figure 5.30). Name the newly created TCP layers *tcp* and *tcpx*, as shown in Figure 5.36. Note that the Netualizer controller automatically assigns port numbers 4000 and 4001 to the *tcp* and *tcpx* layers, respectively. These ports follow the same sequential assignment convention as UDP ports, with increasing values for subsequent layers.

To ensure the correct transmission of segments from the *FIRST* to the *SECOND* stack, which is essential for TCP's connection-oriented nature, the destination port of the *tcp* layer must be explicitly configured. Right-click the *tcp* layer. From the context menu, select the *Dest Port* option. Set the destination port to 4001. This configuration ensures that segments originating from the *tcp* layer are directed specifically to the *tcpx* layer on the *SECOND* stack, aligned with TCP's requirement for established connections between communicating entities.

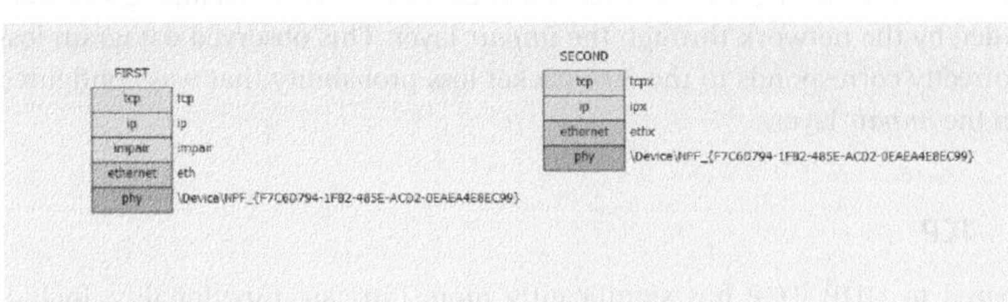

Figure 5.36 TCP Stacks.

Initiate the suite in the standard way and start traffic capture on the *NT* interface within Wireshark. Disregard any script execution failures prompted by the absence of the *udp* layer, as the focus is exclusively on deploying and operating the configuration. Right-click the *tcp* layer and select the *Inject Raw Data* option. Follow the instructions outlined in Section 5.2.4.1 to propel TCP messages down the stack. When prompted for the message to send, enter *Hello World* (or another pertinent message). Click OK twice to bypass the queries regarding the destination L3 address and destination L4 port. The default address and port associated with the *ip* and *tcp* layers will be employed automatically.

The following is a breakdown of the captured TCP traffic shown in Figure 5.37, with a focus on clarity and technical accuracy. Frame Numbers 1 and 2 facilitate the mutual discovery of MAC and IPv4 address mappings through the exchange of ARP packets. This process is essential for establishing communication between devices on an Ethernet network. Frame 3 holds the initial TCP SYN segment, which initiates the connection setup process. Frame 4 contains the corresponding TCP SYN ACK response, signaling acceptance of the connection request. Both frames exhibit sequence numbers of 0, as expected in the initial connection phase. This two-packet exchange typically consumes approximately one RTT. While primarily a SYN ACK response, frame 4 also acknowledges sequence number 1, following the flow outlined in Figure 2.14. Frame Number 5 encapsulates the actual 11-byte *Hello World* payload, as indicated by the *Len = 11* field. It carries sequence number 1 and acknowledges sequence number 1, maintaining synchronization within the TCP communication. Frame Number 6 transmits an empty TCP ACK segment, solely acknowledging sequence number 12, which is associated with the successful delivery of the *Hello World* message.

Figure 5.37 presents a detailed view of the TCP segment header within frame number 5, conforming to the structure described in Figure 2.13. This header

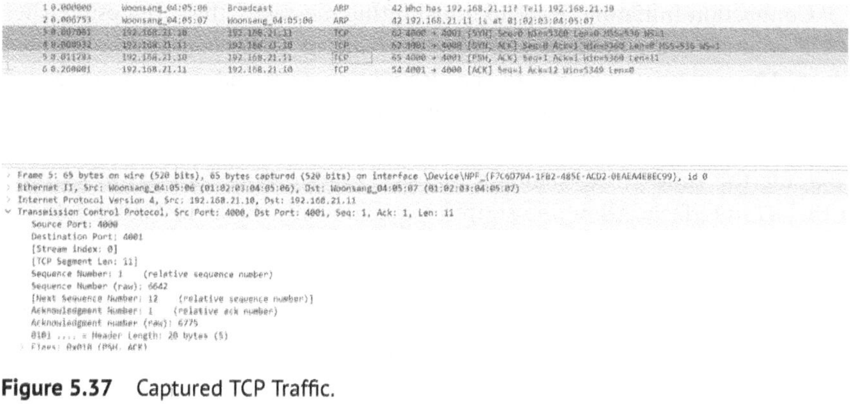

Figure 5.37 Captured TCP Traffic.

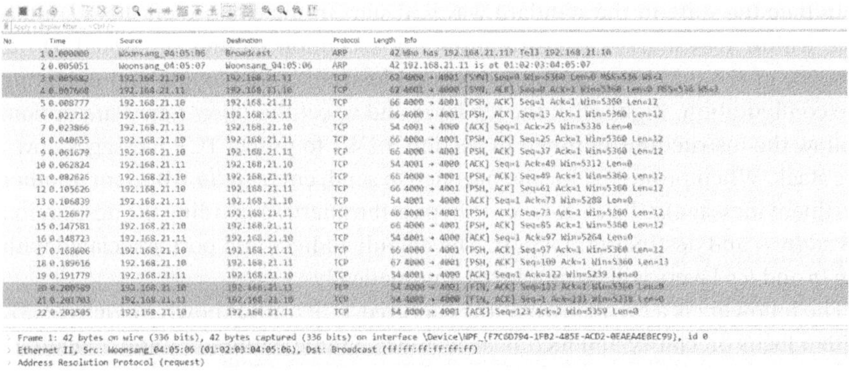

Figure 5.38 Captured 10-message TCP Traffic.

provides valuable insights into the segment's properties and its role within the TCP communication process. To proceed with the examination of loss effects on TCP streams, locate line 7 within the script. Within the *send* function call on this line, substitute the first argument from *udp* to *tcp*. The resulting script is shown below:

```
-- Send ten TCP messages

function main()
    clearOutput();

    for i=1,10 do
        send(tcp, "Hello World" .. i);
    end
end
```

The script produces the same sequence of messages as the one in Section 5.2.4.1. TCP Connection Initiation: Prior to transmitting messages, the script establishes a TCP connection, a fundamental characteristic of TCP's connection-oriented nature. Messages are directed to the destination port and address associated with the *tcpx* and *ipx* layers, respectively, aligning with TCP's requirements for established endpoints.

Execute the suite in the standard manner. Initiate Wireshark traffic capture on the *NT* interface. Allow the script to execute fully. Refer to Figure 5.38 for a representation of the captured traffic. Observe the initial TCP handshake that establishes the connection. Track the transmission of the *Hello World 1* through *Hello World 10* messages. Witness the bidirectional exchange of TCP FIN segments, which terminates the connection. For a detailed illustration of TCP connection teardown, refer to Figure 2.15.

The overall transmission process consumes approximately 200 milliseconds. Approximately 12 RTTs are involved, implying an average RTT of roughly 16 milliseconds. Each TCP segment carries a single message. Right-click on any TCP segment within the connection. Select the *Follow* option from the context menu. Choose *TCP Stream* to reveal the complete content of the TCP stream, as illustrated in Figure 5.39. This functionality offers a comprehensive view of the data exchanged within the stream.

Figure 5.40 shows the content of the TCP stream. The ten *Hello World* messages are displayed as a single ASCII string, with no carriage return characters between

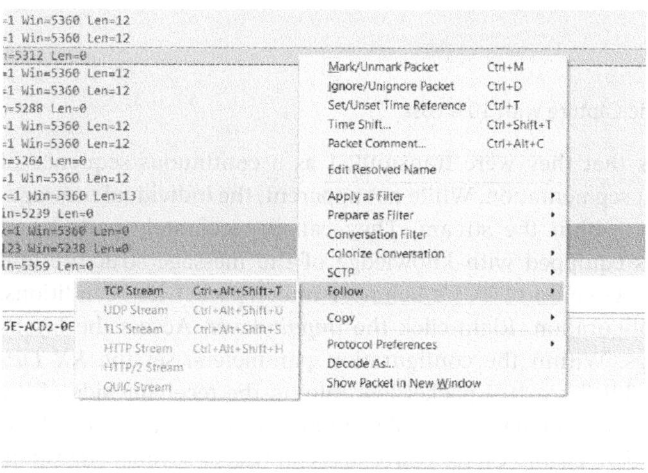

Figure 5.39 TCP Stream Follows Wireshark Option.

Figure 5.40 Stream Dump.

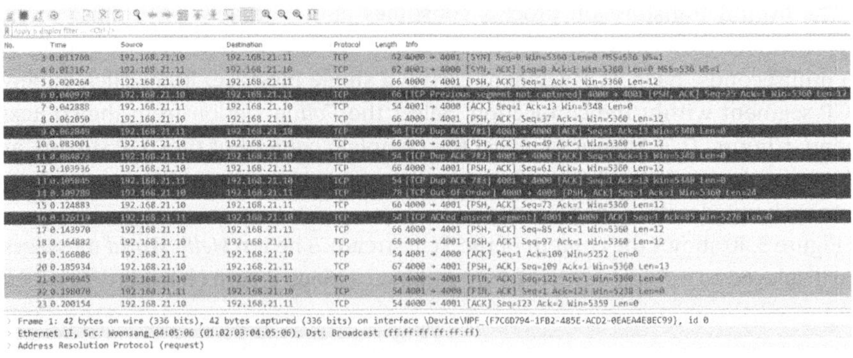

Figure 5.41 TCP Traffic Capture with 10% Loss.

them. This indicates that they were transmitted as a continuous sequence of bytes without explicit segmentation. While not apparent, the individual messages retain their integrity within the stream. They can be accurately parsed and extracted by systems equipped with knowledge of the message structure and length. To initiate observation of TCP's behavior under packet loss conditions, halt the current configuration. Right-click the *impair* layer. Access the layer's configuration settings. Within the configuration parameters, set the *RX Loss* value to 10. This establishes a 10% packet loss rate on the receiving side of the *impair* layer. Resist introducing more substantial loss percentages (such as the 50% used in the UDP example). Excessive loss can trigger severe congestion, potentially leading to connection resets within TCP. This configuration deviates from the UDP example in Section 5.2.4.1 by directly modifying the default settings of the *impair* layer, as opposed to introducing loss through a temporary rule.

In Figure 5.41, the overall transmission time remains consistent at approximately 200 milliseconds, indicating that TCP successfully adapts its transmission strategies to accommodate packet loss. Retransmissions prompt the TCP layer to consolidate multiple messages (both original messages and their retransmissions) within individual segments to enhance efficiency. While retransmissions ensure reliable delivery, they temporarily increase instantaneous transmission rates, potentially leading to suboptimal channel utilization. This is because some bandwidth is devoted to retransmitting data that has already been sent but not successfully received.

5.3 RTC

This section examines the transmission of media utilizing the RTP and SIP protocols. SIP is employed for establishing media sessions, whereas RTP encapsulates the media frames themselves.

5.3.1 SIP and RTP

SIP and RTP, presented in Sections 2.3 and 3.2 respectively, form a pair of session layer protocols that play an important role in the transmission of media over IP networks (IETF RFC 3261, Schooler et al. 2002; IETF RFC 3550, Schulzrinne et al. 2003). SIP functions as a REST mechanism that controls media sessions between endpoints, while RTP handles the packetization of the actual media into speech, audio, and video frames. In traditional telephony contexts, SIP facilitates the establishment and termination of calls. Note that the paths taken by SIP messages and RTP packets from source to destination may not always align. Network and transport layer parameters essential for establishing the RTP media path are conveyed within the SDP payload of the SIP INVITE and 200 OK messages.

Within this section, construct a Netualizer project that incorporates a pair of stacks designed for the transmission of audio from one stack to the other. While SIP offers support for both UDP and TCP transport, opt for encapsulating SIP over UDP to streamline the project. Similarly, RTP, which inherently functions solely over UDP transport, will also be encapsulated over UDP. To initiate this process, open the UDP transport project *transportLayerImp air* that was originally created in Section 5.2.4.1 and proceed to save it under the name *sipRtp*.

As each stack requires two distinct UDP layers (one dedicated to SIP and the other to RTP), proceed to create two additional UDP layers, aptly named *udp2* and *udp2x*. Position *udp2* directly above the *ip* layer and *udp2x* directly above the *ipx* layer, as shown in Figure 5.42. It is important to note that the UDP layers within each stack exhibit a pair of small buttons labeled + and – situated on their left sides. These buttons serve the purpose of enabling to seamlessly cycle through the *udp* and *udp2* layers within the *FIRST* stack, as well as through the *udpx* and *udp2x* layers within the *SECOND* stack.

Initiate the creation of two SIP layers, labeled *sip* and *sipx*, by clicking the beige-colored *SIP* layer button as illustrated in Figure 5.43. Position the *sip* layer directly above the *udp* layer, and likewise, position the *sipx* layer directly above the *udpx* layer, as shown in Figure 5.44.

Proceed to replicate this procedure to construct the RTP stacks. To achieve this, click the *RTP* layer button, as previously illustrated in Figure 5.43, to initiate the

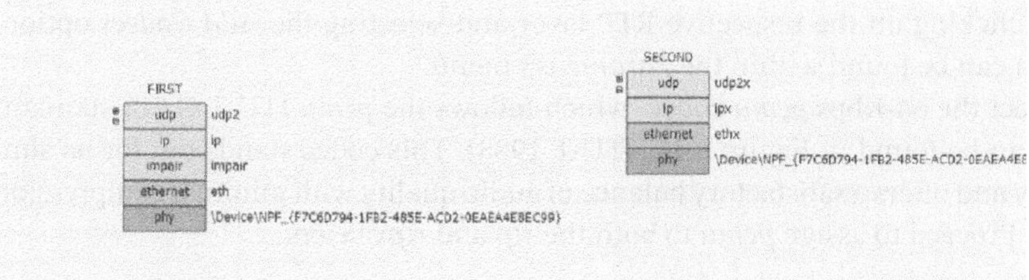

Figure 5.42 UDP Protocol Stacks.

Figure 5.43 Creating the SIP Layer.

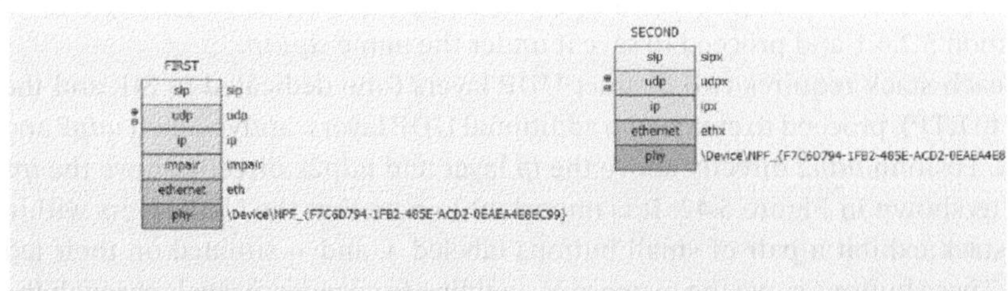

Figure 5.44 SIP Protocol Stacks.

creation of two RTP layers, designated as *rtp* and *rtpx*. Position the *rtp* layer directly above the *udp2* layer, and likewise, position the *rtpx* layer directly above the *udp2x* layer. The representation of these two layers can be observed in Figure 5.45.

A speech codec refers to a computational algorithm that performs the encoding of packetized raw speech into compressed frames, as well as the subsequent decoding of those compressed frames back into raw speech. To ensure proper functionality, a codec must be assigned to each RTP layer. This can be achieved by right-clicking on the respective RTP layer and selecting the *Add Codecs* option, which can be found within the *Parameters* menu.

Select the 64-Kbps *pcmu* codec, which follows the *pcmu* ITU-T G.711 standard and can be found in Figure 5.46 (ITU-T 1988). This codec stands out for its simplicity and offers a satisfactory balance of audio quality with minimal compression rates. Proceed to assign *pcmu* to both the *rtp* and *rtpx* layers.

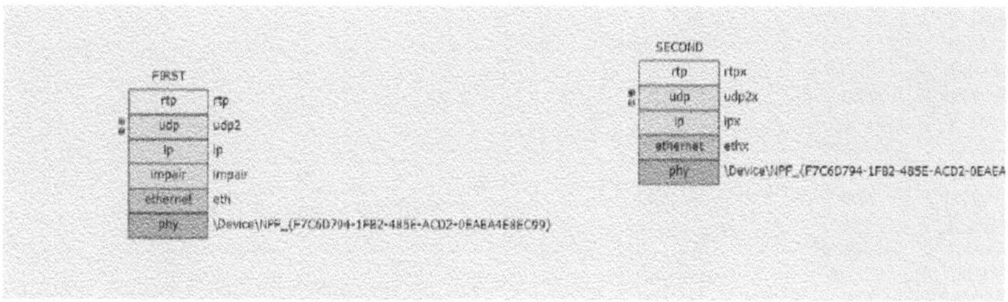

Figure 5.45 RTP Protocol Stacks.

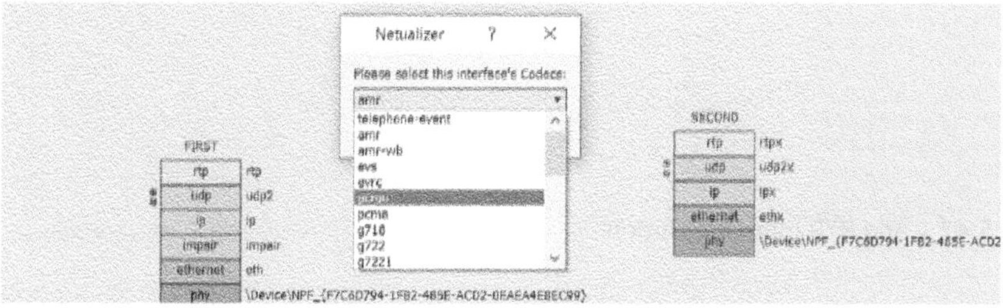

Figure 5.46 Selecting Audio Codec.

Due to the fact that SIP and RTP constitute distinct layers, it is imperative to establish an explicit link between the *sip* layer and the *rtp* layer to ensure the successful facilitation of media sessions. To achieve this, right-click on the *sip* layer and select the *Add Rtp Interfaces* option, which can be found within the *Parameters* menu, as shown in Figure 5.47. Proceed to select the *rtp* layer to associate it with the *sip* layer. Replicate this process for the *sipx* layer within the *SECOND* stack, this time linking it to the *rtpx* layer.

To initiate the generation of RTP packets, the inclusion of an audio generation layer is essential. While Netualizer offers a diverse range of options in this regard, Text-To-Speech (TTS) stands out as a particularly straightforward and effective solution. To employ TTS, click the *tts* button as illustrated in Figure 5.48 to create a TTS layer. Assign the name *tts* to the newly created TTS layer and position it directly above the *rtp* layer within the *FIRST* stack. It is important to note that TTS technology facilitates the playback of audio that has been defined in textual format. In essence, the TTS layer functions by *reading* the provided text and subsequently generating the corresponding raw audio stream.

To guarantee the successful arrival of the audio at the *SECOND* stack, proceed to create an *audio* layer. This can be achieved by clicking the beige-colored *audio* button as illustrated in Figure 5.49. The audio layer serves the purpose of playing the incoming packetized raw audio on the physical sound interface of the agent.

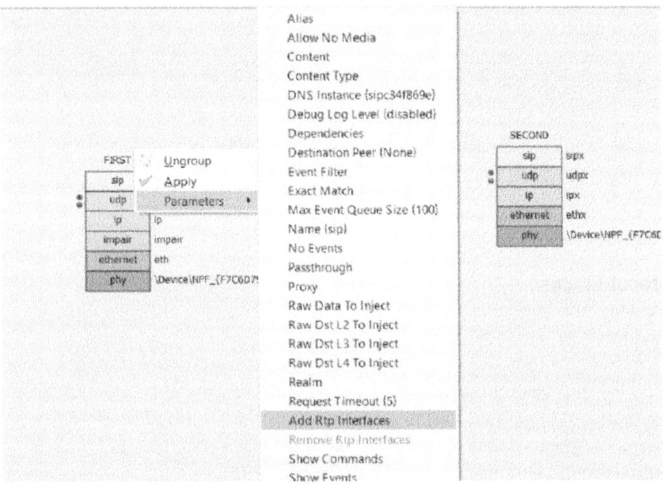

Figure 5.47 Add RTP Interface Option.

Figure 5.48 Creating the TTS Layer.

Provide the name *audio* to the recently created audio layer. Place the audio layer directly above the *rtpx* layer within the *SECOND* stack, as shown in Figure 5.50. With these actions, the configuration process is finalized, and the two stacks are now fully prepared to facilitate a media session between them. To establish a media session between the *sip* layer and the *sipx* layer, ensure that the *sip* layer's destination URL is correctly configured to target the IP address of the *SECOND* stack.

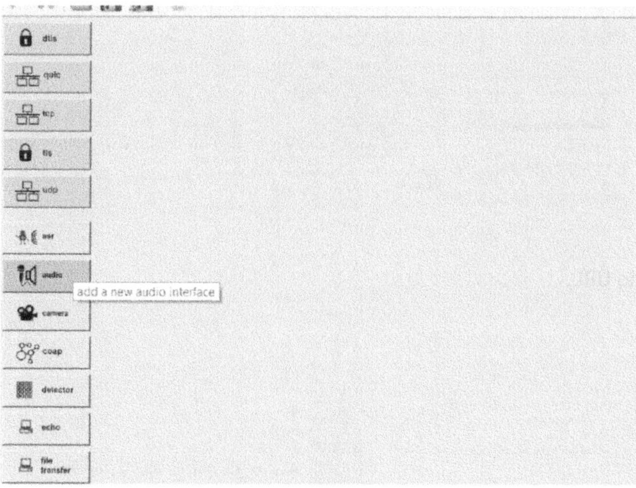

Figure 5.49 Creating Audio Layer.

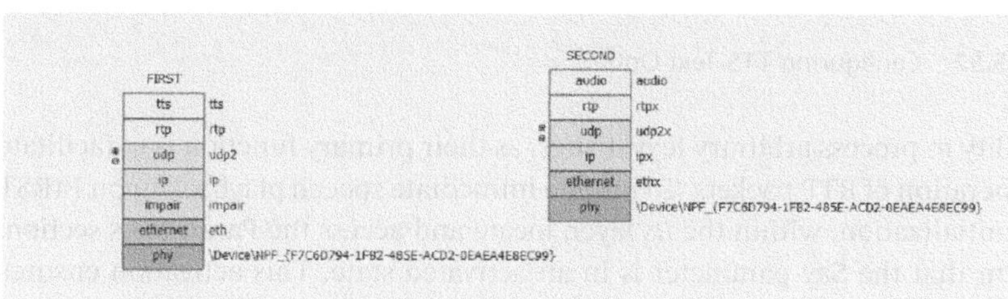

Figure 5.50 Laying down the Audio Layer.

Right-click the sip layer and select the *Parameters* menu. Within the *Parameters* menu, identify the *URL* option. Enter the IP address of the *SECOND* stack as the destination *URL*.

Within the Netualizer's agent configuration panel, locate the field designated for URL entry. Visual guidance for this process can be found in Figure 5.51. Enter the IP address *192.168.21.11* into the URL field. It is important to note that this specific address is exclusively associated with the *ipx* layer within the *SECOND* stack. To start TTS configuration, navigate to the *tts* layer's Parameters section. This section can be accessed by employing right-click functionality. Within the *Parameters* section, locate and select the *Text* option.

Identify the designated text field within the TTS configuration interface, as shown in Figure 5.52. Within the identified text field, input the text string *This is a test* for the purpose of TTS processing. While this specific string serves as a demonstration example, it is essential to note that the TTS system possesses the

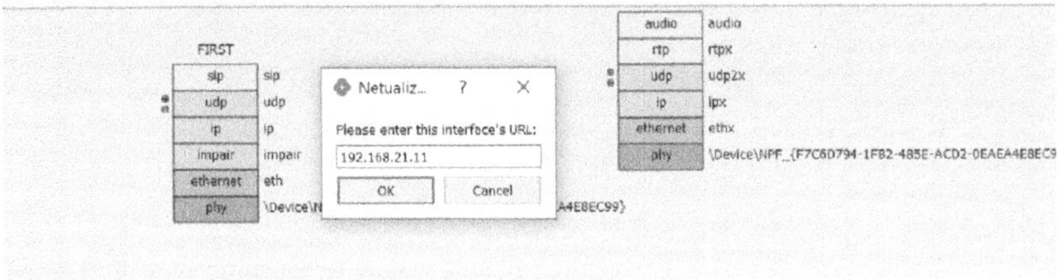

Figure 5.51 Configuring URL.

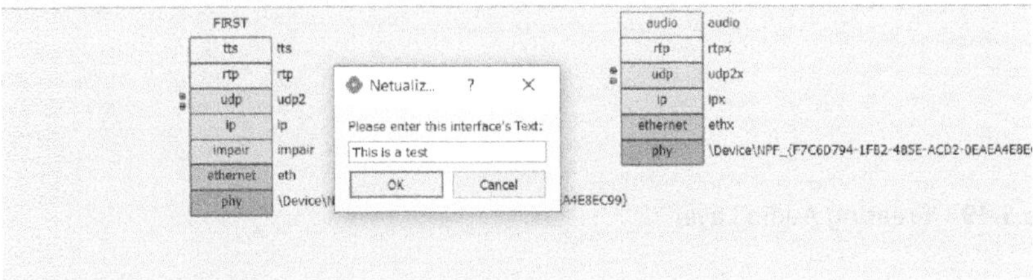

Figure 5.52 Configuring TTS Text Option.

capability to process arbitrary text strings, as their primary function is to facilitate the generation of RTP packets. To initiate immediate speech playback upon FIRST stack initialization, within the *tts* layer, locate and access the Parameters section. Confirm that the Say parameter is in an activated state. This activation ensures prompt speech playback upon stack initialization. In scenarios where manual activation is required subsequent to configuration deployment and execution, the *tts* layer provides a supplementary activation mechanism.

The audio layer is inactive by default, meaning it will not play or record audio without specific configuration. To enable audio playback of incoming media frames, follow these steps. Activate the Play option: Right-click the audio layer and select the *Play* option under *Parameters.* Set Wireshark filter: In Wireshark, configure a filter to capture both SIP and RTP traffic using the expression *sip —— rtp.* Start traffic capture: Begin capturing network traffic on the *NT* interface. Then ensure the project script is altered to perform no actions as follows:

```
-- SIP/RTP Example

function main()
clearOutput();
end
```

Start the suite's execution to ensure the configured settings are successfully deployed to the agent. Employ right-click functionality on the *sip* layer and select

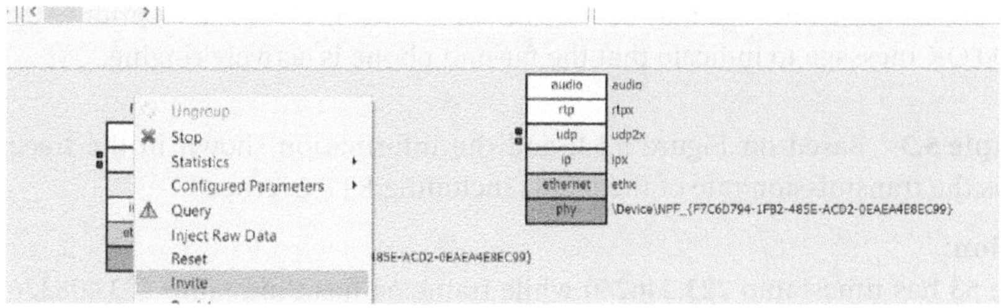

Figure 5.53 Invite Command.

the *Invite* command. This action, as shown in Figure 5.53, establishes an SIP *dialog* that facilitates the media session. Upon successful establishment of the SIP dialog, the *audio* layer within the *SECOND* stack should proceed to play the pre-configured speech sequence, *This is a test*.

Figure 5.54 provides shows the actual Wireshark trace, capturing the following sequence of events. The sip layer initiates the communication by transmitting an SIP INVITE message, which is encapsulated within frame number 46 of the Wireshark trace. The *SECOND* stack responds to the INVITE message with two distinct SIP messages. In frame 47, the SIP 100 Trying message, indicating that the *SECOND* stack is processing the request is sent. Similarly, in frame 51, the SIP 200 OK message, signifying successful acceptance of the session establishment request is transmitted. In frame 52, upon receiving the SIP 200 OK message, the FIRST stack transmits an SIP ACK message in frame number 52 to acknowledge the successful reception of the 200 OK response. This aforementioned sequence of SIP messages, collectively referred to as a transaction, is responsible for establishing the media session between the communicating entities. In certain scenarios,

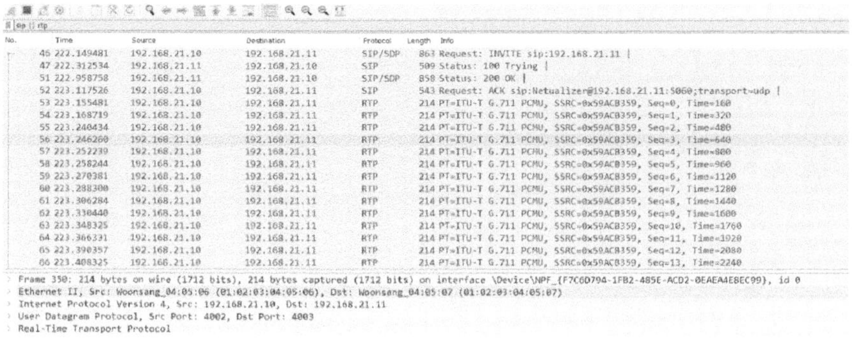

Figure 5.54 SIP and RTP Session Setup.

an SIP 180 Ringing message may be transmitted by the *SECOND* stack prior to the SIP 200 OK message to indicate that the far-end phone is actively ringing.

Example 5.3 Based on Figure 5.54, and the information shown in the trace, what is the transmission rate of the media including RTP overhead?

Solution:
Frame 53 has timestamp 223.246260 while frame 66 has timestamp 223.408326. That is 13 packets transmitted in $408.326 - 246.260 = 162.06$ ms. Because each packet is $214 \times 8 = 1712$ bits long, the overall transmission rate is $R = \frac{1712}{0.162} = 10.56$ Kbps.

Packetized audio is transmitted as encoded speech from the *FIRST* to the *SECOND* stack starting in frame number 53. Note the payload type (PT) is set to *ITU-T G.711 PCMU*, the SSRC is constant for the session, the sequence number increments starting at 0, and the timestamp accumulates in units of 160 samples every 20 milliseconds. Since the codec transmission rate is 64 Kbps, every second $\frac{64000}{8} = 8000$ bytes are generated. Moreover, every 20 milliseconds $8000 \times \frac{20}{1000} = 160$ bytes (or samples), that are encapsulated in RTP packets, are generated.

Figure 5.55 presents a detailed view of the 863-byte SIP INVITE message captured within frame number 46. The SIP INVITE is encoded as an ASCII text message, which can be considered inefficient for transmission over constrained devices due to its relatively high overhead. The message incorporates a substantial number of headers, including the CSeq header, which denotes a sequence number of 1. These headers serve various purposes in managing the SIP communication. The INVITE request additionally contains a body section that encloses an SDP payload. This SDP carries information pertaining to the media session including

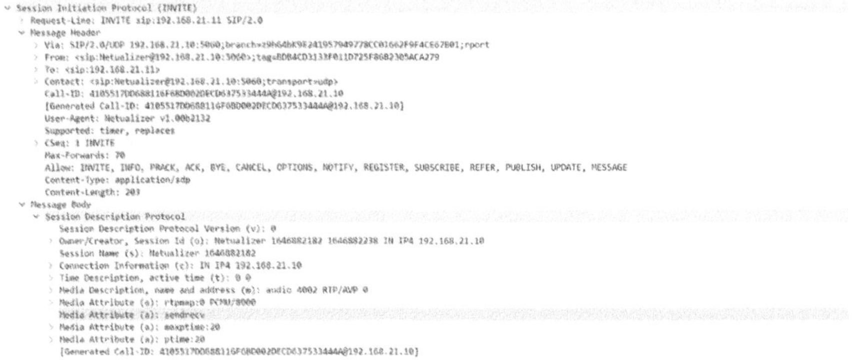

Figure 5.55 INVITE Request.

Figure 5.56 INVITE 200 OK Response.

the IP address designated for the transmission of RTP packets that convey the media frames generated by the *FIRST* stack as well as the specific UDP port number employed for RTP packet delivery.

Figure 5.56 offers a representation of the 858-byte SIP 200 OK message, captured within frame number 47. Similar to the SIP INVITE message, the 200 OK response incorporates a body section that encloses an SDP payload. The SDP within the 200 OK message carries essential details regarding the media session, specifically, the IP address designated for the transmission of RTP packets originating from the *SECOND* stack. The specific UDP port number employed for RTP packet delivery is initiated by the *SECOND* stack.

Figure 5.57 presents a detailed view of a single RTP packet extracted from the trace in Figure 5.54. The packet incorporates a header that follows the format shown in Figure 3.2. It includes a version field within the header that is set to 2, a CSRC count field is set to 0, a sequence number field holds a value of 0 and a timestamp field is set to 160. The SSRC field, which identifies the synchronization source associated with the stream, is set to 0x59acb359. The packet additionally encloses a 160-byte payload, with the initial bytes being displayed in the figure.

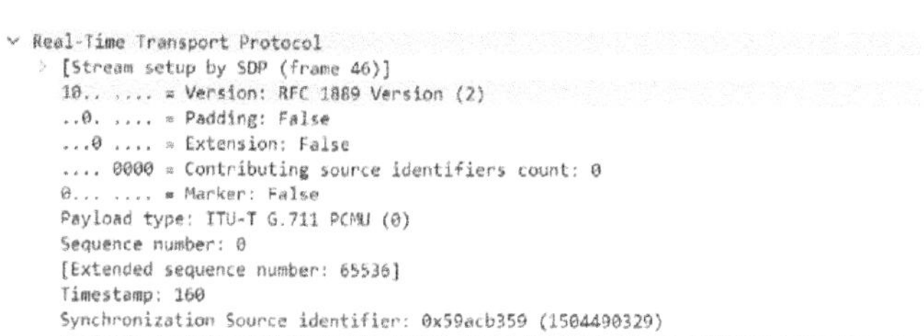

Figure 5.57 RTP Media Packet.

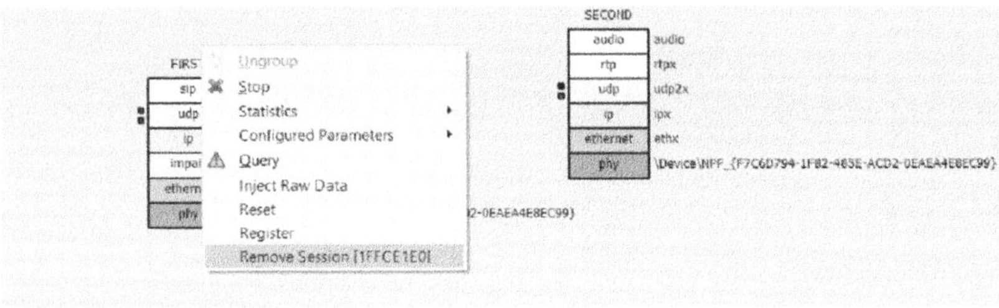

Figure 5.58 Session Removal.

To terminate the session, right-click directly on the *sip* layer. From the options presented, select the option labeled *Remove Session [...]*. A depiction of this process can be found in Figure 5.58. The session identifier, which is displayed within square brackets, is assigned upon session initialization and serves to uniquely distinguish each session.

The following is description of the session termination process, as shown in Figure 5.59. The *FIRST* stack initiates session termination by transmitting an SIP BYE message, which is captured within frame 8685 of the trace. The *SECOND* stack acknowledges the termination request by sending an SIP 200 OK message in frame 8692. It is important to note that RTP packets, responsible for carrying media data, continue to flow even during the session termination process. This shows the independent nature of SIP and RTP stacks. The coordination of session termination between the SIP and RTP stacks is facilitated by mechanisms external to these protocols.

Figure 5.60 provides a detailed view of the SIP BYE message captured within frame 8685. The message incorporates a variety of headers, including the CSeq

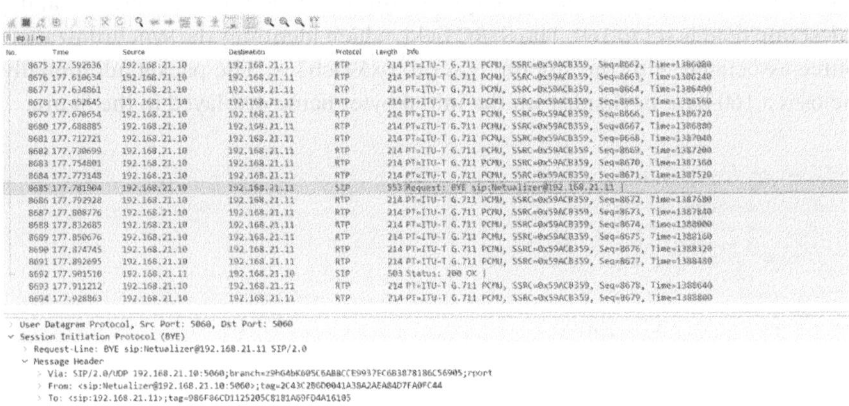

Figure 5.59 SIP and RTP Session Termination.

```
v Session Initiation Protocol (BYE)
  > Request-Line: BYE sip:Netualizer@192.168.21.11 SIP/2.0
  v Message Header
    > Via: SIP/2.0/UDP 192.168.21.10:5060;branch=z9hG4bK605C6AB8CCE9937EC6B3878186C56905;rport
    > From: <sip:Netualizer@192.168.21.10:5060>;tag=2C43C2B6D0041A38A2AEA84D7FA0FC44
    > To: <sip:192.168.21.11>;tag=986F86CD1125205C8181A69FD4A16105
    > Contact: <sip:Netualizer@192.168.21.10:5060;transport=udp>
      Call-ID: 28E1F887CEBEC4FFA6482713EDC67E991EAC8@192.168.21.10
      [Generated Call-ID: 28E1F887CEBEC4FFA6482713EDC67E991EAC8@192.168.21.10]
      User-Agent: Netualizer v1.00b2132
      Supported: timer, replaces
    > CSeq: 2 BYE
      Max-Forwards: 70
      Content-Length: 0
```

Figure 5.60 BYE Request.

header, which has been incremented to a value of 2, indicating a subsequent request within the dialog. The Content-Length header denotes a payload size of zero bytes. This is expected as there is no need to exchange additional SDPs during session termination. The absence of a payload aligns with the logical reasoning that SDP exchange is unnecessary during the session termination process.

Figure 5.61 presents the SIP 200 OK message, which serves as a direct response to the SIP BYE request. The CSeq header value within the SIP 200 OK message matches that of the SIP BYE message (Figure 5.60). This correlation establishes a direct connection between the two messages within the dialog. The Call-ID header value, present in all messages belonging to the same session, remains consistent across Figures 5.55, 5.56, 5.60 and 5.61. This consistency shows the message's association with a specific session. As mentioned in Section 5.5, the combination of the Call-ID, From, and To header values collectively serves to uniquely identify an SIP dialog.

To evaluate the impact of network impairments on media quality begin by stopping the suite's execution. Within the agent configuration panel, initiate the creation of a QoS layer. This can be accomplished by selecting the beige-colored *QoS* layer. Designate the newly created layer as *qos* and position it directly beneath the existing *audio* layer, as illustrated in Figure 5.62. The primary function of the QoS layer is to conduct a comparison between the incoming speech stream and

```
v Session Initiation Protocol (200)
  > Status-Line: SIP/2.0 200 OK
  v Message Header
    > Via: SIP/2.0/UDP 192.168.21.10:5060;branch=z9hG4bK605C6AB8CCE9937EC6B3878186C56905
      User-Agent: Netualizer v1.00b2132
    > From: <sip:Netualizer@192.168.21.10:5060>;tag=2C43C2B6D0041A38A2AEA84D7FA0FC44
    > To: <sip:192.168.21.11>;tag=986F86CD1125205C8181A69FD4A16105
      Call-ID: 28E1F887CEBEC4FFA6482713EDC67E991EAC8@192.168.21.10
      [Generated Call-ID: 28E1F887CEBEC4FFA6482713EDC67E991EAC8@192.168.21.10]
    > CSeq: 2 BYE
    > Contact: <sip:Netualizer@192.168.21.11:5060;transport=udp>
      Supported: timer, replaces
      Content-Length: 0
```

Figure 5.61 BYE 200 OK Response.

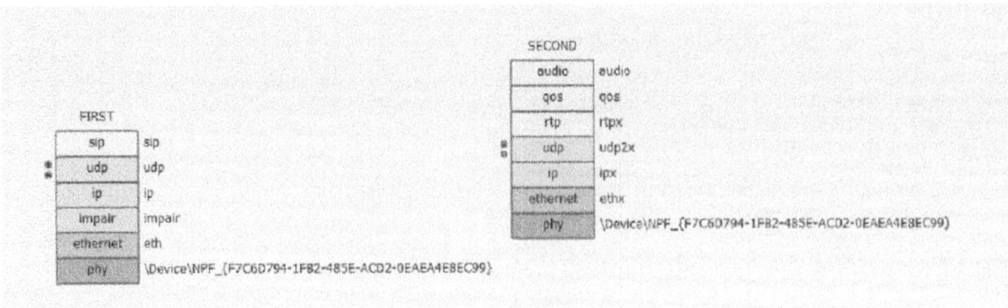

Figure 5.62 QoS Layer Creation.

the stream generated within the tts layer. This comparison yields a PESQ score, which serves as a quantitative measure of speech quality. It is essential to note that the PESQ algorithm has been standardized as ITU-T Recommendation P.862. The PESQ score is a numerical value that falls within the range of −0.5 to 4.5. This score directly corresponds to the perceived quality of the received speech, with higher scores signifying superior speech quality (Herrero and Cadirola 2014).

To establish a connection between the *qos* layer and the *tts* layer, employ right-click functionality directly on the qos layer. Access the *Parameters* section within the context menu. Choose the option labeled *Add PESQ Player* to set a reference for comparison. A depiction of this process can be found in Figure 5.63. The reference can be either a specific media file containing audio content or the name of a layer within the configuration that actively plays audio.

To configure the reference and enable PESQ score display, as shown in Figure 5.64, choose *tts* as the reference for the PESQ comparison. This is because

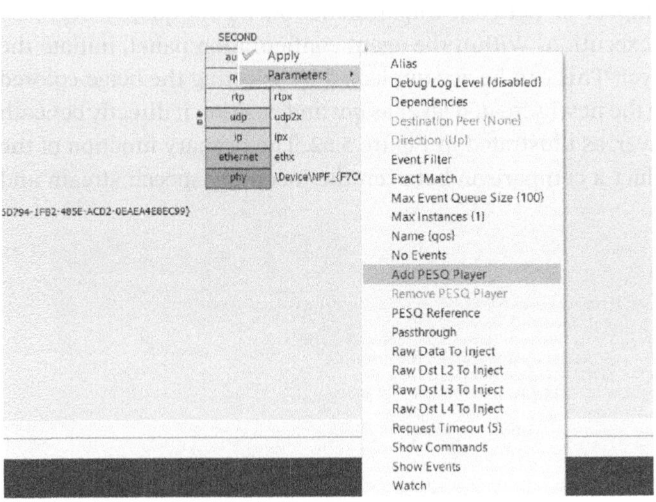

Figure 5.63 QoS Reference Configuration.

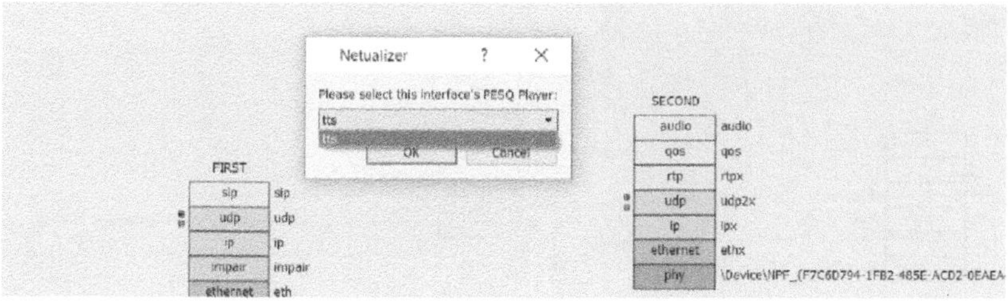

Figure 5.64 Configuring *tts* as Reference.

it is the only layer in the current configuration that generates audio content. Activate the *Show Events* option, located within the Parameters section of the qos layer (Figure 5.63). This will allow real-time display of PESQ scores as they are calculated. The PESQ scores will be displayed in the *Events* window of the Netualizer controller.

To initiate PESQ analysis, start capturing traffic on the *NT* interface. Run the SIP and RTP project suite once again. Facilitate the establishment of a media session between the FIRST and SECOND stacks. Allow the *This is a test* speech sequence to be played by the audio layer. Right-click directly on the *qos* layer and select the *Start PESQ* option, as shown in Figure 5.65. This action will trigger the PESQ analysis process.

Following the initiation of PESQ analysis, allow approximately 30 seconds to elapse. This duration is intended to ensure the completion of a substantial number of speech sequence playouts. Terminate the PESQ analysis process by selecting the *Stop PESQ* option, as shown in Figure 5.66. The selection of this option will prompt the *qos* layer to compute the final PESQ score.

Figure 5.67 provides a representation of how the Netualizer controller displays the PESQ score, accompanied by a timestamp, within the *Events* window. The computed PESQ score in this instance is 4.48. This score signifies exceptionally high speech quality when employing the ITU-T G.711 codec. It is important

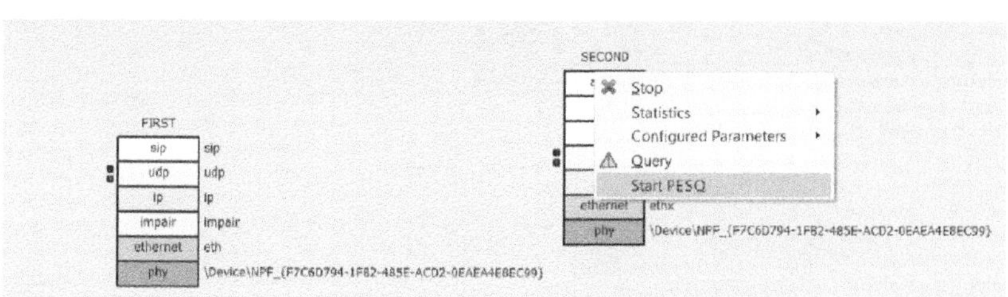

Figure 5.65 Start PESQ Option.

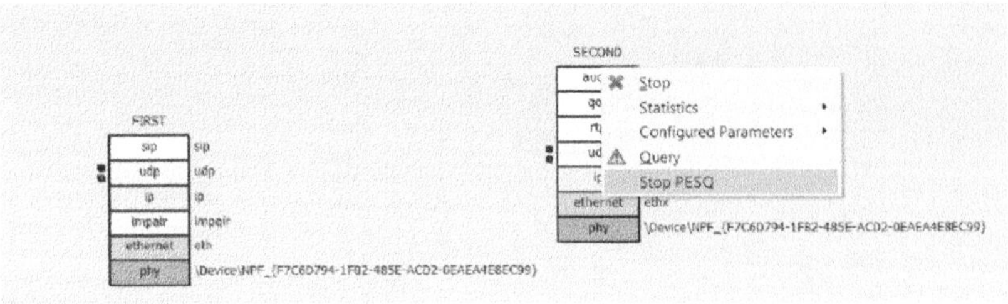

Figure 5.66 Stop PESQ Option.

Figure 5.67 PESQ Score on Events Window.

to note that different codecs, particularly those with higher compression rates, typically exhibit lower maximum PESQ scores.

To investigate next how network impairments such as packet loss affect speech quality in this setup, set the ***impair*** layer's *TX Loss* option to 50% and repeat the experiment. As shown in Figure 5.68, the PESQ score plummets to approximately 1.06, indicating a significant degradation in speech quality. It is important to note that the relationship between packet loss and PESQ scores is not linear. Even negligible levels of packet loss can drastically impact quality scores, making this particularly relevant for IoT Lossy Low-Power Networks (LLNs) that may chronically experience packet loss. This experiment shows the substantial impact of packet loss on speech quality. Even low levels of loss can significantly degrade

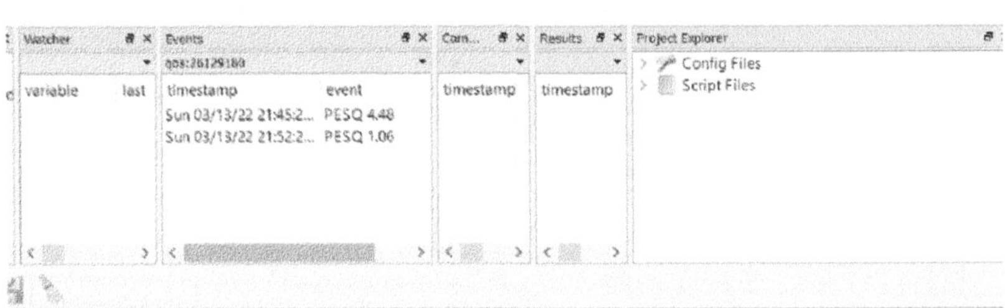

Figure 5.68 PESQ Score on Events Window (50% Packet Loss).

perceived speech quality. This is particularly relevant for LLNs where reliable network connectivity is important for maintaining acceptable call quality.

5.3.2 Secure RTC

Secure RTC is enabled by the protocols SIP over TLS and SRTP. These protocols, in turn, depend on TLS and DTLS, respectively. This section provides a comprehensive discussion of these essential protocols.

5.3.3 TLS

Load the *transportTcpLayerImpair* project created earlier in Section tcpLayer. Create a duplicate of this project and save it under the name *tls*. To add TLS layers to the *tcp* and *tcpx* layers in the *FIRST* and *SECOND* stacks, respectively, locate and click the TLS layer button (shown in Figure 5.30). Create a TLS layer named *tls* and position it directly above the *tcp* layer. Create another TLS layer named *tlsx* and place it above the *tcpx* layer. The resulting *FIRST* and *SECOND* stacks, now equipped with their corresponding TLS layers, are illustrated in Figure 5.69.

To successfully establish a TLS session, three specific files are required: (1) a private key file that holds a secret key that must be kept confidential, (2) a certificate file that contains a public key and serves as a form of digital identification, and (3) a CA certificate file that holds the public key of a trusted entity that verifies the authenticity of certificates.

A private key and its corresponding public key are generated together as a matched pair. Information encrypted with the public key can only be decrypted using the private key, and vice versa. The public key is used to create a Certificate Signing Request (CSR), which is a request for a certificate to be issued by a CA. Upon receiving a CSR, a CA verifies the identity associated with the public key and, if valid, signs a certificate that binds the public key to that identity. Each TLS layer needs a private key and a certificate for its own authentication. Additionally, a CA certificate is used to validate the certificates received from

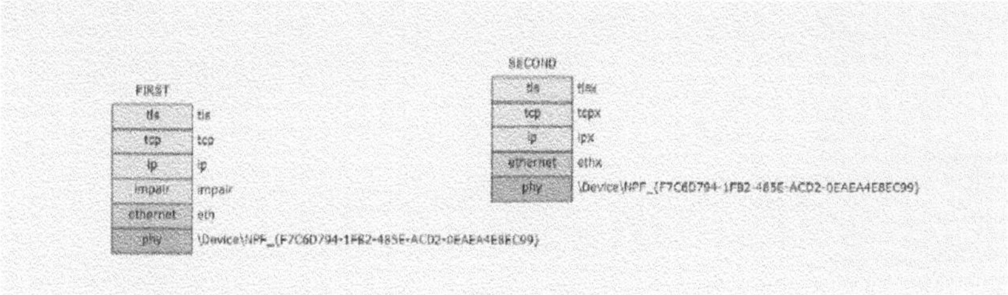

Figure 5.69 TLS Protocol Stacks.

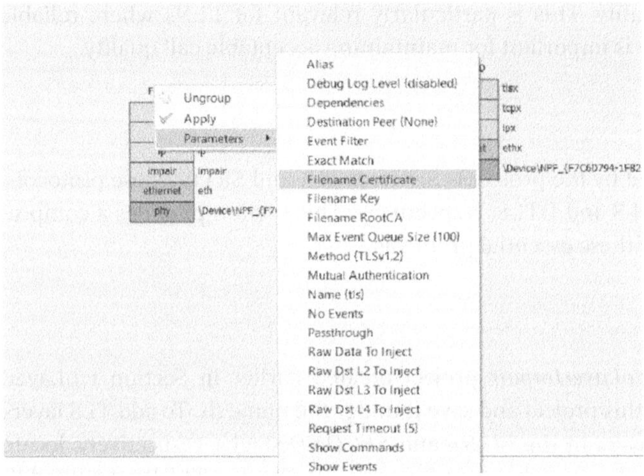

Figure 5.70 Configuration of Certificates and Private Key.

other parties, ensuring trust in their identities. To configure these files for a TLS layer, right-click on the layer and choose the *Parameters* option (as shown in Figure 5.70).

Netualizer provides pre-installed certificate files for convenience. Upon its initial installation, Netualizer stores the following files: (1) a private key file named *server.key*, (2) its corresponding certificate file named *server.pem* and a CA certificate file named *CA.pem* that validates the *server.pem* certificate. These files can be found within the *certs* directory, situated within Netualizer's home directory. There is the option to utilize these files whenever configuring TLS or DTLS layers. To configure these files for a TLS layer, right-click on the layer and choose the *Parameters* option. As illustrated in Figure 5.71, select the *server.pem* file by clicking on the *Filename Certificate* option within the TLS layer. Locate and choose the *server.key* file by clicking the *Filename Key* option within the

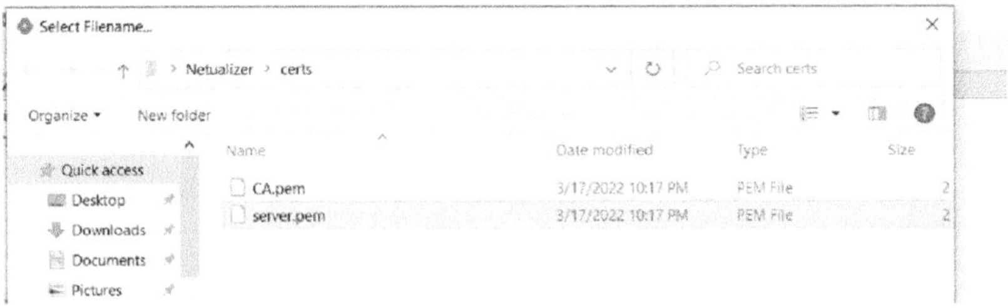

Figure 5.71 *server.pem* Selection.

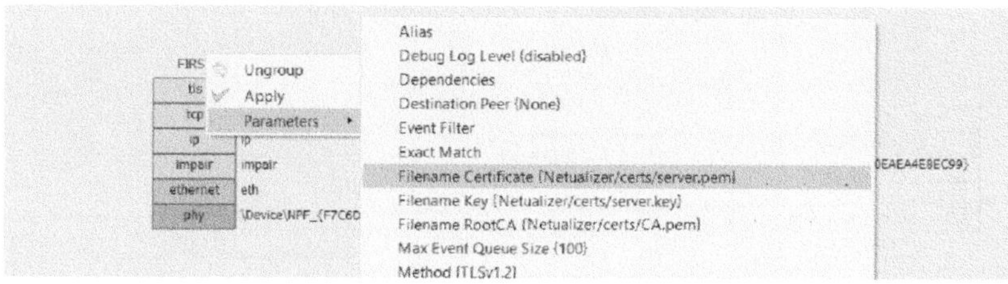

Figure 5.72 TLS Certificates and Private Key.

Parameters menu. Similarly, select the *CA.pem* file by clicking the *Filename RootCA* option within the same *Parameters* menu.

Refer to Figure 5.72 for a representation of the configured TLS security parameters on the *tls* layer. Apply the same configuration steps to the *tlsx* layer. Right-click on the *tlsx* layer and select *Parameters*. Set the private key file, certificate file, and CA certificate file using the same process as it was done for the *tls* layer. More importantly, set the script to do nothing to facilitate direct packet injection:

```
-- TLS Example

function main()
clearOutput();
end
```

Begin capturing traffic on the Wireshark NT interface. Set the filter to *tcp* to specifically capture TLS packets that are transmitted over TCP. Launch the suite. Inject the *Hello World* message into the *tls* layer by choosing the *Inject Raw Data* option, closely following the injection procedure outlined in Section udpLayer. Navigate to the *tlsx* layer. Click on *Upstream Last Packet* within the *Statistics* section (as shown in Figure 5.73). This will display the received message.

Figure 5.74 shows the Netualizer controller displaying the content of the most recently received packet at the *tlsx* layer. It is important to note that the data displayed in this figure has undergone encryption within the *tls* layer and decryption within the *tlsx* layer. This demonstrates the successful capture and decryption of the TLS traffic.

Figure 5.75 shows the TLS packets exchanged between the *FIRST* and *SECOND* stacks, captured in Wireshark. Frames 11 and 12 establish the underlying TCP connection. Specifically, frame 11 initiates the TCP connection with a TCP SYN message and frame 12 confirms the connection with a TCP SYN ACK message. Frames 13 through 16 initiate and complete the TLS handshake. In frame 13, the *tls* layer in the *FIRST* stack sends a Client Hello message to start the TLS handshake. In frame 14, the *tlsx* layer in the *SECOND* stack responds with a Server Hello message. Encryption and message authentication parameters are negotiated during this exchange. In frame 16, the *SECOND* stack sends its

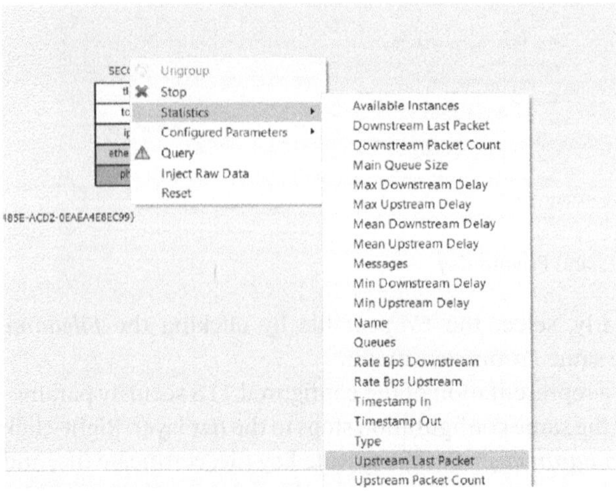

Figure 5.73 Upstream Last Packet Option.

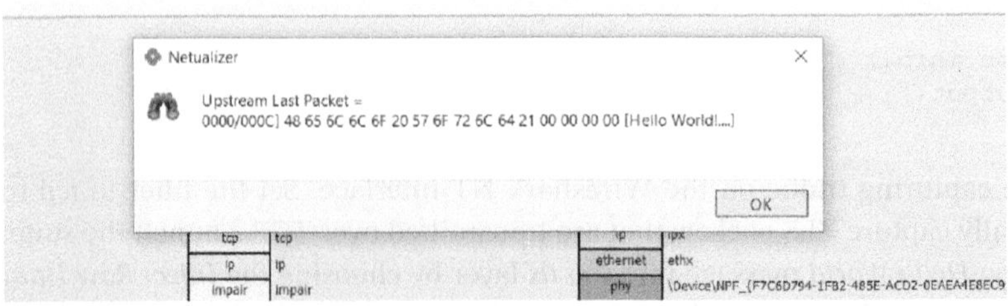

Figure 5.74 Content of Upstream Last Packet.

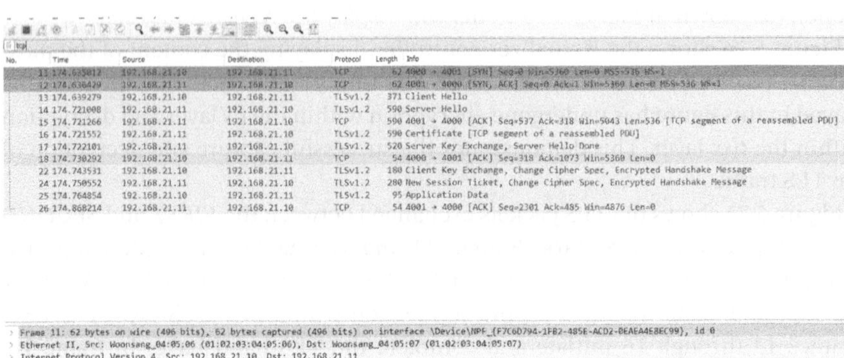

Figure 5.75 Server Authentication.

certificate as part of the default server-side authentication process. The *FIRST* stack verifies the received certificate against the CA certificate. Upon successful verification, a secure TLS session is established. Encrypted Application Data Transmission (Frame 25 onwards). All subsequent application messages, such as the one in frame 25, are encrypted and authenticated to ensure confidentiality and integrity.

Mutual authentication (also known as mTLS), requiring both parties to verify each other's identity, is often necessary for enhanced security. To enable mutual authentication in the TLS layers (*tls* and *tlsx*), halt the suite's execution. Access the agent configuration. Right-click on the TLS stack. Locate and select the *Mutual Authentication* option situated within the *Parameters* menu (shown in Figure 5.76). Restart the suite. Initiate traffic capture on the *NT* interface using Wireshark. Inject the *Hello World* message for encryption on the *tlsx* layer within the *SECOND* stack.

Figure 5.77 presents the Wireshark trace resulting from the mutual authentication scenario. The following are the key events in the trace. In frame 8, the server

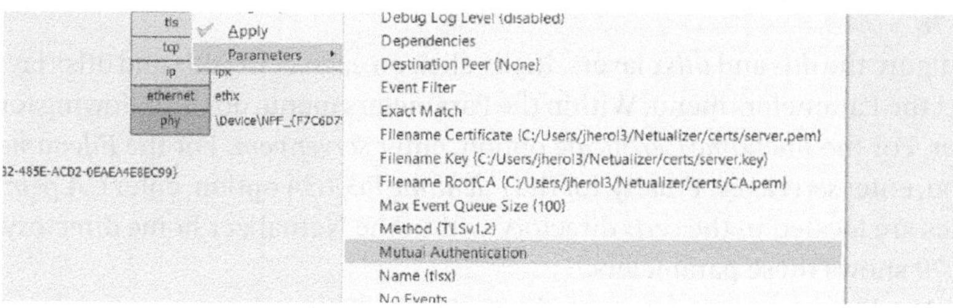

Figure 5.76 Mutual Authentication Option.

Figure 5.77 Captured TLS Packets with Mutual Authentication.

(*tlsx* layer in the *SECOND* stack) sends its certificate to the client (*tls* layer in the *FIRST* stack). In frame 12, the server initiates mutual authentication by requesting a certificate from the client, using a *Certificate Request* message. In frame 17, the client responds by transmitting its certificate to the server. Only after both the server and client certificates are successfully verified against the CA certificate by the server and client, respectively, is the secure session established.

5.3.4 DTLS

For session layer protocols like CoAP that utilize UDP transport, DTLS serves as an alternative to TLS. To begin, open the *transportLayerImpair* project created in Section 5.2.4.1 and save it as *dtls*. Follow the steps outlined in Section 5.3.3 for TLS, but construct a DTLS stack instead. Locate the DTLS layer button on the left side of the configuration panel (illustrated in Figure 5.30) and click it to generate a DTLS layer. Assign the name *dtls* to this layer and position it above the *udp* layer. To create a second DTLS layer, replicate this process. Name this layer *dtlsx* and place it above the *udpx* layer. Both the FIRST and SECOND stacks are shown in Figure 5.78.

To configure the *dtls* and *dtlsx* layers. Right-click on each of the dtls and dtlsx layers. Select the Parameters menu. Within the Parameters menu, do the following for each layer. For the *Filename Certificate* option, enter *server.pem*. For the *Filename Key* option, enter *server.key*. Finally, for the *Filename RootCA* option, enter *CA.pem*. These files are located in the *certs* directory within the Netualizer home directory. Figure 5.79 shows these parameters.

Verify that the script is inactive as shown below. This is important for direct packet injection, as any active script commands could interfere with the process:

```
-- DTLS Example

function main()
clearOutput();
end
```

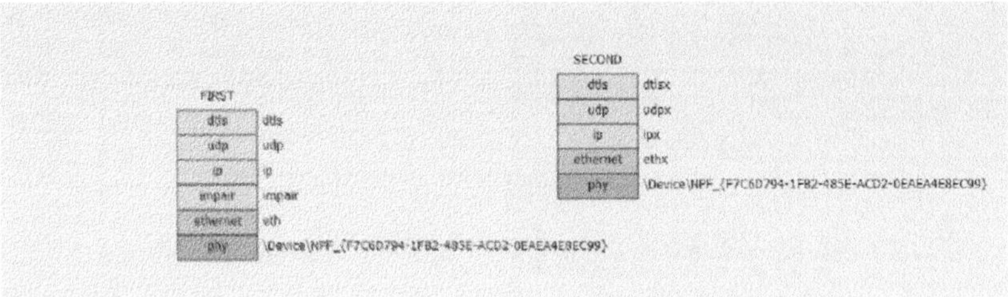

Figure 5.78 DTLS Protocol Stacks.

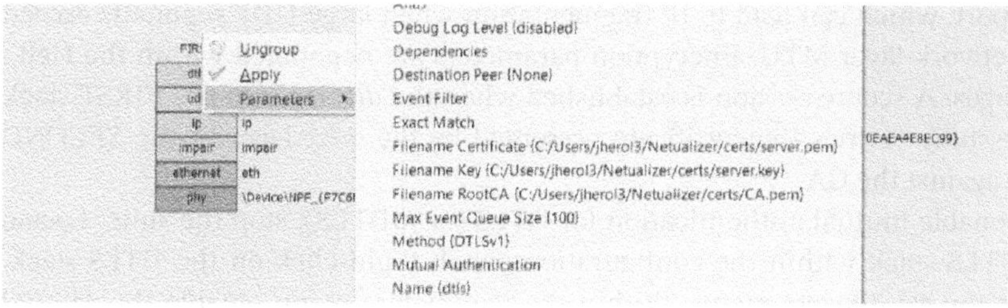

Figure 5.79 DTLS Certificates and Private Key.

To capture and analyze DTLS traffic, initiate traffic capture by opening Wireshark. Begin packet capture on the *NT* interface. Apply a filter to specifically display DTLS traffic: *dtls*. Run the suite. Locate the *Inject Raw Data* option within the suite's interface. Use this option to introduce the *Hello World* message into the *dtls* layer. Follow the transmission procedure outlined in Section 5.2.4.1. (This section should provide specific instructions). Access the *dtlsx* layer. Open the *Statistics* menu. Select the *Upstream Last Packet* option to examine the content of the most recently received message.

Figure 5.80 shows the DTLS packets exchanged between the *FIRST* and the *SECOND* stacks. These are key observations. Frame 13 contains the client Hello message, originating from the *dtls* layer in the *FIRST* stack. Frame 15 contains the server Hello message, sent by the *dtlsx* layer in the *SECOND* stack. Server-side authentication is the default behavior, as evidenced by the transmission of a certificate within this frame. Frame 16 carries the encrypted and authenticated *Hello World* message. Note that under TLS Messages are transmitted over a TCP stream, ensuring reliable, in-order delivery while DTLS employs UDP for

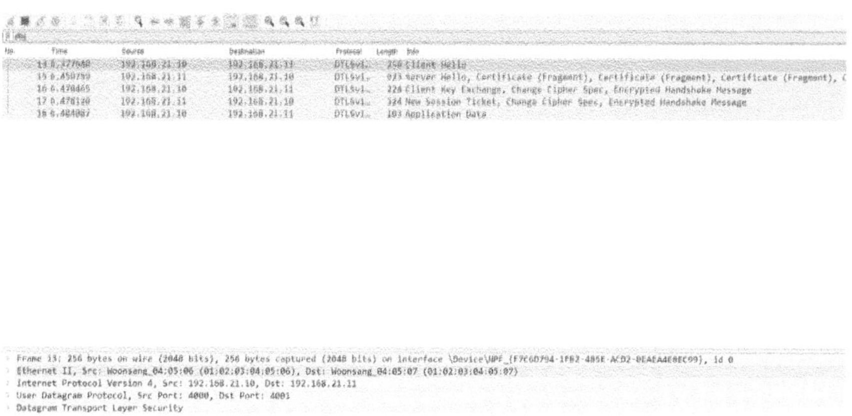

Figure 5.80 Captured DTLS Packets with Server Authentication.

transport, which can lead to IP fragmentation when large UDP segments exceed the network layer MTU. Encryption parameters are negotiated within the Hello messages. A secure session is established when the *dtls* layer in the FIRST stack successfully verifies the certificate presented by the *dtlsx* layer in the *SECOND* stack against the CA.

To enable mutual authentication for DTLS (or mDTLS), stop the suite. Locate the DTLS stack within the configuration panel. Right-click on the DTLS stack. Select the *Parameters* menu. Within the Parameters menu, enable the *Mutual Authentication* option. Restart the suite for the changes to take effect. To capture and analyze DTLS traffic with mutual authentication, open Wireshark. Begin packet capture on the *NT* interface. Following the same steps as before, insert the *Hello World* message into the *dtls* layer within the FIRST stack. The message will be encrypted upon transmission. Upon receipt at the *dtls* layer in the *SECOND* stack, the message will be decrypted.

Figure 5.81 shows the transmitted DTLS messages between the two stacks when mutual authentication is enabled. Frame number 15 carries the server certificate generated by the *dtlsx* layer in the *SECOND* stack while frame number 17 carries the client certificate generated by the *dtls* layer in the *FIRST* stack. The session is established after server and client certificates are respectively verified against the CA by the client and the server. Frame number 20 carries the actual encrypted and authenticated *Hello World* message. Figure 5.81 shows the DTLS messages exchanged between the two stacks with mutual authentication enabled. These are some key observations. Frame 15 contains the server certificate, generated by the *dtlsx* layer in the SECOND stack. Frame 17 contains the client certificate, generated by the *dtls* layer in the FIRST stack. Session establishment occurs after both server and client certificates are successfully verified against the CA by their

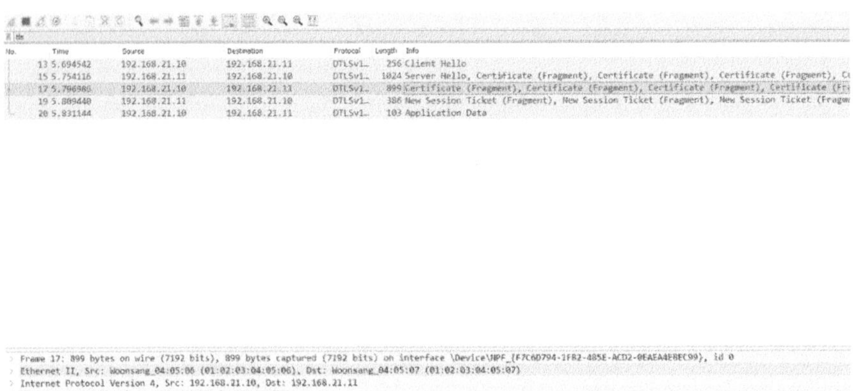

Figure 5.81 Captured DTLS Packets with Mutual Authentication.

respective parties. Frame 20 carries the actual encrypted and authenticated *Hello World* message. This figure demonstrates the secure communication established through mutual authentication in DTLS.

5.3.5 SRTP

SRTP is a protocol designed to safeguard RTP traffic through encryption and authentication (Carrara et al. 2004). It protects the confidentiality and integrity of RTP data. It is negotiated and used alongside SIP for real-time multimedia sessions. Master keys for encryption and authentication are exchanged within the SDP of the SIP INVITE and 200 OK messages. Typically, SIP messages are themselves encrypted and authenticated using TLS (over TCP) or DTLS (over UDP). However, for this section, use unencrypted SIP to clearly observe the SRTP negotiation process. Reload the *sipRtp* project that was created in Section 5.3.1. Save this project with a new name, sipSrtp. Proceed with configuring SRTP within this project.

To implement SRTP, replace the existing RTP layers with SRTP layers. Begin by detaching the following layers: *tts*, *audio*, and *qos*. Proceed to remove the *rtp* and *rtpx* layers entirely. Figure 5.82 illustrates these steps. Refer to the figure for guidance during the removal process.

To create and position the SRTP layers, click the beige-colored SRTP layer button (refer to Figure 5.83 for reference). Create two SRTP layers. Name the layers *srtp* and *srtpx*, respectively. Position the srtp layer above the *udp2* layer. Place the *srtpx* layer above the *udp2x* layer. Reattach the following layers to their original positions: *tts*, *audio*, and *qos*. Figure 5.84 shows the *FIRST* and *SECOND* stacks with the newly added SRTP layers correctly positioned. Refer to the figure to ensure accurate placement.

To establish the connection between SRTP and SIP layers, access the SIP layer configuration. Locate the *sip* layer within the configuration panel. Right-click on the *sip* layer. Select the *Parameters* menu. Choose the *Add Rtp Interfaces* option. In the list of RTP layers, add the *srtp* layer. Repeat the process for the *sipx* layer:

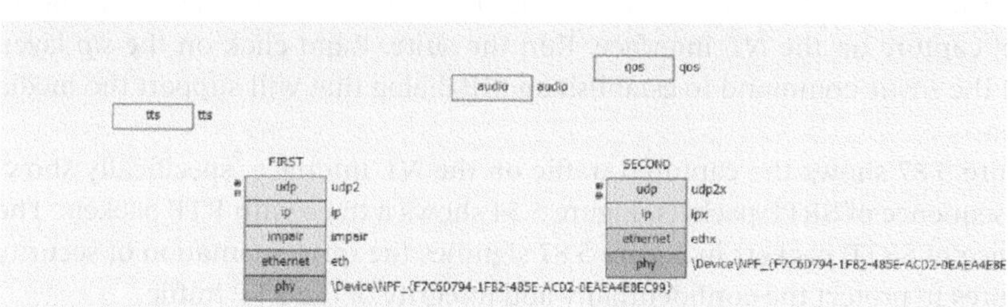

Figure 5.82 Detaching and Removing Protocol Layers.

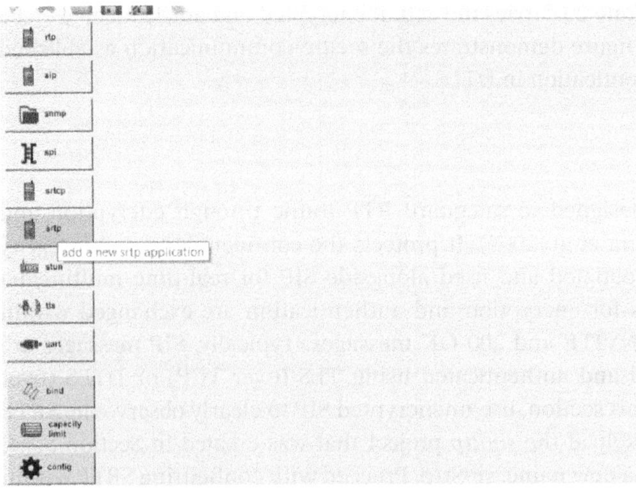

Figure 5.83 SRTP Layer Creation.

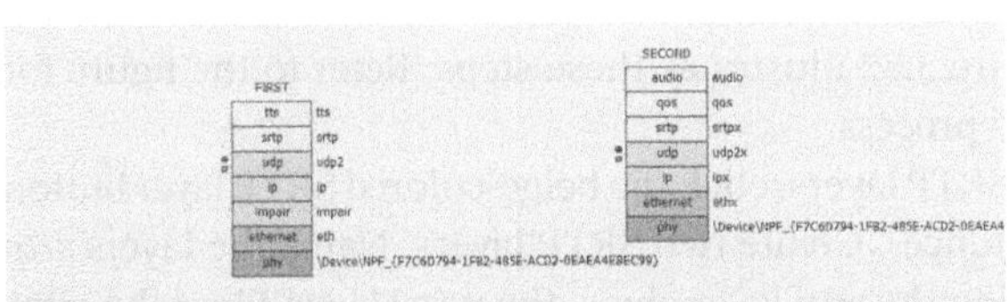

Figure 5.84 SRTP Protocol Stacks.

Right-click on the *sipx* layer. Access the *Parameters* menu. Select *Add Rtp Interfaces*. Add the *srtpx* layer to the list of RTP layers associated with the *sipx* layer. Figure 5.85 provides a representation of these steps. Refer to the figure for guidance during the association process.

To assign a codec to the RTP stacks and initiate an SIP dialog access the *Parameters* (shown in Figure 5.86) menu within the configuration panel. Right-click on the *Add Codecs* option. Choose the 64-Kbps *pcmu* (or ITU-T G.711 μ-law) codec. Open Wireshark. Set the filter to *sip* —— *rtp* to capture SIP and RTP traffic. Begin packet capture on the *NT* interface. Run the suite. Right-click on the *sip* layer. Select the *Invite* command to establish an SIP dialog that will support the media session.

Figure 5.87 shows the captured traffic on the *NT* interface, specifically showing a sequence of SRTP packets. Figure 5.54 shows a trace with RTP packets. The presence of SRTP packets in Figure 5.87 signifies the implementation of security measures to protect the confidentiality and integrity of the RTP traffic.

Figure 5.88 demonstrates the SDP contained within the SIP INVITE message. It incorporates a crypto media attribute that designates the cipher suite

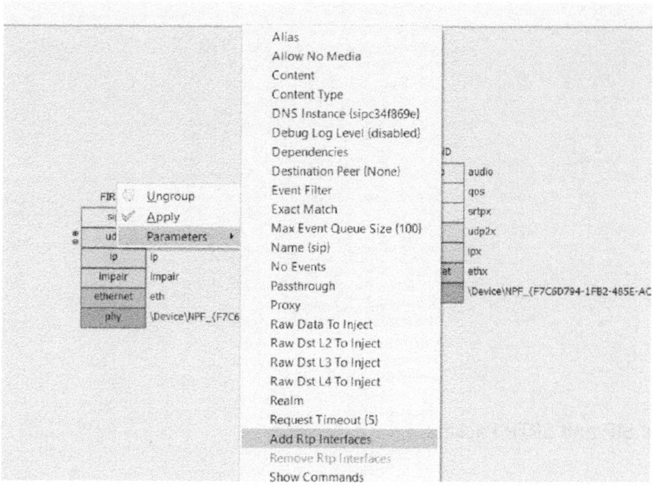

Figure 5.85 Add RTP Interfaces Option.

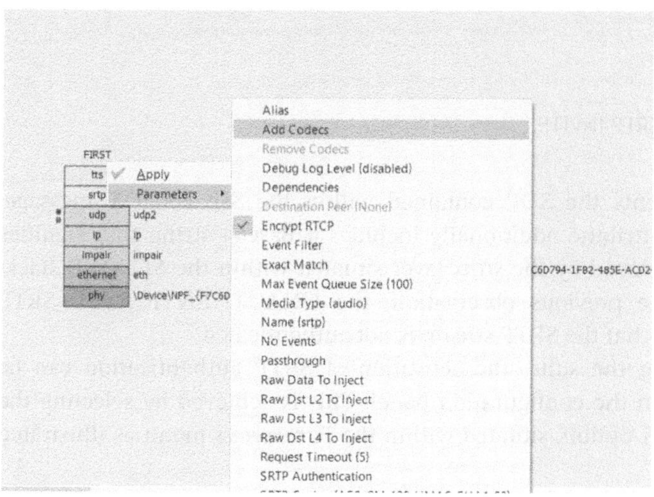

Figure 5.86 Add Codecs Option.

AES_CM_128_HMAC_SHA1_80, which offers AES encryption in Counter Mode (CM) and SHA1 authentication. It is important to note that the master key of the *srtp* layer within the *FIRST* stack is encoded as a Base64 string. The presence of the keyword UNAUTHENTICATED_SRTP signifies that the SRTP stream lacks authentication. This is attributable to the fact that the *SRTP Authentication* option within the *Parameters* menu, as shown in Figure 5.86, is not enabled.

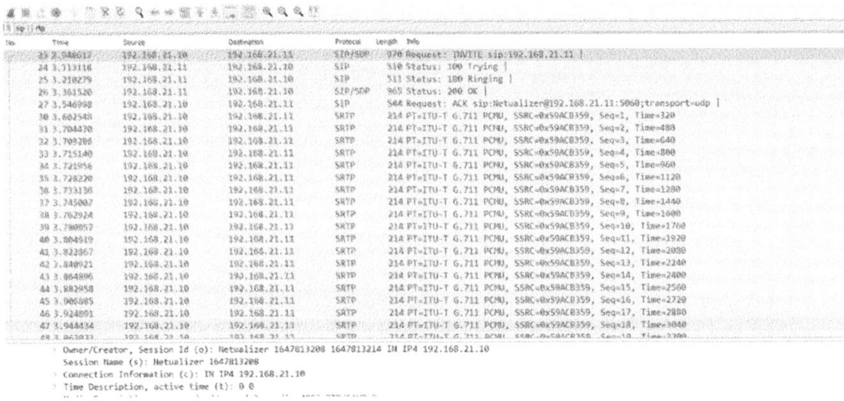

Figure 5.87 Captured SIP and SRTP Packets.

Figure 5.88 SDP of SRTP INVITE.

Figure 5.89 presents the SDP contained within the SIP 200 OK message. The *crypto* media attribute additionally includes a Base64 string that signifies the master key employed by the *srtpx* layer situated within the *SECOND* stack. Consistent with the previous observation, the UNAUTHENTICATED_SRTP keyword designates that the SRTP stream is not authenticated.

Upon deactivating the suite, the activation of SRTP authentication can be accomplished within the configuration panel. This is achieved by selecting the *SRTP Authentication* option, situated within the *Parameters* menu as illustrated in Figure 5.90.

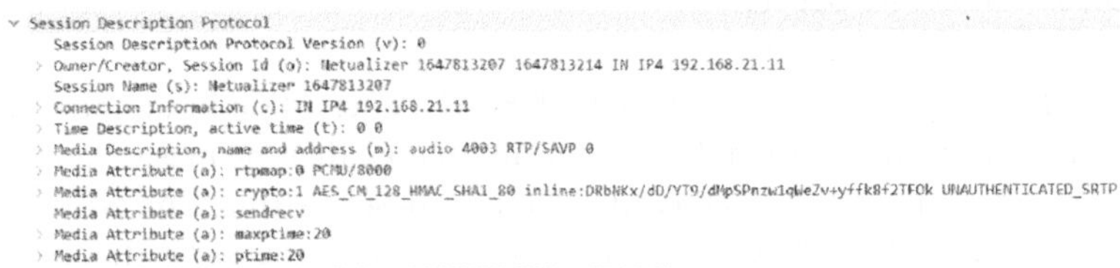

Figure 5.89 SDP of SRTP 200 OK.

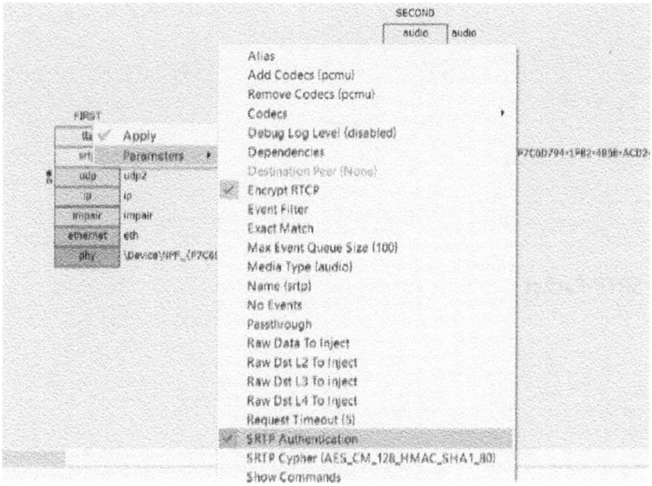

Figure 5.90 SRTP Authentication Option.

Restart the suite and establish a new session between the *FIRST* and *SECOND* stacks. In this case, the captured traffic shows that the SIP INVITE and 200 OK SDPs, respectively shown in Figures 5.91 and 5.92, do not exhibit the UNAUTHENTICATED_SRTP keyword anymore. Reinitiate the communication session and establish a new session between the *FIRST* and *SECOND* stacks. Upon examination of the captured traffic, it is evident that the SIP INVITE and 200 OK SDPs,

Figure 5.91 SDP of SRTP INVITE with Authentication.

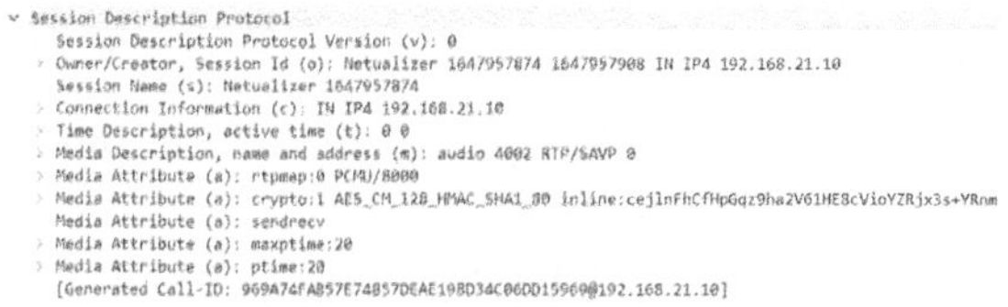

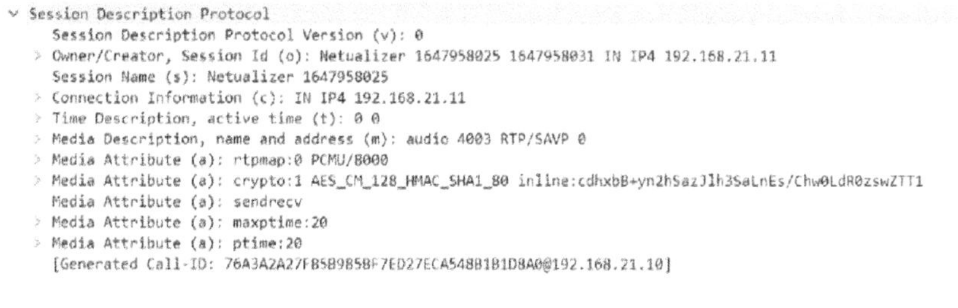

Figure 5.92 SDP of SRTP 200 OK with Authentication.

```
˅ Real-Time Transport Protocol
  > [Stream setup by SDP (frame 3)]
    10.. .... = Version: RFC 1889 Version (2)
    ..0. .... = Padding: False
    ...0 .... = Extension: False
    .... 0000 = Contributing source identifiers count: 0
    0... .... = Marker: False
    Payload type: ITU-T G.711 PCMU (0)
    Sequence number: 7897
    [Extended sequence number: 73433]
    Timestamp: 1263680
    Synchronization Source identifier: 0x59acb359 (1504490329)
    SRTP Encrypted Payload: d150eb5768e7cb4d5fcd735de02939a74137f9ba0617aa215f1398efae6396bebe6d6d27…
    SRTP Auth Tag: b8349fbe87e7b6233cfa
```

Figure 5.93 Captured SRTP Packet.

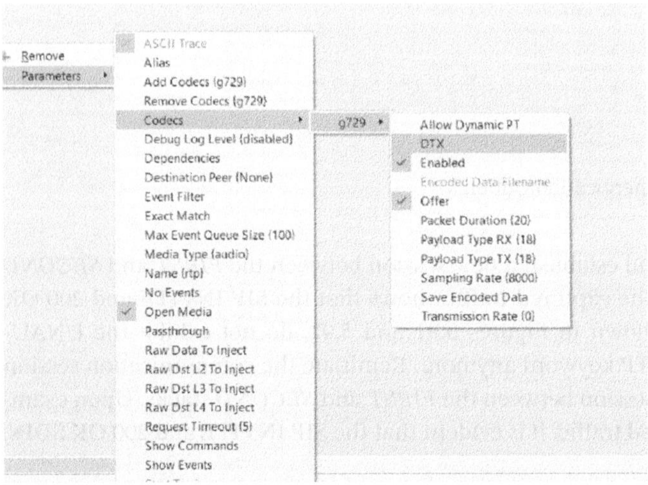

Figure 5.94 ITU-T Recommendation G.729 Configuration.

respectively shown in Figures 5.91 and 5.92, no longer exhibit the UNAUTHENTICATED_SRTP keyword.

The SRTP packets incorporate encrypted payloads in conjunction with authentication tags. Figure 5.93 offers an illustrative example, showcasing an SRTP packet that possesses a 10-byte authentication tag. It is important to note that the length of the authentication tag, expressed in units of bits, is explicitly indicated within the final numerical component of the cipher suite *AES_CM_128_HMAC_SHA1_80*, as specified within the SDP.

5.4 Codecs

Netualizer possesses the capability to accommodate a multiplicity of codecs, the configuration of which can be undertaken as shown within Figure 5.86.

The enablement of audio and speech codecs requires verification that the *Media Type* option has been designated as *audio*. It is generally the case that any codec that has been introduced within Section 1.4 can be selected, as shown by Figure 5.46.

In the instance of ITU-T Recommendation G.729 being designated as the codec of choice (ITU-T 1996), the process of accessing and subsequently modifying its configuration parameters requires the following sequence of actions shown in Figure 5.94: Firstly, one must navigate to the *Parameters* menu, whereupon the *Codecs* option should be selected. This action will grant the user the capability to configure a predetermined set of parameters, inclusive of DTX, packet duration, sampling rate, and payload types.

Figure 5.95 presents a scenario in which AMR-WB has been selected as the codec of choice (3GPP 2008). It is important to note that this particular codec, by virtue of its heightened complexity, offers a more expansive array of options pertaining to fine-tuning. These options include, but are not restricted to, IETF RFC 4867 standard support, sampling rate selection, robust-sorting, and additional parameters that are elucidated within Section 3.5.1.1.

To enable the utilization of video codecs, one must first execute a modification of the *Media Type* field, designating it as *Video*. Upon completion of this action, a comprehensive enumeration of the available video codecs will be presented within the graphical interface, as illustrated in Figure 5.96. This enumeration includes a substantial proportion of the video codecs that have been previously introduced within Section 1.4.3.

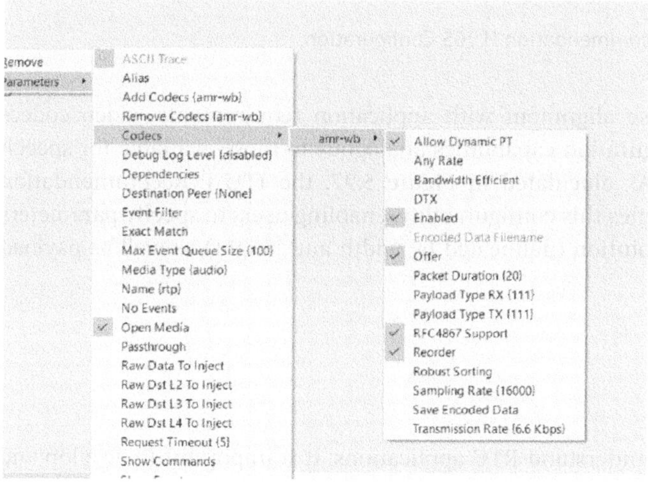

Figure 5.95 AMR-WB Configuration.

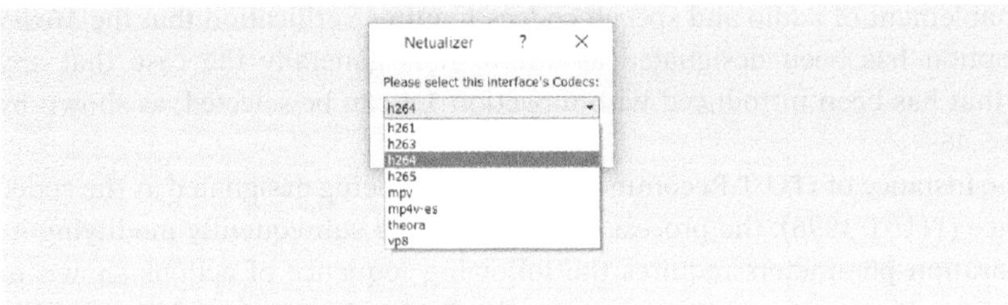

Figure 5.96 Selection of Video Code.

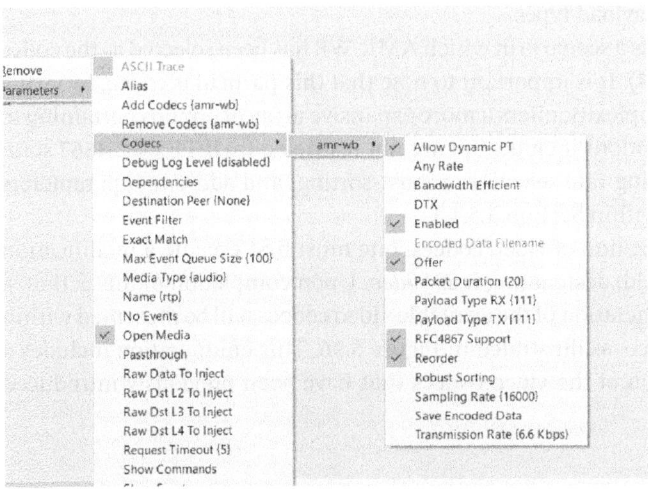

Figure 5.97 ITU-T Recommendation H.265 Configuration.

To facilitate precise alignment with application requirements, video codecs offer granular configuration capabilities analogous to those available for speech and audio codecs. As elucidated in Figure 5.97, the ITU-T Recommendation H.265 codec exemplifies this configurability, enabling users to specify parameters including image resolution (delineated by width and height) as well as payload types (ITU-T 2019).

5.5 Summary

To better build and understand RTC applications, it is important to develop and deploy protocol stacks. Netualizer is a tool that enables developers to develop and deploy protocol stacks (or alternative phrasing as suggested). Wireshark, in turn, can be used to analyze the traffic generated by these protocol stacks. In non-3GPP networks, RTC relies on protocols of the Internet suite that have been traditionally used in the context of LAN topologies. Ethernet supports encapsulating IPv4 and IPv6 datagrams at its link layer. In turn, RTP and SIP protocols can be encapsulated over UDP segments such that codecs like ITU-T Recommendation G.711 can be negotiated. With RTC mechanisms in place, media quality can be evaluated to obtain MOS metrics by means of the PESQ algorithm.

To support security in non-3GPP networks, both DTLS and TLS can be used respectively in the context of UDP and TCP transport. For encryption and authentication in RTP sessions, SRTP can be used instead. SRTP relies on SDES signaling to exchange master and salting keys. In addition to speech, other media types like audio and video can be encapsulated over RTP and SRTP to support RTC.

5.6 Homework Problems and Questions

Problem 5.1 What is the average RTT obtained from the ICMP datagrams that are seen in Figure 5.20? What is the effective transmission rate in each direction?

Problem 5.2 Repeat Problem 5.1 above but look at the ICMPv6 datagrams instead. Specifically, refer to Figure 5.23.

Problem 5.3 What is the actual payload of the UDP segment in Figure 5.33? How long is it?

Problem 5.4 How long does it take for the TCP connection in Figure 5.38 to get established?

Problem 5.5 How long does it take for the SIP session in Figure 5.54 to get established? What is the effective transmission rate in each direction?

Problem 5.6 From the receiver perspective, what is the throughput of the incoming RTP traffic in Figure 5.54? What is actual goodput?

Problem 5.7 How long does it take for the TCP connections and the TLS sessions in Figures 5.75 and 5.77 to get established?

Problem 5.8 How long does it take for the SIP session in Figure 5.87 to get established? What is the effective transmission rate in each direction?

Problem 5.9 From the receiver perspective, what is the throughput of the incoming RTP traffic in Figure 5.87? What is actual goodput?

5.7 Lab Exercises

Problem 5.10 Build a scenario in Netualizer with the following characteristics:

- Two Stacks: *EP1* and *EP2*
- Physical Layer/Link Layer: Ethernet (NT Interface)
- Client IP address: 192.168.21.10
- Server IP address: 192.168.21.12
- Session Layer: SIP/RTP (over UDP)
- Codec: ITU-T Recommendation G.711
- Text-To-Speech on EP1: *Hello Everyone*
- Audio Player on EP2

Capture traffic using Wireshark and place a call from EP2 to EP1. Let the media run for 1 minute. Answer the following:

1. How many RTP packets can be seen?
2. How many SIP packets can be seen?
3. What is the SIP throughput and goodput in each direction?
4. What is the RTP throughput and goodput in each direction?

Problem 5.11 Introduce an impairment layer between the UDP and the RTP layer in the EP1 stack. Set the transmission loss to 20%. Let the media run for 1 minute.
Answer the following:

1. How many RTP packets can be seen?
2. How many SIP packets can be seen?
3. What is the SIP throughput and goodput in each direction?
4. What is the RTP throughput and goodput in each direction?

Bibliography

3GPP. (2008). TS 26.190: Speech codec speech processing functions; adaptive multirate-wideband (AMR-WB) speech codec; transcoding functions. Technical report TS 26.190, 3rd Generation Partnership Project.

Amazon Corporation. AWS: Amazon web services. Available from: https://aws.amazon.com. (Accessed March 15th, 2024)

Carrara, E., Norrman, K., McGrew, D., Naslund, M., and Baugher, M. (2004). The secure real-time transport protocol (SRTP). RFC 3711, March. Available from: https://www.rfc-editor.org/info/rfc3711. (Accessed March 15th, 2024)

Dr., Deering, S.E. and Hinden, B. (2017). Internet protocol, version 6 (IPv6) specification. RFC 8200, July. Available from: https://rfc-editor.org/rfc/rfc8200.txt. (Accessed March 15th, 2024)

Graziani, R. (2012). IPv6 fundamentals: a straightforward approach to understanding IPv6. Pearson Education.

Herrero, R. and Cadirola, M. (2014). Effect of FEC mechanisms in the performance of low bit rate codecs in lossy mobile environments. In Proceedings of the Conference on Principles, Systems and Applications of IP Telecommunications, IPTComm '14, New York, NY, USA: Association for Computing Machinery. ISBN 978-1-450-32124-2. DOI: 10.1145/2670386.2670387. Available from: https://doi.org/10.1145/2670386.2670387. (Accessed March 15th, 2024)

IEEE. (2018). IEEE standard for Ethernet. IEEE Std 802.3-2018 (Revision of IEEE Std 802.3-2015), pp. 1–5600.

Ierusalimschy, R. (2006). *Programming in Lua*.

ITU-T. (1988). G.711: Pulse code modulation (PCM) of voice frequencies.

ITU-T. (1996). G.729: Coding of speech at 8 kbit/s using conjugate-structure algebraic-code-excited linear prediction (CS-ACELP).

ITU-T. (2019). H.265: High efficiency video coding.

L7TR. Netualizer: network virtualizer. Available from: https://www.l7tr.com. (Accessed March 15th, 2024)

Microsoft Corporation. Azure. Available from: https://azure.microsoft.com. (Accessed March 15th, 2024)

RFC. (1980). User datagram protocol. RFC 768, August. Available from: https://www.rfc-editor.org/info/rfc768. (Accessed March 15th, 2024)

RFC. (1981a). Internet protocol. RFC 791, September. Available from: https://www.rfc-editor.org/info/rfc791. (Accessed March 15th, 2024)

RFC. (1981b). Internet control message protocol. RFC 792, September. Available from: https://www.rfc-editor.org/info/rfc792. (Accessed March 15th, 2024)

RFC. (1981c). Transmission control protocol. RFC 793, September. Available from: https://www.rfc-editor.org/info/rfc793. (Accessed March 15th, 2024)

Schooler, E., Rosenberg, J., Schulzrinne, H., Johnston, A., Camarillo, G., Peterson, J., Sparks, R., and Handley, M.J. (2002). SIP: session initiation protocol. RFC 3261, July. Available from: https://rfc-editor.org/rfc/rfc3261.txt. (Accessed March 15th, 2024)

Schulzrinne, H., Casner, S.L., Frederick, R., and Jacobson, V. (2003). RTP: a transport protocol for real-time applications. RFC 3550, July. Available from: https://rfc-editor.org/rfc/rfc3550.txt. (Accessed March 15th, 2024)

Sethi, P. and Sarangi, S.R. (2017). Internet of things: architectures, protocols, and applications. *Journal of Electrical and Computer Engineering*, 2017:9324035, January. ISSN 2090-0147. DOI: 10.1155/2017/9324035. Available from: https://doi.org/10.1155/2017/9324035. (Accessed March 15th, 2024)

Wireshark. Wireshark: Network analyzer. Available from: https://www.wireshark.org. (Accessed March 15th, 2024)

6

3GPP Networks

6.1 Introduction

LTE and 5G are two widely used cellular network technologies, defined by the 3GPP standard [Section 4.3 (Cox 2014; 2020; Herrero 2021)]. LTE and 5G utilize VoLTE and VoNR, respectively, to enable RTC applications. Both technologies encapsulate SIP and RTP traffic over their respective link layers. The LTE link layer is divided into three sublayers: the MAC sublayer, the RLC sublayer, and the PDCP sublayer. 5G introduces an additional upper sublayer called SDAP.

Similar to other technologies covered in this book, Netualizer (L7TR) supports the SIM7600 hardware network adapter showcased in Figure 6.1 (SIMCom). This module features an M.2 connector that, coupled with a USB adapter, enables LTE network support and integration with embedded devices, Raspberry Pi, and PC-based agents. Connecting a Netualizer agent to a SIM7600G network interface via USB allows the *SIM7600G+LTE* option to appear in the Netualizer agent configuration panel, alongside the physical layer selection. Choosing this option activates LTE connectivity for the agent.

To streamline the process and generate LTE traffic without a physical network interface, enable the *Virtual Hardware Support* option located within the *Agents* menu (refer to Figure 7.3). The virtual hardware support facilitates the deployment of specialized scenarios without requiring physical network adapters. This capability enables the initial setup of virtual networking scenarios, which can then be seamlessly transitioned to hardware adapters like SIM7600G when physical LTE network interfaces become available. Note that the instructions provided for LTE network deployment in this chapter are equally applicable to 5G network deployment.

Figure 6.2 illustrates the topology central to this chapter's discussion. Due to LTE's native support for IP networking, the scenario employs a protocol stack where RTC protocols, namely SIP and RTP, handle media transport. Both SIP and RTP are specifically conveyed over UDP and IPv4. The overarching objective of

Hands-On Real Time Communications: A Practical Guide to RTC Protocols in Non-3GPP, 3GPP, 4G/5G/6G and IoT Networks, First Edition. Rolando Herrero.

Companion website: www.wiley.com/go/herrero/RTCProtocols

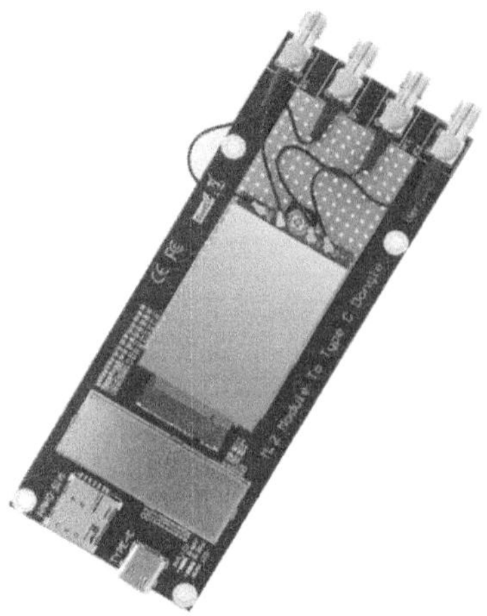

Figure 6.1 SIM7600G-based Radio.

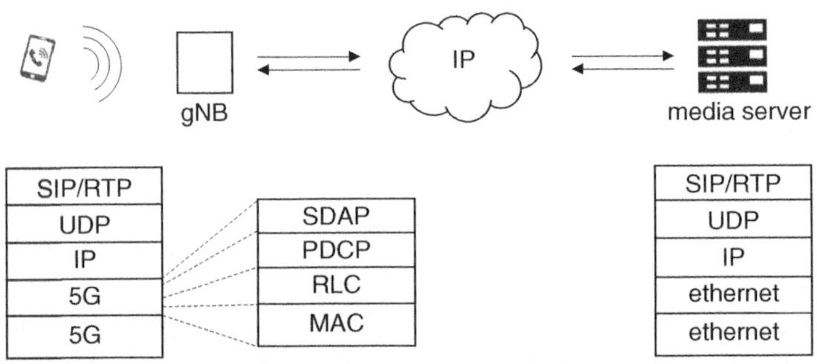

Figure 6.2 LTE Topology under Consideration.

this chapter is to establish UDP protocol stacks within Netualizer, forming the foundation for enabling RTC protocols that can function seamlessly in both LTE networks and traditional non-3GPP networks, as outlined in Chapter 5.

6.2 Lower Layers

6.2.1 Physical/Link

The following are the steps to create a new LTE Netualizer project. Access the *File* menu and select the *New Project* option. Enter *lte* as the project's name. Ensure the checkbox for attaching the local agent is selected, as shown in Figure 6.3. Then proceed with the project creation process. Upon successful project creation, the agent configuration will be automatically incorporated into the project and displayed within the Netualizer configuration panel.

Figure 6.3 4G Project.

To configure the physical layer within the Netualizer configuration panel, initiate Physical Layer Creation: Locate and click the green *phy* button. Select Virtual 4G Layer: A window showcasing available network interfaces will emerge. From this list, carefully select the *Virtual 4G* layer, as illustrated in Figure 6.4. By following these concise steps, establish the virtual 4G layer as the foundation for the LTE network configuration within Netualizer.

Refer to the following steps to configuring the 4G link layer within Netualizer. First, position the Virtual 4G Layer: place the virtual 4G layer within the configuration layout, ensuring its alignment with Figure 6.5. Create the 4G Link Layer: Access the orange *4G* layer button and click it to establish the corresponding 4G link layer. Bear in mind that this link layer includes the functionalities associated

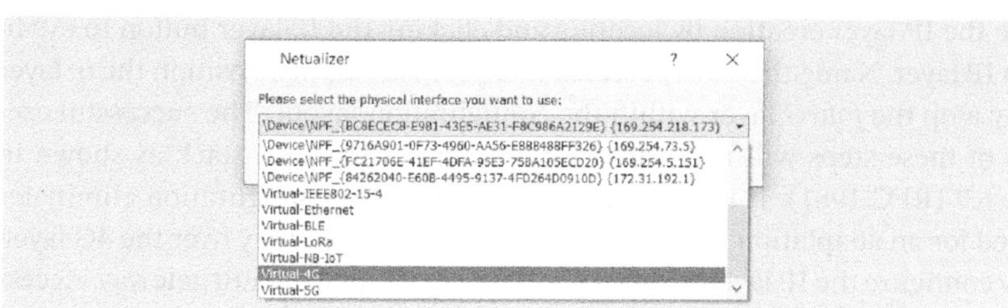

Figure 6.4 The Virtual 4G Interface.

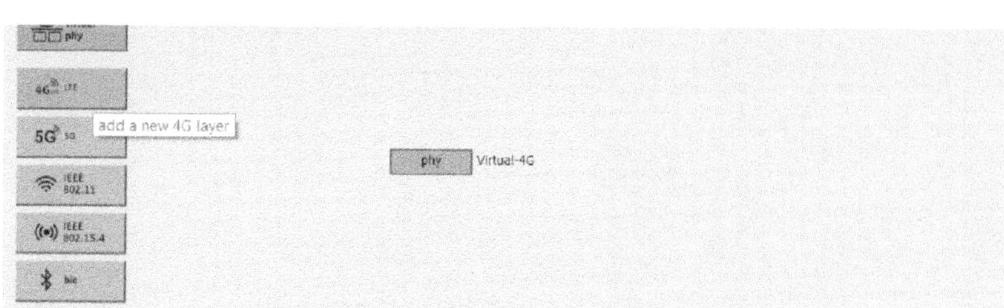

Figure 6.5 4G Link Layer Selection.

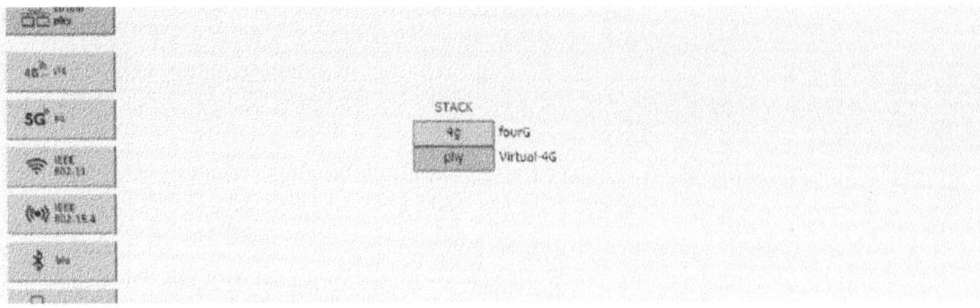

Figure 6.6 4G 2-Layer Stack.

with the sublayers shown in Figure 6.2, namely, MAC, RLC, and PDCP. Note that the virtual 4G layer can be seamlessly substituted with a physical SIM7600G network adapter to facilitate 4G deployment in real-world scenarios.

Continue by assigning the link layer name. Specifically, set the name *fourG* to the 4G link layer. Position the *fourG* layer directly atop the physical layer within the configuration layout. The successful execution of these steps will result in the formation of the 2-layer stack as shown in Figure 6.6. The subsequent section will focus on the configuration of an IP layer, which can be seamlessly positioned atop the 4G layer without requiring intermediary adaptation mechanisms.

6.2.2 Network/Transport

Initiate the IP layer creation by locating and clicking the *IP* layer button to establish an IP layer. Name the newly created IP layer *ip* for clarity. Position the *ip* layer directly atop the *fourG* layer within the configuration layout. The successful execution of these steps will result in the formation of the IPv4 stack as shown in Figure 6.7 (RFC 1981). It is important to note that this configuration eliminates the need for an adaptation layer, enabling direct IP connectivity over the 4G layer.

Now configure the IP layer's static IP address, mask, and default gateway. Access the IP layer configuration by right-clicking the *ip* layer to reveal its configuration options. Very importantly, ensure that DHCP is disabled prior to proceeding with

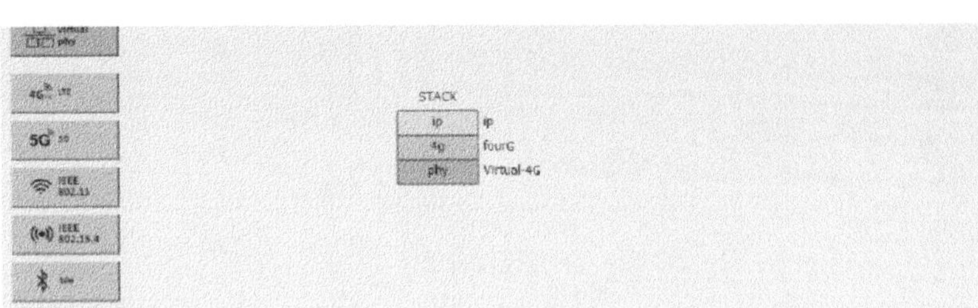

Figure 6.7 IP Stack.

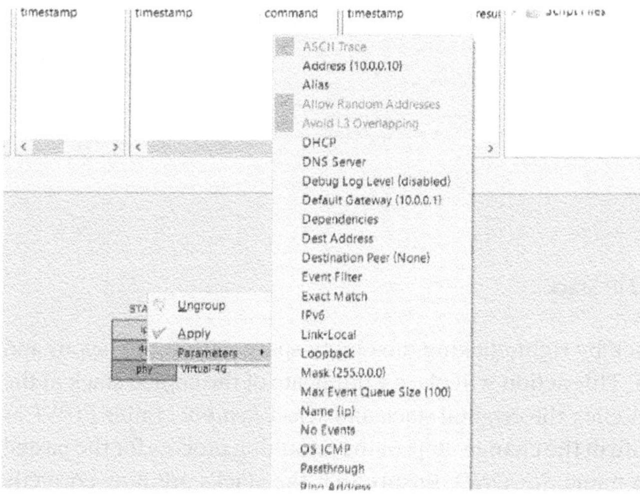

Figure 6.8 IP Stack Configuration.

manual configuration. Set static IP address by entering *10.0.0.10* as the IP address. Set the subnet mask by assigning *255.0.0.0* as the subnet mask. Set the default gateway by inputting *10.0.0.1* as the default gateway address. Refer to Figure 6.8 for the confirmation of these settings. Upon successful completion of these steps, the IP layer will be equipped with a static IP configuration, enabling direct communication within the network.

To create additional copies of the network stack to establish new endpoints, identify the stack to duplicate, referring to Figure 6.9 for guidance. Select the stack by clicking directly on the stack to show it. Locate the *Copy* option within the floating menu and click it. The execution of these steps will generate a precise duplicate of the entire network stack, including all layers and their respective configurations. This replicated stack, shown in Figure 6.10, can then be positioned at a distinct location within the layout to represent a separate endpoint within the network.

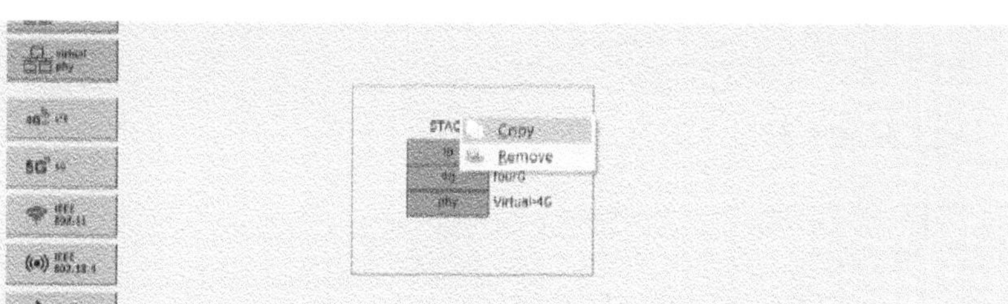

Figure 6.9 IP Stack Copy Selection.

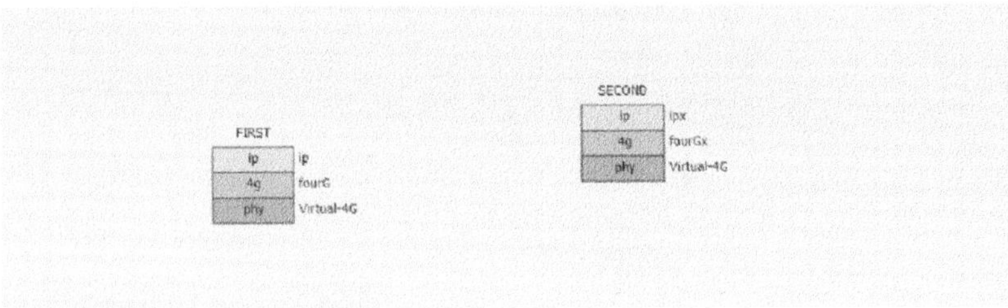

Figure 6.10 Duplicated IP Stack.

Paste the copied stack by right-clicking the empty space within the layout and select the *Paste* option. This action will place a duplicate of the copied stack at the chosen location. Right-click the original stack and select *Rename*. Enter *FIRST* as the new name and confirm the change. Repeat the renaming process for the pasted stack, assigning it the name *SECOND*. Ensure that the stacks are now correctly labeled as *FIRST* and *SECOND* within the layout. This process creates two distinct network stacks, each representing a separate endpoint within the topology.

To modify the IP address of the second stack, access the IP layer configuration by right-clicking the *ip* layer within the *SECOND* stack to reveal its configuration options. Navigate to the *Parameters* menu and find the *Address* field. Enter *10.0.0.11* as the new IP address. Apply the modification to establish the updated IP address for the *SECOND* stack. Refer to Figure 6.11 for confirmation of the correct configuration. By following these precise steps, differentiate the IP address of the *SECOND* stack, enabling distinct communication paths within the network.

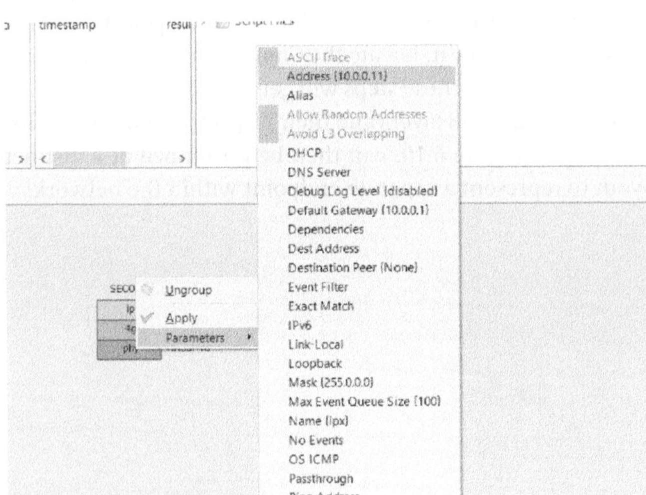

Figure 6.11 IP Address Changes.

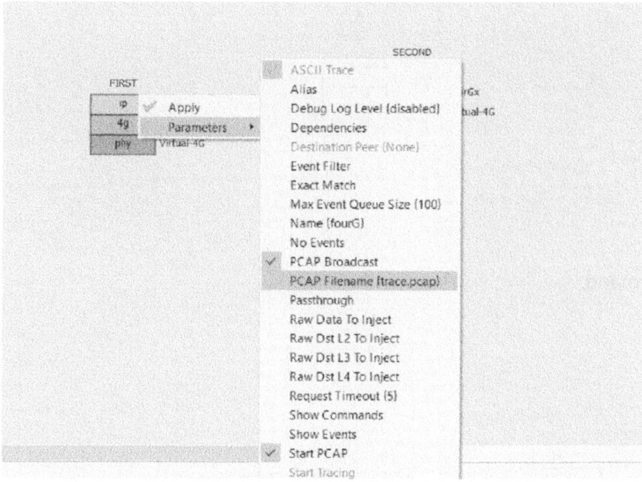

Figure 6.12 PCAP Capture Configuration.

Capture PCAP traces within the virtual 4G layer, as Wireshark is not compatible with virtual traffic (Wireshark). Choose the stack where to capture traffic. For this example, focus on the *FIRST* stack. Right-click the *ip* layer within the selected stack to reveal its configuration options. Navigate to the *Parameters* menu and find the section related to PCAP settings. Activate the *Start PCAP* option to initiate traffic capture. Specify *trace.cap* as the desired filename for the captured trace. This file will store the captured traffic data for subsequent analysis. Upon completion of these steps, the PCAP capture process will start within the virtual 4G layer, recording network traffic for the selected stack. Figure 6.12 shows this configuration menu within the *FIRST* stack, offering a reference for the process.

Execute the project and initiate traffic generation by locating the *Scripts* menu and click the *Run Suite* option. This action will deploy the crafted configuration onto the designated agent, enabling the network stacks to function as intended. Right-click the *ip* layer within the *FIRST* stack to reveal its available actions. Select the *Ping* option to initiate traffic generation, as shown in Figure 6.13. This action will trigger a series of ping packets to be transmitted across the network, enabling to observe and analyze network behavior. By following these concise steps, bring the configured network to life and establish the foundation for further experimentation and analysis within the virtual LTE environment.

Run the ping command and observe RTT measurements by specifying the destination address. Specifically, within the ping command prompt, enter *10.0.0.11* as the target address. This directs the ping packets specifically toward the *ip* layer within the *SECOND* stack. The *FIRST* stack will start the generation and transmission of ICMP echo requests, traversing the virtual 4G network adapter to reach

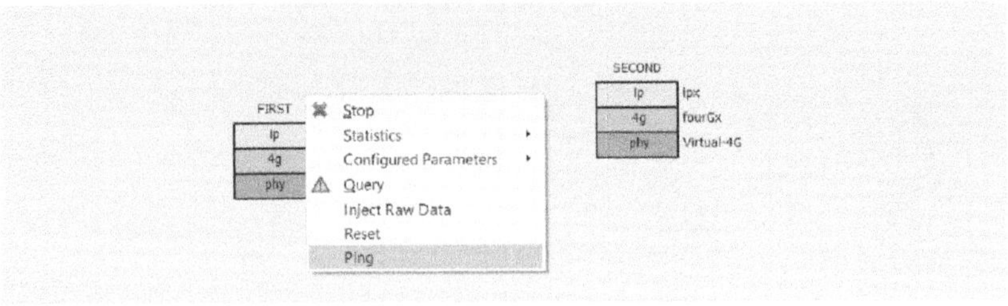

Figure 6.13 Ping Command.

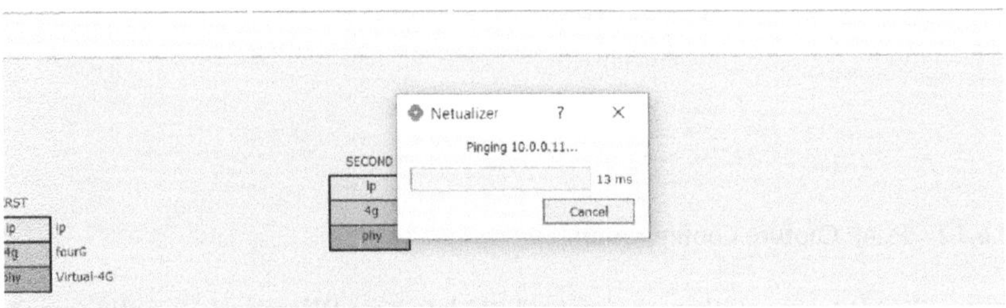

Figure 6.14 Ping Execution on *SECOND* Stack.

their intended destination. Upon successful arrival of the ICMP echo requests at the *ip* layer within the *SECOND* stack, the layer will respond by transmitting ICMP echo replies in return. Netualizer will measure and display the RTT associated with the ping communication, providing valuable insights into network latency and performance. Refer to Figure 6.14 for a representation of this RTT display. It is important to note that this ICMP traffic exchange can be replicated between any of the configured stacks, enabling comprehensive RTT estimations for various network paths within the virtual LTE environment.

Give the traffic a duration of approximately 30 seconds to traverse the network, enabling sufficient data collection for analysis. Navigate to the agent configuration panel and click the red cross button to halt the configuration's execution. Access the project folder, specifically the *Netualizer/Projects/4g* directory, and identify the *trace.cap* file, which contains the captured network traffic. Ensure that Wireshark, a powerful network analysis tool, is installed and readily accessible on the system. Double-click the *trace.cap* file to initiate its opening within Wireshark, granting the ability to show and inspect the captured network data in detail. Figure 6.15 offers a representation of the captured traffic within Wireshark, showcasing a sequence of ICMP requests and replies that have traversed the virtual 4G network. By carefully examining this trace, it is possible to get valuable insights into network behavior, performance characteristics, and potential issues, gaining a deeper understanding of the virtual LTE environment.

Example 6.1 Consider Figure 6.15, what is the transmission rate of the traffic originated at *10.0.0.11*?

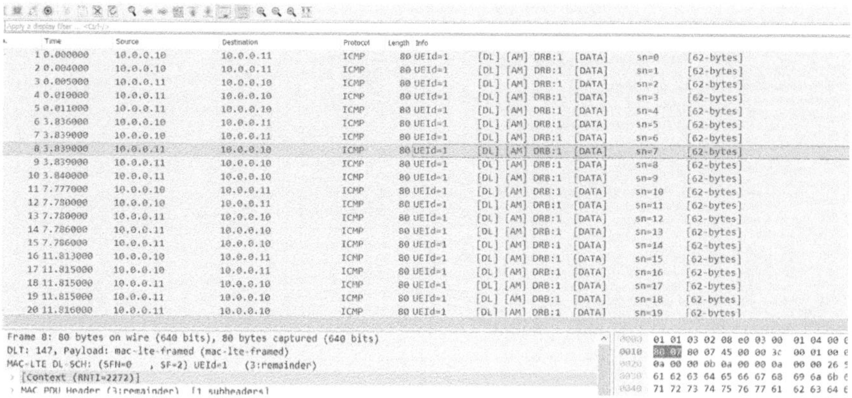

Figure 6.15 Captured ICMP Packets.

Solution:

There are 12 ICMP packets originated at *10.0.0.11* between timestamps 11.816 and 0.005 corresponding to 11.811 seconds. Since each packet is 80 bytes, the transmission rate is $R = \frac{80\times8}{11.811} = 54.19$ bps.

To configure Wireshark to accurately interpret 4G PCAP files, navigate to the *Edit* menu within Wireshark and select the *Preferences* option. Within the Preferences window, locate the *Protocols* option and click it. Choose *DLT_USER* from the available options. Access the *Encapsulation Table* and proceed to modify its settings. Locate the entry for *User 0 (DLT=147)* and assign the *mac-lte-framed* Wireshark dissector to it. This configuration ensures that Wireshark correctly interprets the structure of 4G PCAP files, enabling accurate analysis of their contents. Figure 6.16 provides a representation of a captured ICMP echo request packet within Wireshark, showcasing its transmission over various layers. The ICMP Echo Request is the core message initiating the ping request. It is encapsulated over IPv4 that, in turn, is encapsulated over PDCP, a key component of LTE for data handling. Wireshark also shows the RLC layer, tasked with managing data transmission reliability and the MAC layer, responsible for regulating access to the shared wireless medium. It is important to remember that these three LTE sublayers (PDCP, RLC, and MAC) collectively form the 4G link layer, providing the foundation for IP-based communication within the virtual LTE environment.

Halt the current configuration by navigating to the agent configuration panel and click the red cross button to pause the ongoing configuration. Click the *UDP* layer button to create two separate UDP layers (RFC 1980). Assign the names *udp*

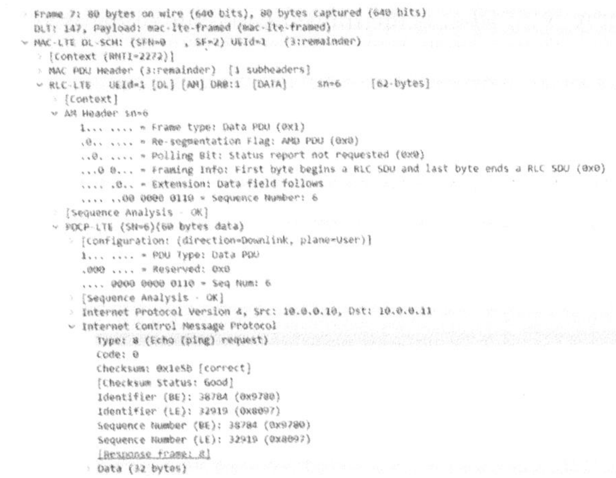

Figure 6.16 ICMP Echo Packets.

and *udpx* to these newly created layers for clarity. Position the *udp* layer directly atop the *ip* layer, and position the *udpx* layer directly on top of the *ipx* layer within the configuration layout. Access the configuration options for both UDP layers and set their respective transport ports to 4000. This ensures that they communicate on the same port, enabling seamless UDP-based data exchange. Navigate to the *Scripts* menu and click the *Run Suite* option to redeploy the updated configuration onto the agent. This action will activate the newly introduced UDP layers, preparing the stacks for UDP communication. Figure 6.17 likely shows a representation of this configuration, aiding in its accurate execution.

Proceed and access the injection option by right-clicking the *udp* layer within the *FIRST* stack to reveal its available actions. Choose the *Inject Raw Data* option, enabling to manually introduce a message into the UDP layer for transmission. Enter the text *Hello World!* as the content of the message to be sent. Set the destination IP Address to *10.0.0.11* (corresponding to the *ip* layer within the *SECOND* stack). Set the destination UDP port to 4000 (aligned with the configured port on the *udpx* layer). Upon completion of these steps, click the *Send*

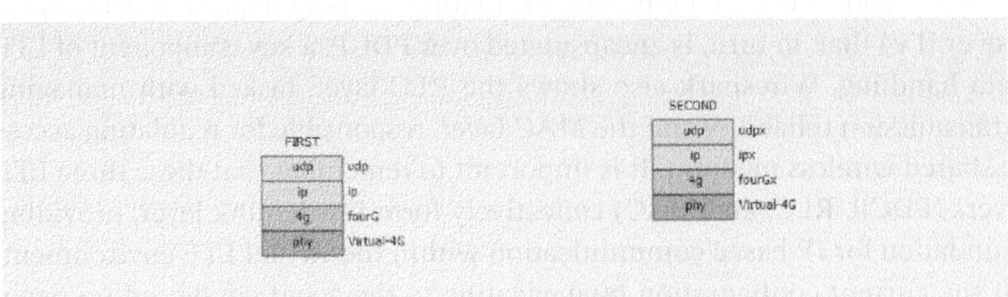

Figure 6.17 UDP Protocol Stacks.

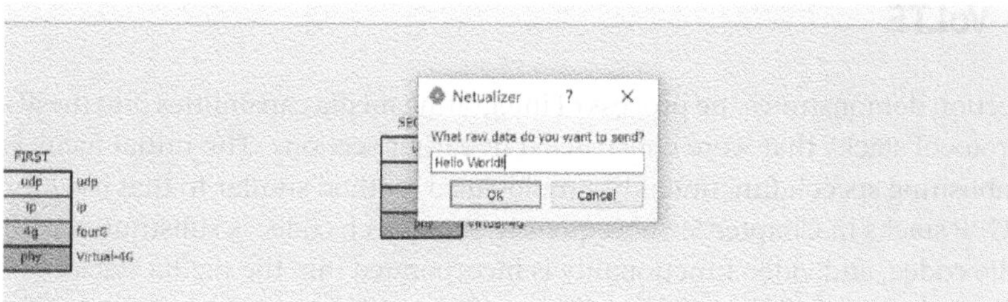

Figure 6.18 Raw Data Injection.

button (or equivalent action) to trigger the transmission of the *Hello World!* message across the virtual LTE network to the specified destination. Figure 6.18 likely provides a representation of this injection process, offering guidance for its execution.

Begin by halting the current configuration. This is typically achieved by clicking the red cross button within the agent configuration panel. Open the file named *trace.cap*, which contains the recorded network traffic. This file can be analyzed using tools such as Wireshark. Within the trace, identify the specific UDP segment that encapsulates the *Hello World!* message. Observe the layered structure of the captured traffic. The UDP segment, responsible for carrying the actual message, is contained within an IPv4 packet. IPv4, in turn, is encapsulated within the 4G link layer, which consists of the PDCP, RLC, and MAC sublayers. Note that IPv4 functions as a unifying interface, enabling the seamless integration of diverse protocols like UDP within the broader Internet suite. This is a key concept that has been emphasized throughout the book. Figure 6.19 provides a representation of this encapsulation structure, aiding in its comprehension.

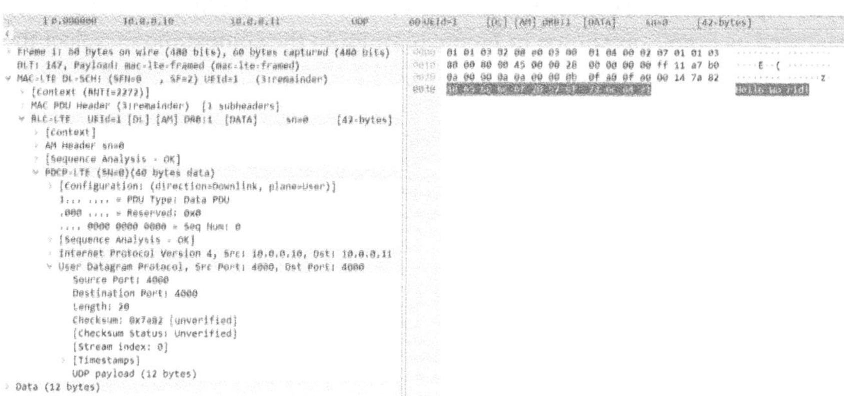

Figure 6.19 Captured UDP Packet.

6.3 VoLTE

This section demonstrates the process of integrating media capabilities into the 4G LTE protocol stacks that were constructed in earlier sections. The initial focus is on establishing speech functionality, employing a method similar to that used for non-3GPP stacks in Chapter 5. Subsequently, the speech codec is substituted with an audio codec, and video functionality is incorporated into the media stacks.

6.3.1 Speech

Follow the instructions outlined in Section 5.3 to incorporate SIP and RTP (Schulzrinne et al. 2003) layers into both stacks, enabling them to support RTC functionality. The objective is to establish a call from the *FIRST* stack to the *SECOND* stack. Start by adding two SIP (Schooler et al. 2002) layers, designated as *sip* and *sipx*, positioned directly above the existing *udp* and *udpx* layers, respectively. Proceed by creating two new UDP layers, named *udp2* and *udp2x*, and position them above the ip and ipx layers, respectively. Construct two RTP layers, designated as *rtp* and *rtpx*, configured with the *pcmu* ITU-T G.711 codec. Place these layers above the *udp2* and *udp2x* layers, respectively. Establish a link between the *sip* and *rtp* layers, and similarly, link the *sipx* and *rtpx* layers together. Build a TTS layer, named *tts*, and position it above the *rtp* layer. Configure the text associated with the *tts* layer as *Hello World!* and enable playback by selecting the Say option within the layer. Create an *audio* layer, designated as *audio*, and position it above the *rtpx* layer. Enable playback on the *audio* layer by selecting the *Play* option. Refer to Figure 6.20 for a representation of the completed RTC stacks.

To initiate traffic generation within the two configured stacks, it is imperative to first transfer the configuration from the configuration panel to the agent. Navigate to the *Scripts* menu and select the option labeled *Run Suite*. Right-click on the stack designated as *FIRST*, as shown in Figure 6.21. From the context menu, select the option labeled *New Dialpad*.

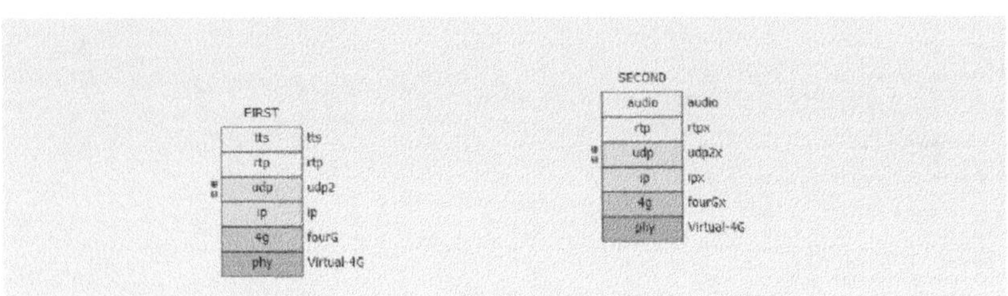

Figure 6.20 RTC 4G Stacks.

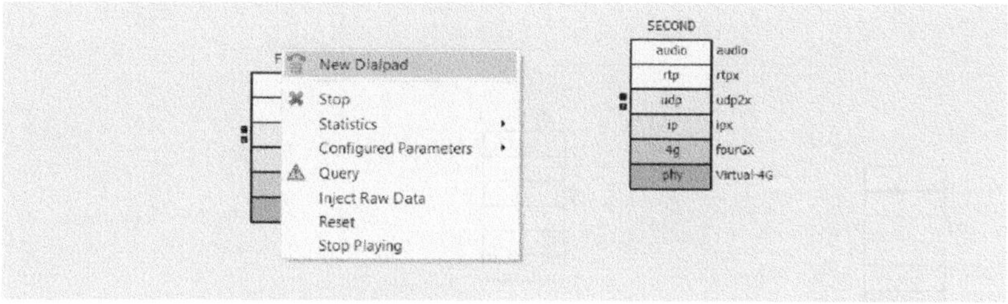

Figure 6.21 Netualizer Dialpad Option.

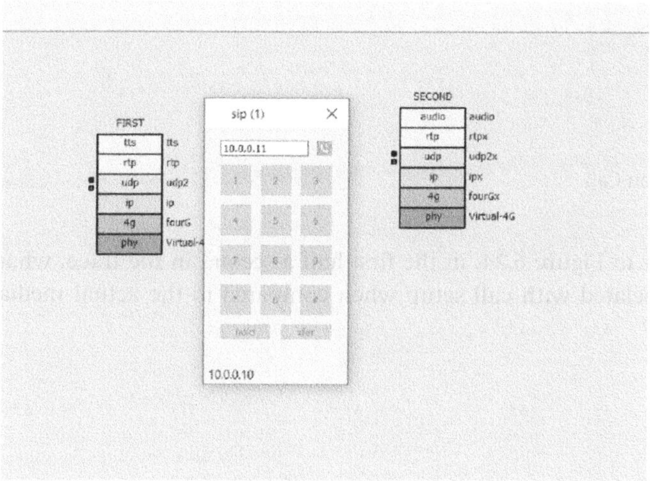

Figure 6.22 Idle Dialpad.

Figure 6.22 shows the dialpad that appears upon selecting the New Dialpad option. The Netualizer component automatically populates the dialpad edit box with the IP address corresponding to the *SECOND* stack.

Initiate a call from the *FIRST* stack to the *SECOND* stack by clicking the green dial button. Figure 6.23 shows the dialpad after the call is successfully established. Maintain the call for approximately 30 seconds, then press the red hang-up button to terminate the call.

Conclude the configuration process and proceed to open the *trace.cap* PCAP trace file. Refer to Figure 6.24 for a representation of the captured trace, which includes UDP segments responsible for transporting both SIP and RTP traffic. It is important to note that all traffic is encapsulated within IPv4, which is subsequently encapsulated within the 4G link layer.

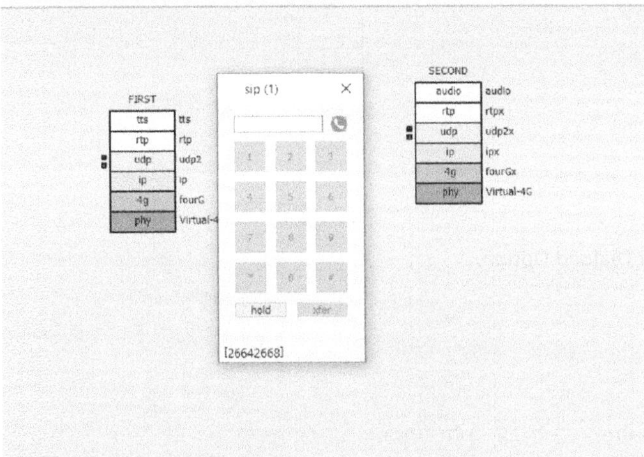

Figure 6.23 Dialpad on Call.

Example 6.2 Refer to Figure 6.24, in the first half a second in the trace, what is the overhead associated with call setup when compared to the actual media packetization traffic?

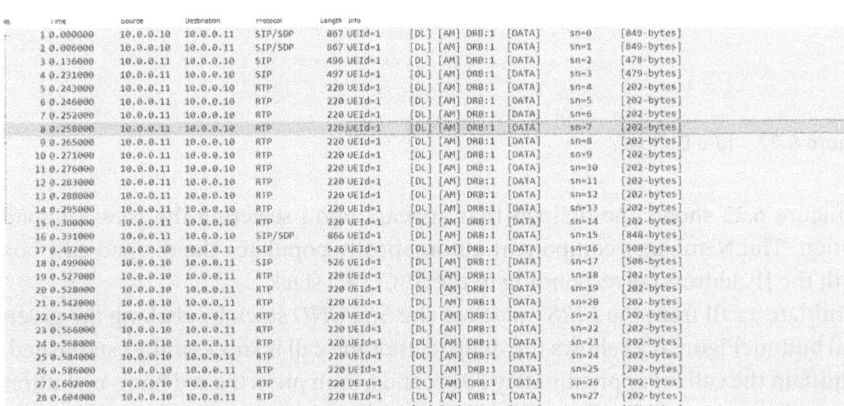

No.	Time	Source	Destination	Protocol	Length	Info
1	0.000000	10.0.0.10	10.0.0.11	SIP/SDP	867	UEId=1 [DL] [AM] DRB:1 [DATA] sn=0 [849-bytes]
2	0.006000	10.0.0.10	10.0.0.11	SIP/SDP	867	UEId=1 [DL] [AM] DRB:1 [DATA] sn=1 [849-bytes]
3	0.136000	10.0.0.11	10.0.0.10	SIP	496	UEId=1 [DL] [AM] DRB:1 [DATA] sn=2 [478-bytes]
4	0.231000	10.0.0.11	10.0.0.10	SIP	497	UEId=1 [DL] [AM] DRB:1 [DATA] sn=3 [479-bytes]
5	0.243000	10.0.0.11	10.0.0.10	RTP	220	UEId=1 [DL] [AM] DRB:1 [DATA] sn=4 [202-bytes]
6	0.246000	10.0.0.11	10.0.0.10	RTP	220	UEId=1 [DL] [AM] DRB:1 [DATA] sn=5 [202-bytes]
7	0.252000	10.0.0.11	10.0.0.10	RTP	220	UEId=1 [DL] [AM] DRB:1 [DATA] sn=6 [202-bytes]
8	0.258000	10.0.0.11	10.0.0.10	RTP	220	UEId=1 [DL] [AM] DRB:1 [DATA] sn=7 [202-bytes]
9	0.265000	10.0.0.11	10.0.0.10	RTP	220	UEId=1 [DL] [AM] DRB:1 [DATA] sn=8 [202-bytes]
10	0.271000	10.0.0.11	10.0.0.10	RTP	220	UEId=1 [DL] [AM] DRB:1 [DATA] sn=9 [202-bytes]
11	0.276000	10.0.0.11	10.0.0.10	RTP	220	UEId=1 [DL] [AM] DRB:1 [DATA] sn=10 [202-bytes]
12	0.283000	10.0.0.11	10.0.0.10	RTP	220	UEId=1 [DL] [AM] DRB:1 [DATA] sn=11 [202-bytes]
13	0.288000	10.0.0.11	10.0.0.10	RTP	220	UEId=1 [DL] [AM] DRB:1 [DATA] sn=12 [202-bytes]
14	0.295000	10.0.0.11	10.0.0.10	RTP	220	UEId=1 [DL] [AM] DRB:1 [DATA] sn=13 [202-bytes]
15	0.300000	10.0.0.11	10.0.0.10	RTP	220	UEId=1 [DL] [AM] DRB:1 [DATA] sn=14 [202-bytes]
16	0.331000	10.0.0.11	10.0.0.10	SIP/SDP	866	UEId=1 [DL] [AM] DRB:1 [DATA] sn=15 [848-bytes]
17	0.497000	10.0.0.10	10.0.0.11	SIP	526	UEId=1 [DL] [AM] DRB:1 [DATA] sn=16 [508-bytes]
18	0.499000	10.0.0.10	10.0.0.11	SIP	526	UEId=1 [DL] [AM] DRB:1 [DATA] sn=17 [508-bytes]
19	0.527000	10.0.0.10	10.0.0.11	RTP	220	UEId=1 [DL] [AM] DRB:1 [DATA] sn=18 [202-bytes]
20	0.528000	10.0.0.10	10.0.0.11	RTP	220	UEId=1 [DL] [AM] DRB:1 [DATA] sn=19 [202-bytes]
21	0.542000	10.0.0.10	10.0.0.11	RTP	220	UEId=1 [DL] [AM] DRB:1 [DATA] sn=20 [202-bytes]
22	0.543000	10.0.0.10	10.0.0.11	RTP	220	UEId=1 [DL] [AM] DRB:1 [DATA] sn=21 [202-bytes]
23	0.566000	10.0.0.10	10.0.0.11	RTP	220	UEId=1 [DL] [AM] DRB:1 [DATA] sn=22 [202-bytes]
24	0.568000	10.0.0.10	10.0.0.11	RTP	220	UEId=1 [DL] [AM] DRB:1 [DATA] sn=23 [202-bytes]
25	0.584000	10.0.0.10	10.0.0.11	RTP	220	UEId=1 [DL] [AM] DRB:1 [DATA] sn=24 [202-bytes]
26	0.586000	10.0.0.10	10.0.0.11	RTP	220	UEId=1 [DL] [AM] DRB:1 [DATA] sn=25 [202-bytes]
27	0.602000	10.0.0.10	10.0.0.11	RTP	220	UEId=1 [DL] [AM] DRB:1 [DATA] sn=26 [202-bytes]
28	0.604000	10.0.0.10	10.0.0.11	RTP	220	UEId=1 [DL] [AM] DRB:1 [DATA] sn=27 [202-bytes]

Figure 6.24 Captured RTC Traffic.

Solution:

In the trace, the call setup SIP traffic accounts for 849 + 849 + 478 + 479 + 848 + 508 + 202 = 4213 bytes. Similarly, the media packetization traffic accounts for eleven 202-byte packets. The overhead is about $\frac{4213}{202\times11} = 1.896$ or 189.6.

```
> Frame 8: 220 bytes on wire (1760 bits), 220 bytes captured (1760 bits)
  DLT: 147, Payload: mac-lte-framed (mac-lte-framed)
v MAC-LTE DL-SCH: (SFN=0    , SF=2) UEId=1    (3:remainder)
  > [Context (RNTI=2272)]
  v MAC PDU Header (3:remainder)  [1 subheaders]
    > Sub-header (lcid=3, length is remainder)
  v RLC-LTE   UEId=1 [DL] [AM] DRB:1  [DATA]       sn=7       [202-bytes]
    > [Context]
    > AM Header sn=7
    > [Sequence Analysis - OK]
    > PDCP-LTE (SN=7)(200 bytes data)
v Real-Time Transport Protocol
  > [Stream setup by SDP (frame 1)]
    10.. .... = Version: RFC 1889 Version (2)
    ..0. .... = Padding: False
    ...0 .... = Extension: False
    .... 0000 = Contributing source identifiers count: 0
    0... .... = Marker: False
    Payload type: ITU-T G.711 PCMU (0)
    Sequence number: 3
    [Extended sequence number: 65539]
    Timestamp: 480
    Synchronization Source identifier: 0x95c32b85 (2512595845)
    Payload: ffffffffffffffffffffffffffffffffffffffffffffffffffffffffffffffffffffff83
```

Figure 6.25 Captured RTP Packet.

Figure 6.25 shows a sample RTP with sequence number 3. This packet is transported over a UDP layer, which is in turn encapsulated over IPv4. The IPv4 layer is further encapsulated over the 4G link layer, which comprises the PDCP, RLC, and MAC sublayers. It is important to note that the payload contains the actual ITU-T G.711 speech frames.

6.3.2 Audio and Video

To enable audio functionality on the *SECOND* RTP layer, right-click on the *SECOND* RTP layer. Select the option to add the EVS codec, as demonstrated in Figure 6.26. It is important to note that this stack now supports two distinct media codecs. The speech ITU-T G.711 codec (ITU-T 1988) and the audio EVS codec (3GPP 2018). The objective is to configure the EVS codec on the *FIRST* stack as well. This ensures that EVS, being the only codec common to both stacks, becomes the negotiated codec during call establishment.

Figure 6.27 illustrates the EVS codec configuration that has been optimized to deliver the highest possible audio quality. This optimization involves setting the *Transmission Rate* parameter to its maximum value of 96 Kbps and adjusting the *Sampling Rate* parameter to 48000 sps (3GPP 2018).

To configure the FIRST stack to utilize the EVS codec, locate the *rtp* layer within the *FIRST* stack. Right-click on the *rtp* layer to reveal its configuration options. Remove the currently assigned *pcmu* ITU-T G.711 codec. Select the *evs* codec as its replacement. Refer to Figure 6.28 for guidance throughout this process.

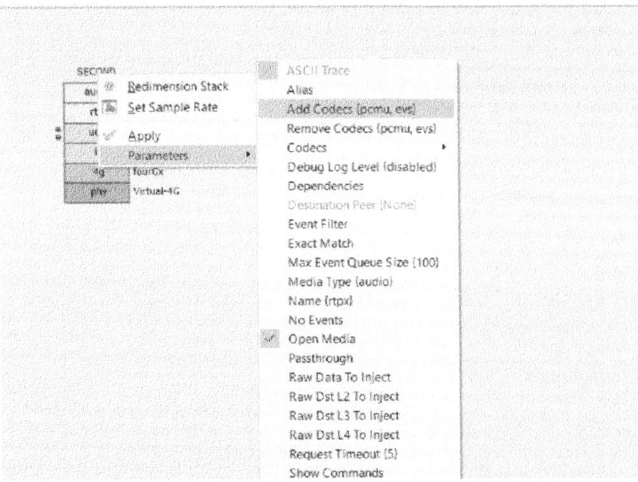

Figure 6.26 *SECOND* Stack EVS Codec Support.

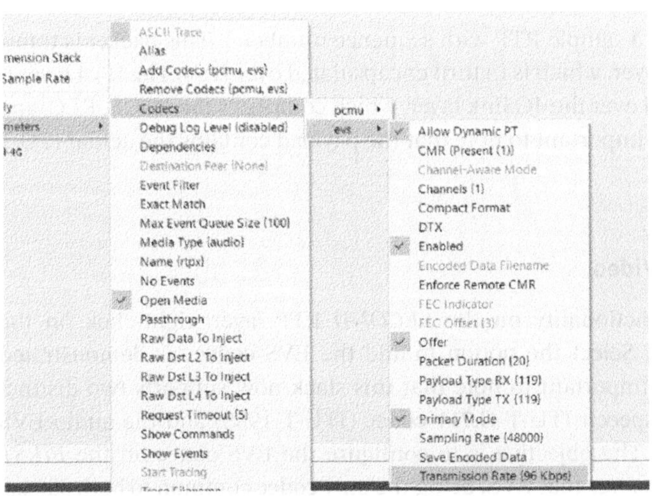

Figure 6.27 EVS Codec Configuration.

To incorporate video codec functionality into the stacks, create two new UDP layers, designated as *udp3* and *udp3x*. Position *udp3* on top of the ip layer and also position *udp3x* on top of the ipx layer. Construct two RTP layers, named *rtp2* and *rtp2x*. Place *rtp2* on top of *udp3* and then place *rtp2x* on top of *udp3x*. Refer to Figure 6.29 for a representation of the correct layer placement.

The newly created RTP stacks are initially configured to handle audio streams. To enable video support, right-click on the respective RTP layers and modify the

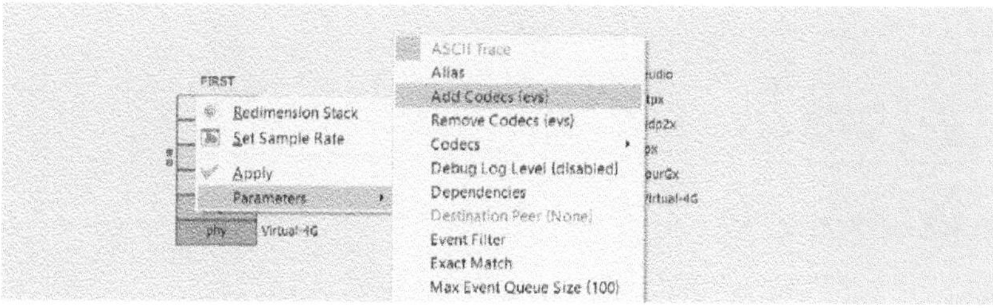

Figure 6.28 *FIRST* Stack EVS Codec Support.

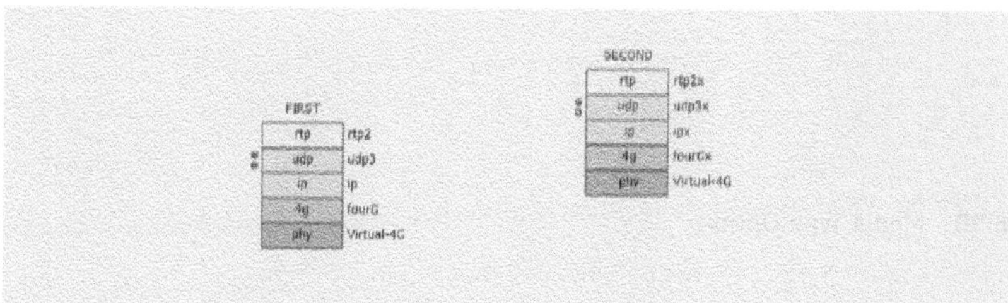

Figure 6.29 RTP Stacks with Audio and Video Support.

media type from audio to video. Upon setting the media type to video, as illustrated in Figure 6.30, there will be the option to select a video codec. Choose *h265* to activate the ITU-T Recommendation H.265 codec.

To enable the playback of video clips, it is necessary to replace the existing *tts* layer with a media player layer. This is because TTS layers are solely capable of playing audio streams and the player layers, on the other hand, possess the ability to play both audio and video. To proceed, remove the current *tts* layer and then create a new player layer by clicking on the beige-colored button labeled *player* (refer to Figure 6.31 for guidance).

Assign the name *player* to the newly created layer. Position the player layer directly above the *rtp* layer, replacing the previously removed *tts* layer. Refer to Figure 6.32 for a representation of the correct layer placement.

Right-click on the newly created *player* layer to access its configuration options. Within the configuration, specify a valid video clip as the filename. In this particular case, the *hp.mov* clip has been chosen. Ensure that both the *Play* and *Play on Media* options are enabled within the configuration as shown in Figure 6.33.

Navigate to the *Scripts* menu and select the option labeled *Run Suite* to initiate the configuration process. Right-click on the stack designated as *FIRST* to activate the dialpad. Click on the green button labeled *Place call* to start a media session with the *SECOND* stack. Refer to Figure 6.34 for guidance.

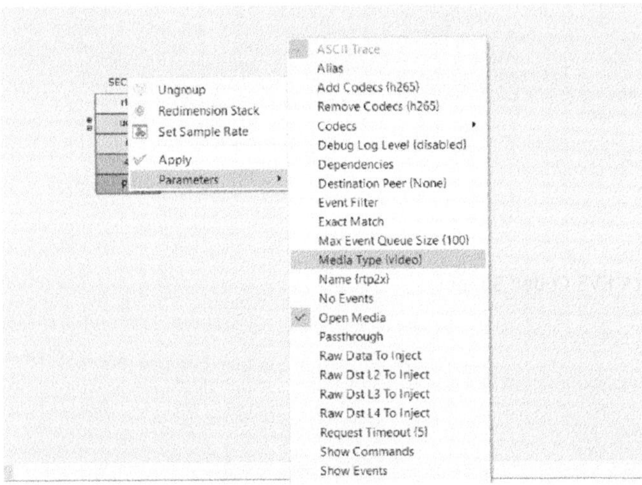

Figure 6.30 Media Type Option.

Figure 6.31 Player Layer Creation.

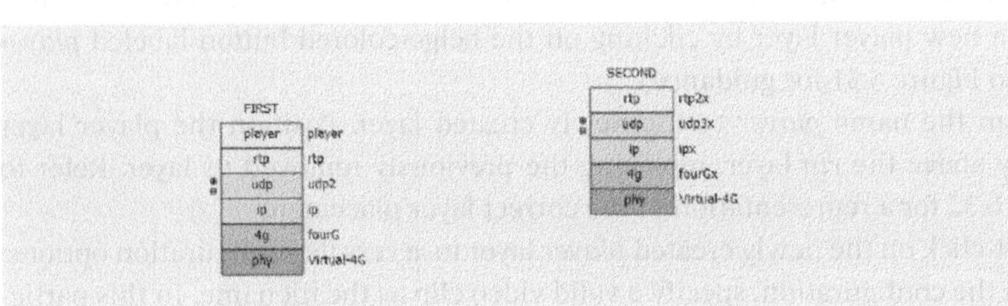

Figure 6.32 RTP Stack with Player.

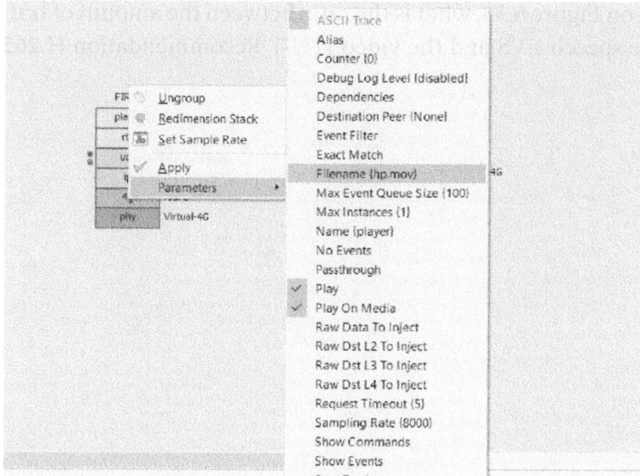

Figure 6.33 Player Layer Configuration.

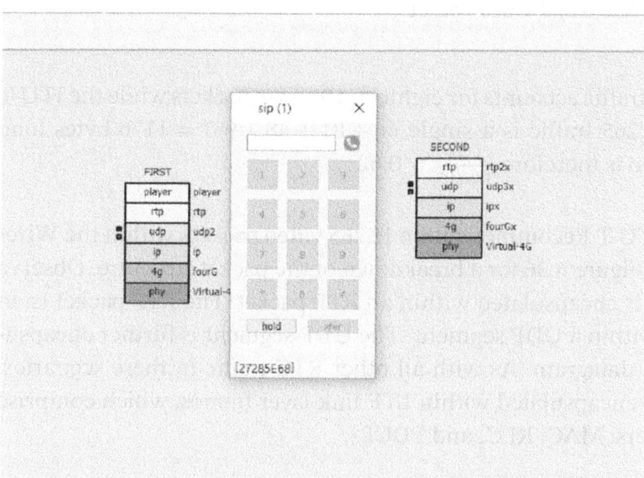

Figure 6.34 Call in Progress.

Maintain the call for approximately 30 seconds. Click the red button labeled *Hang up* to terminate the call. Conclude the configuration process. Open the *trace.cap* file using Wireshark to examine the captured network traffic. Refer to Figure 6.35 for a representation of the trace content, which includes SIP traffic, audio RTP packets utilizing the EVS codec, and video RTP packets employing the ITU-T Recommendation H.265 codec (ITU-T 2019).

Example 6.3 Based on Figure 6.35, what is the ratio between the amount of traffic associated with the speech EVS and the video ITU-T Recommendation H.265 codecs?

1	0.000000	10.0.0.10	10.0.0.11	SIP/SDP	1031	UEId=1	[DL]	[AM]	DRB:1	[DATA]	sn=0 [1013-bytes]
2	0.017000	10.0.0.10	10.0.0.11	SIP/SDP	1031	UEId=1	[DL]	[AM]	DRB:1	[DATA]	sn=1 [1013-bytes]
3	0.123000	10.0.0.11	10.0.0.10	SIP	496	UEId=1	[DL]	[AM]	DRB:1	[DATA]	sn=2 [478-bytes]
4	0.222000	10.0.0.11	10.0.0.10	SIP	497	UEId=1	[DL]	[AM]	DRB:1	[DATA]	sn=3 [479-bytes]
5	0.262000	10.0.0.11	10.0.0.10	H.265	84	UEId=1	[DL]	[AM]	DRB:1	[DATA]	sn=4 [66-bytes]
6	0.271000	10.0.0.11	10.0.0.10	H.265	99	UEId=1	[DL]	[AM]	DRB:1	[DATA]	sn=5 [81-bytes]
7	0.272000	10.0.0.11	10.0.0.10	H.265	66	UEId=1	[DL]	[AM]	DRB:1	[DATA]	sn=6 [48-bytes]
8	0.273000	10.0.0.11	10.0.0.10	H.265	999	UEId=1	[DL]	[AM]	DRB:1	[DATA]	sn=7 [981-bytes]
9	0.275000	10.0.0.11	10.0.0.10	EVS	123	UEId=1	[DL]	[AM]	DRB:1	[DATA]	sn=8 [105-bytes]
10	0.276000	10.0.0.11	10.0.0.10	EVS	123	UEId=1	[DL]	[AM]	DRB:1	[DATA]	sn=9 [105-bytes]
11	0.294000	10.0.0.11	10.0.0.10	EVS	123	UEId=1	[DL]	[AM]	DRB:1	[DATA]	sn=10 [105-bytes]
12	0.312000	10.0.0.11	10.0.0.10	EVS	123	UEId=1	[DL]	[AM]	DRB:1	[DATA]	sn=11 [105-bytes]
13	0.324000	10.0.0.11	10.0.0.10	SIP/SDP	1030	UEId=1	[DL]	[AM]	DRB:1	[DATA]	sn=12 [1012-bytes]
14	0.335000	10.0.0.11	10.0.0.10	EVS	123	UEId=1	[DL]	[AM]	DRB:1	[DATA]	sn=13 [105-bytes]
15	0.355000	10.0.0.11	10.0.0.10	EVS	123	UEId=1	[DL]	[AM]	DRB:1	[DATA]	sn=14 [105-bytes]
16	0.371000	10.0.0.11	10.0.0.10	EVS	123	UEId=1	[DL]	[AM]	DRB:1	[DATA]	sn=15 [105-bytes]
17	0.395000	10.0.0.11	10.0.0.10	EVS	123	UEId=1	[DL]	[AM]	DRB:1	[DATA]	sn=16 [105-bytes]
18	0.413000	10.0.0.11	10.0.0.10	EVS	123	UEId=1	[DL]	[AM]	DRB:1	[DATA]	sn=17 [105-bytes]
19	0.431000	10.0.0.11	10.0.0.10	EVS	123	UEId=1	[DL]	[AM]	DRB:1	[DATA]	sn=18 [105-bytes]
20	0.457000	10.0.0.11	10.0.0.10	EVS	123	UEId=1	[DL]	[AM]	DRB:1	[DATA]	sn=19 [105-bytes]
21	0.473000	10.0.0.11	10.0.0.10	EVS	123	UEId=1	[DL]	[AM]	DRB:1	[DATA]	sn=20 [105-bytes]
22	0.492000	10.0.0.10	10.0.0.11	SIP	526	UEId=1	[DL]	[AM]	DRB:1	[DATA]	sn=21 [508-bytes]
23	0.492000	10.0.0.11	10.0.0.10	EVS	123	UEId=1	[DL]	[AM]	DRB:1	[DATA]	sn=22 [105-bytes]
24	0.496000	10.0.0.10	10.0.0.11	SIP	526	UEId=1	[DL]	[AM]	DRB:1	[DATA]	sn=23 [508-bytes]
25	0.515000	10.0.0.11	10.0.0.10	EVS	123	UEId=1	[DL]	[AM]	DRB:1	[DATA]	sn=24 [105-bytes]
26	0.526000	10.0.0.10	10.0.0.11	EVS	123	UEId=1	[DL]	[AM]	DRB:1	[DATA]	sn=25 [105-bytes]
27	0.532000	10.0.0.10	10.0.0.11	EVS	123	UEId=1	[DL]	[AM]	DRB:1	[DATA]	sn=26 [105-bytes]
28	0.535000	10.0.0.11	10.0.0.10	EVS	123	UEId=1	[DL]	[AM]	DRB:1	[DATA]	sn=27 [105-bytes]
29	0.541000	10.0.0.10	10.0.0.11	EVS	123	UEId=1	[DL]	[AM]	DRB:1	[DATA]	sn=28 [105-bytes]

Figure 6.35 Captured Audio and Video Packets.

Solution:

In the trace, the EVS traffic accounts for eighteen 105-byte packets while the ITU-T Recommendation H.265 traffic is a single $66 + 81 + 48 + 981 = 1176$ bytes long video frame. The ratio is therefore $\frac{1176}{18 \times 105} = 0.62$.

Select one of the ITU-T Recommendation H.265 video packets within the Wireshark trace. Refer to Figure 6.36 for a breakdown of the packet structure. Observe that the video frame is encapsulated within an RTP packet. The RTP packet is, in turn, encapsulated within a UDP segment. The UDP segment is further encapsulated within an IPv4 datagram. As with all other RTC traffic in these scenarios, the IPv4 datagram is encapsulated within LTE link layer frames, which comprise three distinct sublayers: MAC, RLC, and PDCP.

6.4 Summary

3GPP networks are key to everyday communications. Within the context of LTE and other 4G technologies, VoLTE plays a crucial role. Leveraging tools like Netualizer and Wireshark, it is possible to build an LTE-compatible set of physical and link layers. These layers include MAC, RLC, and PDCP sublayers, which enable the encapsulation of IP traffic. With the IP layer established, transport and application layers can be utilized to support the establishment of speech and video calls.

```
> Frame 5: 84 bytes on wire (672 bits), 84 bytes captured (672 bits)
  DLT: 147, Payload: mac-lte-framed (mac-lte-framed)
v MAC-LTE DL-SCH: (SFN=0   , SF=2) UEId=1   (3:remainder)
  > [Context (RNTI=2272)]
  v MAC PDU Header (3:remainder)  [1 subheaders]
    > Sub-header (lcid=3, length is remainder)
  v RLC-LTE   UEId=1 [DL] [AM] DRB:1  [DATA]     sn=4      [66-bytes]
    > [Context]
    > AM Header sn=4
    > [Sequence Analysis - OK]
    v PDCP-LTE (SN=4)(64 bytes data)
      > [Configuration: (direction=Downlink, plane=User)]
        1... .... = PDU Type: Data PDU
        .000 .... = Reserved: 0x0
        .... 0000 0000 0100 = Seq Num: 4
      > [Sequence Analysis - OK]
      > Internet Protocol Version 4, Src: 10.0.0.11, Dst: 10.0.0.10
      > User Datagram Protocol, Src Port: 4005, Dst Port: 4004
v Real-Time Transport Protocol
  > [Stream setup by SDP (frame 1)]
    10.. .... = Version: RFC 1889 Version (2)
    ..0. .... = Padding: False
    ...0 .... = Extension: False
    .... 0000 = Contributing source identifiers count: 0
    0... .... = Marker: False
    Payload type: H265 (120)
    Sequence number: 0
    [Extended sequence number: 65536]
    Timestamp: 0
    Synchronization Source identifier: 0x7ad8f5b2 (2061039026)
v H.265
  > NAL unit header or first byte of the payload
  > H265 NAL Unit Payload
```

Figure 6.36 Captured ITU-T Recommendation H.265 RTP Packet.

Similar to non-3GPP networks, the VoLTE protocol stack incorporates UDP transport, which encapsulates both SIP and RTP packets. SIP is responsible for establishing, maintaining, and terminating calls, while RTP handles the actual transmission of media streams. In the context of VoLTE, 3GPP-compliant codecs like EVS and ITU-T Recommendation H.265 are essential for transmitting these speech and video streams effectively.

6.5 Homework Problems and Questions

Problem 6.1 Consider Figure 6.15, what is the transmission rate in each direction? What can be inferred from the RTT?

Problem 6.2 Looking at Figure 6.24, what is the RTP traffic transmission rate in each direction? What is the average interpacket RTP delay in each direction?

Problem 6.3 Figure 6.16 shows an ICMP echo packet. What is the overhead due to the LTE headers?

Problem 6.4 Given the captured UDP packet in Figure 6.19, what is the ratio between the IPv4/UDP and the LTE overhead?

Problem 6.5 For the captured RTP packet in Figure 6.26, what component is performing the encapsulation of the actual IPv4 and UDP headers?

Problem 6.6 Consider Figure 6.36, what is the overhead of the RTP header in the PDCP-LTE layer?

6.6 Lab Exercises

Problem 6.7 Build a scenario in Netualizer with the following characteristics:

- Two Stacks: *EP1* and *EP2*
- Physical Layer/Link Layer: 4G/VoLTE (Emulated)
- Client IP address: 192.168.21.10
- Server IP address: 192.168.21.12
- Session Layer: SIP/RTP (over UDP)
- Codec: ITU-T Recommendation G.711
- Text-To-Speech on EP1: *Hello Everyone*
- Audio Player on EP2

Capture traffic using Wireshark and place a call from EP2 to EP1. Let the media run for 1 minute. Answer the following:

1. How many RTP packets can be seen?
2. How many SIP packets can be seen?
3. What is the SIP throughput and goodput in each direction?
4. What is the RTP throughput and goodput in each direction?
5. How these results compare to those of the exercises in Problem 6.10?

Problem 6.8 Introduce an impairment layer between the UDP and the RTP layer in the EP1 stack. Set the transmission loss to 20%. Let the media run for 1 minute.

Answer the following:

1. How many RTP packets can be seen?
2. How many SIP packets can be seen?
3. What is the SIP throughput and goodput in each direction?
4. What is the RTP throughput and goodput in each direction?
5. How these results compare to those of the exercises in Problem 6.11?

Problem 6.9 Repeat Problem 6.7 but on a 5G/VoNR architecture. How does 4G/VoLTE compare to 5G/VoNR?

Problem 6.10 Now repeat Problem 6.8 but on a 5G/VoNR architecture. How does 4G/VoLTE compare to 5G/VoNR?

Bibliography

3GPP. (2018). TS 126.442: Mandatory speech codec speech processing functions; AMR speech codec; general description. Technical report TS 126.442, 3rd Generation Partnership Project.

Cox, C. (2014). *An introduction to LTE: LTE, LTE-advanced, SAE, VoLTE and 4G mobile communications.* 2nd ed. Wiley Publishing. ISBN 1-118-81803-2.

Cox, C. (2020). An introduction to 5G: the new radio, 5G network and beyond. Wiley. ISBN 978-1-119-60269-9. Available from: https://books.google.com/books?id=aJsMEAAAQBAJ. (Accessed March 15th, 2024)

Herrero, R. (2021). Fundamentals of IoT communication technologies. *Textbooks in telecommunication engineering.* Springer International Publishing. ISBN 978-3-030-70079-9. Available from: https://books.google.com/books?id=k70rzgEACAAJ. (Accessed March 15th, 2024)

ITU-T. (1988). G.711: Pulse code modulation (PCM) of voice frequencies.

ITU-T. (2019). H.265: High efficiency video coding.

L7TR. Netualizer: Network virtualizer. Available from: https://www.l7tr.com. (Accessed March 15th, 2024)

RFC. (1980). User datagram protocol. RFC 768, August. Available from: https://www.rfc-editor.org/info/rfc768. (Accessed March 15th, 2024)

RFC. (1981). Internet control message protocol. RFC 792, September. Available from: https://www.rfc-editor.org/info/rfc792. (Accessed March 15th, 2024)

Schooler, E., Rosenberg, J., Schulzrinne, H., Johnston, A., Camarillo, G., Peterson, J., Sparks, R., and Handley, M.J. (2002). SIP: session initiation protocol. RFC 3261, July. Available from: https://rfc-editor.org/rfc/rfc3261.txt. (Accessed March 15th, 2024)

Schulzrinne, H., Casner, S.L., Frederick, R., and Jacobson, V. (2003). RTP: a transport protocol for real-time applications. RFC 3550, July. Available from: https://rfc-editor.org/rfc/rfc3550.txt. (Accessed March 15th, 2024)

SIMCom. Sim7600g: R2 LTE CAT-4 MPCIE. Available from: https://www.simcom.com/product/SIM7600G.html. (Accessed March 15th, 2024)

Wireshark. Wireshark: network analyzer. Available from: https://www.wireshark.org. (Accessed March 15th, 2024)

7

IoT Networks

7.1 Introduction

Networks designed for the IoT provide connectivity within Cyber-Physical Systems to link devices with applications. Certain devices within these networks are capable of recording audio, speech, and video, enabling RTC functionalities. Section 4.4 focuses on the specifics of integrating IoT protocols with established RTP and SIP protocols, leveraging IPv6 networking as the underlying infrastructure (Deering and Hinden 2017). This section focuses on the integration of WPAN and LPWAN IoT protocols with RTC protocols, with the goal of prototyping protocol stacks on the Netualizer platform (Herrero 2021; 2023). Similar to Chapter 5, Wireshark is employed for analyzing network traffic generated by these integrated protocols.

The next two sections provide instructions on constructing protocol stacks that support IPv6 for the network layer and UDP for the transport layer. Upon completion of these layers, the addition of DTLS, RTP, SRTP, and SIP layers will enable support for RTC applications, as illustrated in Figure 4.34. Within the Netualizer platform, the implementation of DTLS is discussed in Section 5.3.5, while the implementation of RTP and SIP is covered in Section 5.3.1. Finally, Section 5.3.5 details the implementation of SRTP (Carrara et al. 2004).

7.2 WPAN

This section focuses on IEEE 802.15.4 WPAN. It explores how the physical and link layers of this standard, in conjunction with the 6LoWPAN adaptation layer, enable support for IPv6 and UDP network communication within WPAN topologies.

Hands-On Real Time Communications: A Practical Guide to RTC Protocols in Non-3GPP, 3GPP, 4G/5G/6G and IoT Networks, First Edition. Rolando Herrero.

Companion website: www.wiley.com/go/herrero/RTCProtocols

7.2.1 Physical and Link Layers

The IEEE 802.15.4 standard (discussed in detail in Section 4.4.1.1) defines the physical and link layer technologies that form the foundation of the ZigBee protocol stack (Zig 2015). These layers play an important role in enabling end-to-end IPv6 support via 6LoWPAN, while simultaneously facilitating wide-ranging WPAN IoT connectivity (Herrero 2021). Building protocol stacks that support IEEE 802.15.4 physical and link layers can be achieved through the functionalities of Netualizer and Wireshark. As described in Sections 5.1.1 and 5.1.2, these tools offer capabilities for creating and analyzing such stacks (L7TR; Wireshark).

Netualizer is compatible with the CC2531 hardware-based network interface. The CC2531 is a popular IEEE 802.15.4 physical layer chip developed and manufactured by Texas Instruments (TI). It includes, among other features, a USB interface that enables integration with Raspberry Pi and other Netualizer agents, such as traditional personal computers and laptops. Figure 7.1 shows a TI CC2531 network interface.

When configuring a physical layer in Netualizer and the agent is physically connected through its USB interface to a CC2531 network interface, a new option appears. As shown in Figure 7.2, the *CC2531+IEEE802.15.4+1* {*COM10*} physical interface becomes available in the list of interfaces on the agent configuration. This option represents the USB-connected CC2531 IEEE 802.15.4 network interface associated with the serial COM10 port. The *+1* in the name indicates that this is the first IEEE 802.15.4 interface detected by this agent. As more physical

Figure 7.1 CC2531-based Radio.

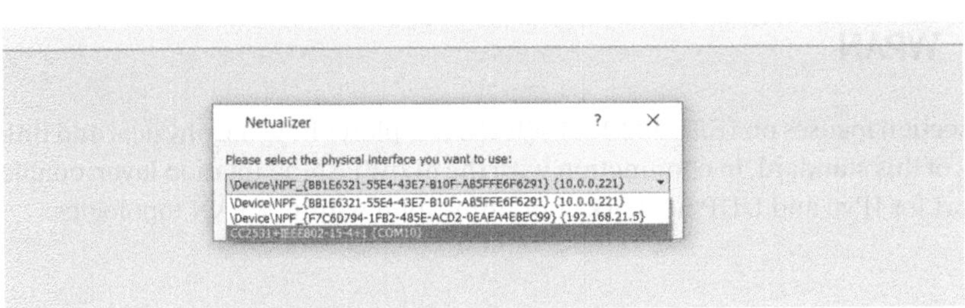

Figure 7.2 Network Interface Selection.

IEEE 802.15.4 interfaces are connected, the number will increment consecutively (IEEE 2020).

To simplify development and testing of IEEE 802.15.4 applications without requiring hardware like the TI CC2531, Netualizer offers virtual hardware support. This feature allows users to deploy and simulate IoT scenarios entirely within the software environment, without needing external hardware devices. Additionally, any Netualizer agent, regardless of its location, can utilize these virtual interfaces. This enables a workflow where initial development and testing occur with virtual interfaces, which can then be seamlessly replaced with physical hardware like the CC2531 once ready. To enable virtual hardware support, open the *Agents* menu, illustrated in Figure 7.3. Locate the *Virtual Hardware Support* option and enable it.

Create a new Netualizer project. Specifically, in the *File* menu, choose *New Project*. Enter a descriptive name, such as *ieee802154*, in the project name field. Ensure the checkbox to attach the local agent is selected, as shown in Figure 7.4. Then click *OK*. Note that for detailed instructions on creating projects, refer to Section 5.1.1 (in Chapter 5). Upon project creation, the agent configuration will be automatically added and displayed open in the Netualizer configuration panel.

With the configuration panel open, click the green *phy* button to begin creating a new physical layer. Netualizer will display a list of available network interfaces (shown in Figure 7.5). This includes the *NT* interface, any connected hardware-based interfaces, and the *Virtual-IEEE802-15-4* interface, which becomes accessible when virtual hardware support is enabled. The

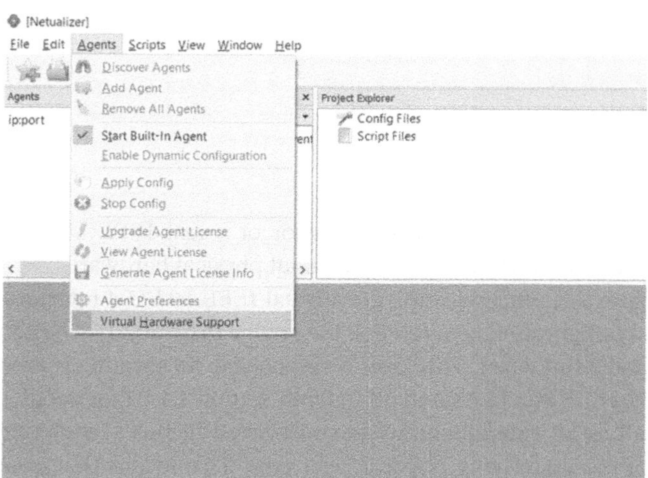

Figure 7.3 Virtual Hardware Support Option.

Figure 7.4 Project Name Selection.

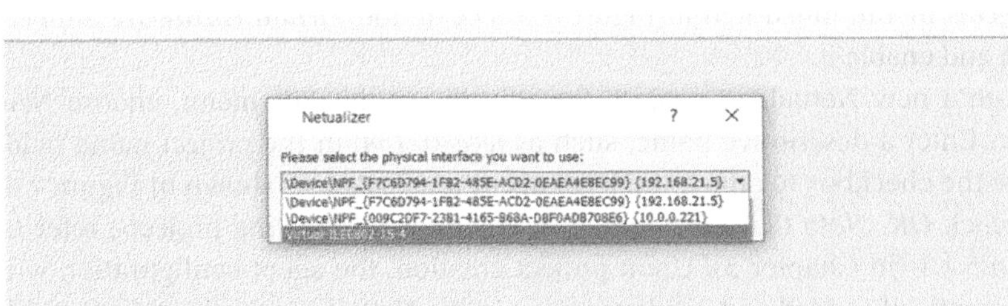

Figure 7.5 The IEEE 802.15.4 Interface.

Figure 7.6 IEEE 802.15.4 Layer Creation.

Virtual-IEEE802-15-4 interface emulates the behavior of a real IEEE 802.15.4 interface, allowing for development and testing without physical hardware.

As indicated in Figure 7.6, after selecting the Virtual *IEEE 802.15.4* interface, position it within the configuration layout. Click the orange *IEEE 802.15.4* button to generate the associated link layer. This layer is responsible for forming frames in compliance with the IEEE 802.15.4 standard (refer to Section 4.4.1.1 for details). The link layer, as well as all subsequent layers constructed in this section, are essential regardless of the underlying physical layer type. This means that even if a hardware-based TI CC2531 physical layer were used, the IEEE 802.15.4 link layer and the layers above it would still be required.

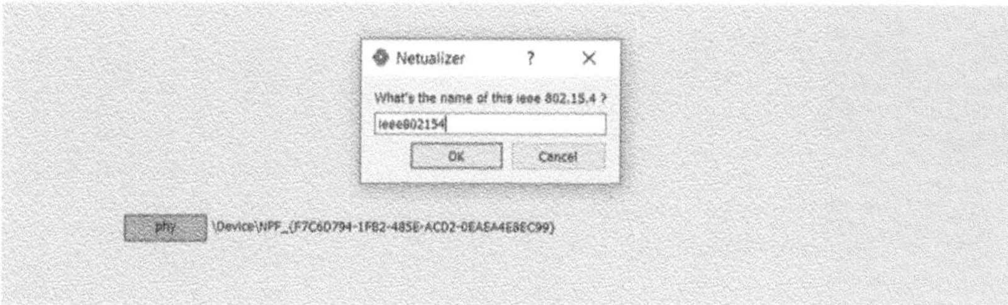

Figure 7.7 IEEE 802.15.4 Layer Name Selection.

Figure 7.8 IEEE 802.15.4 2-Layer Stack.

Referring to Figure 7.7, before positioning the link layer above the *Virtual-IEEE802-15-4* physical layer, assign it a descriptive name, such as *ieee802154*. This name serves as a unique identifier for the layer within Lua scripts that interact with Netualizer APIs ((Ierusalimschy 2006) for more details on Lua). Once named, place the link layer on top of the *Virtual-IEEE802-15-4* physical layer, as illustrated in Figure 7.8.

Arrange the *ieee802154* layer directly above the physical layer within the configuration layout. Rename the stack by double-clicking the default *STACK* label situated above the 2-layer stack. Modify the label to *FIRST* for clear identification. Note that stack labels serve to distinguish individual stacks within configurations that include multiple stacks, as demonstrated in Chapter 4. Figure 7.8 shows the currently constructed, concise *FIRST* stack.

7.2.2 Network Layer

IPv6 is essential for IoT, however, IPv6 datagrams cannot be directly transmitted over IEEE 802.15.4. To address this, 6LoWPAN (described in Section 4.4.1.2) is the recommended mechanism for IPv6 adaptation in IEEE 802.15.4 scenarios (Montenegro et al. 2007). To add a 6LoWPAN layer click the *6LoWPAN* layer button shown in Figure 7.9. Label the layer *sixlow* (Netualizer layer names cannot start with numbers). Place the *sixlow* layer on top of the *ieee802154* layer.

Figure 7.9 6LoWPAN Layer Creation.

6LoWPAN serves as an adaptation mechanism, but a dedicated IPv6 layer is still required (Deering and Hinden 2017; Graziani 2012). To create a new IP layer, click the *IP* layer button shown in Figure 7.9. Follow the steps outlined in Section 2.2 to construct an IP layer, but this time place it over IEEE 802.15.4 instead of Ethernet. Name the newly created IP layer *ip* and position it above the 6LoWPAN *sixlow* layer to activate link layer adaptation. The resulting IPv6 stack is shown in Figure 7.10.

To create the second IPv6 stack, it is recommended to duplicate the existing *FIRST* stack rather than constructing one from the beginning. To duplicate the *FIRST* stack, select the *FIRST* stack by clicking on it. Right-click to open the context menu and choose the *Copy* option, as shown in Figure 7.11.

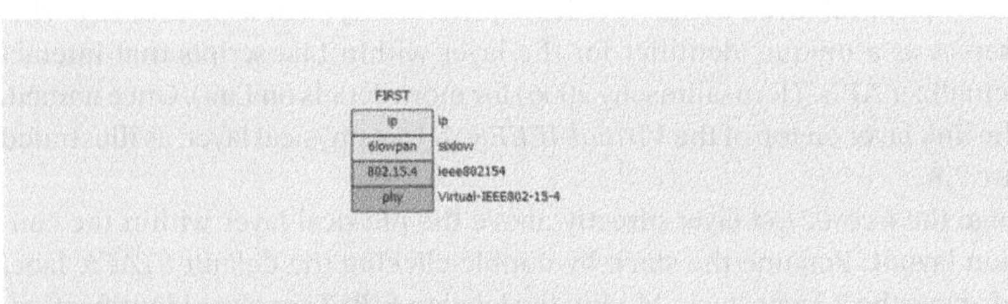

Figure 7.10 IEEE 802.15.4 4-Layer Stack.

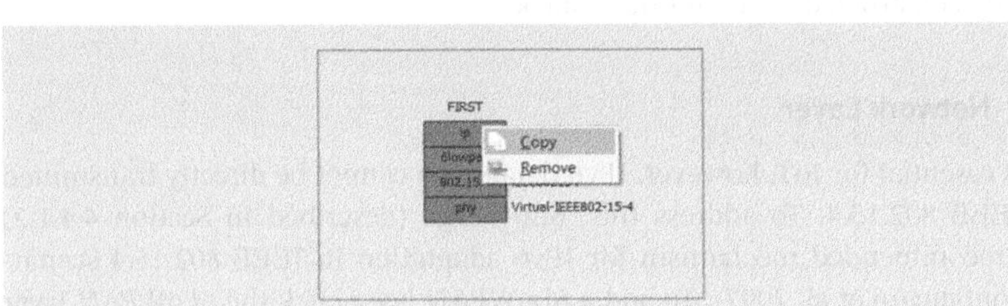

Figure 7.11 IPv6 Stack Copy Selection.

To create the second IPv6 stack, right-click in the agent configuration panel, near the original *FIRST* stack, and choose the *Paste* option. Double-click on the label of the pasted stack to rename it *SECOND*. Figure 7.12 illustrates both the *FIRST* and *SECOND* stacks, complete with their respective layers. The Netualizer agent automatically appends an *x* suffix to the names of copied layers to ensure unique layer names and avert configuration issues that could hinder script execution.

To ensure connectivity, a few modifications are required to the copied layers in the *SECOND* stack. Specifically, modify the network layer address of the *ipx* layer by right-clicking on the *ipx* layer within the *SECOND* stack. Select the *Address* field from the *Parameters* menu option, as shown in Figure 7.13.

To avoid overlapping addresses and potential connectivity issues, modify the network layer address of the *ipx* layer in the *SECOND* stack by rightclicking on the *ipx* layer within the *SECOND* stack. Select the *Address* field from the *Parameters*

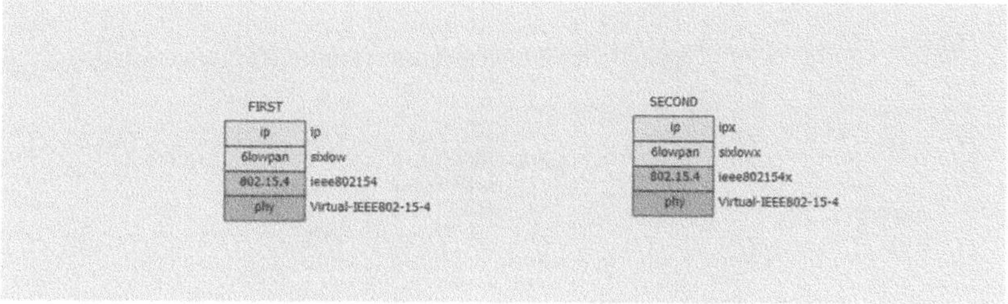

Figure 7.12 Duplicated IPv6 Stack.

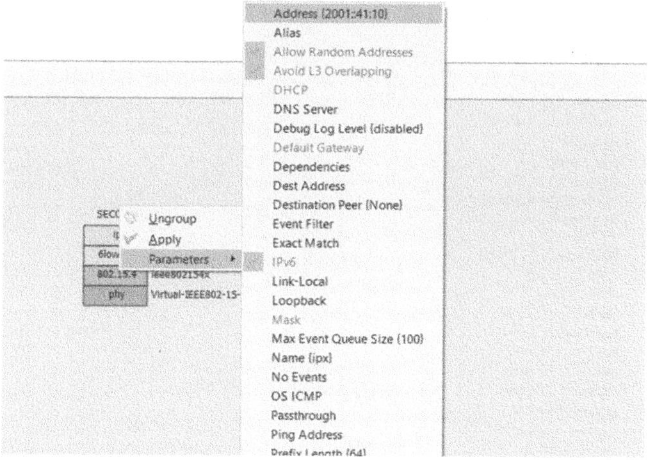

Figure 7.13 Address Field.

menu option. Change the IPv6 address from *2001::41:10* to *2001::41:11* as shown in Figure 7.14.

Configure the link layer address of the *ieee802154x* layer. Right-click on the *ieee802154x* layer within the *SECOND* stack. Select the *Address* field from the *Parameters* menu option, as shown in Figure 7.15.

Modify the link layer address of the *ieee802154x* layer in the *SECOND* stack. Specifically, right-click on the *ieee802154x* layer within the *SECOND* stack. Select the *Address* field from the *Parameters* menu option. Change the address from *01:02:03:04:05:06* to *01:02:03:04:05:07*, as shown in Figure 7.16. Note that IEEE 802.15.4 addresses are typically 64 or 16 bits in length, depending on whether they are manufacturer-assigned or router-assigned. Netualizer automatically converts 48-bit addresses into 64-bit addresses by inserting an *FF:FE* sequence in the middle of the 48-bit address.

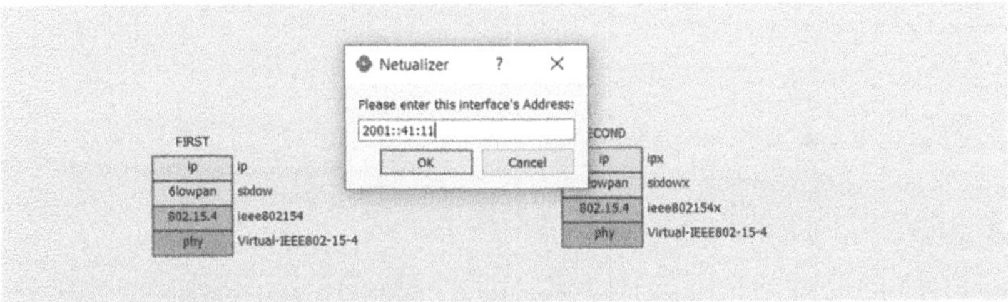

Figure 7.14 IPv6 Address Changes.

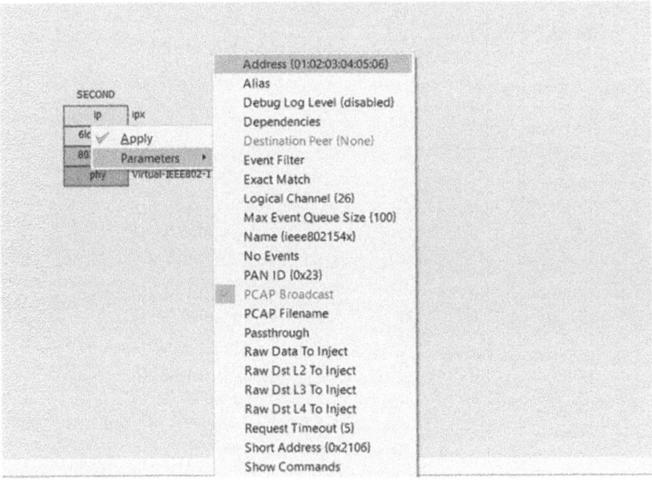

Figure 7.15 IEEE 802.15.4 Address Field.

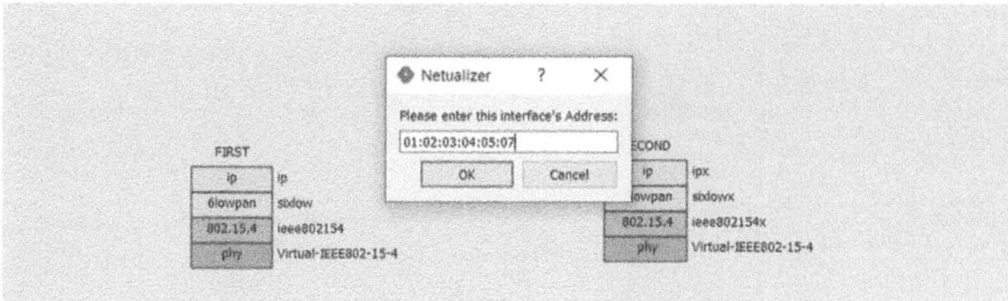

Figure 7.16 IEEE 802.15.4 Address Changes.

Wireshark cannot directly capture traffic generated by virtual interfaces. Therefore, to enable analysis with Wireshark, network traffic must be captured at the underlying IEEE 802.14.5 link layer. To enable link layer capture right-click on the *ieee802154* layer in the Netualizer configuration panel. Select the *Parameter* menu option and set the *Start PCAP* flag as indicated in Figure 7.17. The captured traffic will be saved in the PCAP format, which is widely used for network analysis and is compatible with Wireshark. This means that the captured traffic can be opened in Wireshark for in-depth examination at a later time. To learn more about using Wireshark, please refer to Section 5.1.2.

Specifying the PCAP filename is done by right-clicking on the *ieee802154* layer in the Netualizer configuration panel. Select the *PCAP Filename* option. Enter *ieee802154.cap* as the filename to store the captured IEEE 802.15.4 frames (shown in Figure 7.18). Save the captured file in the current project directory for easy access.

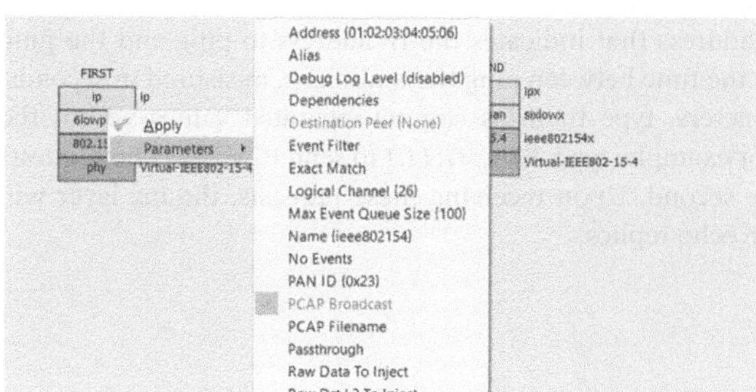

Figure 7.17 Wireshark Capture Setup.

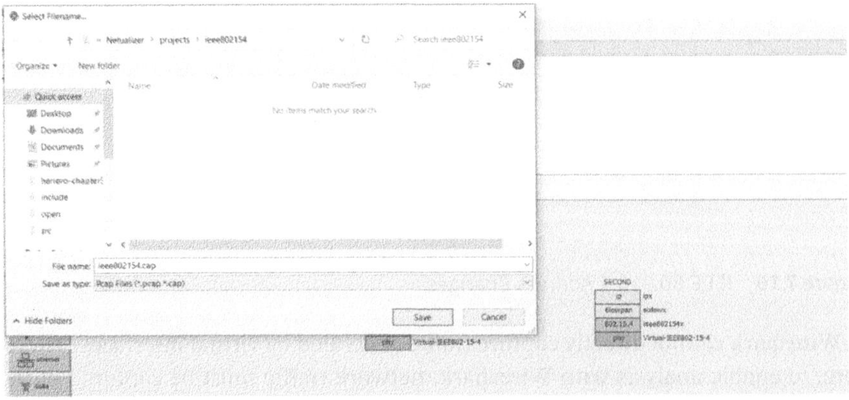

Figure 7.18 PCAP Filename Configuration.

With the configuration complete, everything is now ready to execute the project. Since this project involves a single agent configuration and does not require any functional scripts, the setup of the *FIRST* and *SECOND* stacks is sufficient. To initiate the project, navigate to the *Scripts* menu. Select the *Run Suite* option. Figure 7.19 illustrates the running suite from the perspective of the agent configuration panel.

With both IPv6 stacks operational, make sure to create traffic between them and verify their connectivity by pinging the IP address of the *ipx* layer in the *SECOND* stack. To initiate a ping, right-click on the *ip* layer in the *FIRST* stack. Select the *Ping* command. Figure 7.20 shows the *Ping* command within the agent configuration panel. The *Ping* command functions with two key parameters; the destination network address that indicates the IP address to ping and the ping interval that specifies the time between ping transmissions, measured in seconds. To enter these parameters, type them as comma-separated values within the command prompt. For example, type *2001::41:11,1* to send ICMPv6 echo requests to the *ipx* layer every second. Upon receiving these requests, the *ipx* layer will respond with ICMPv6 echo replies.

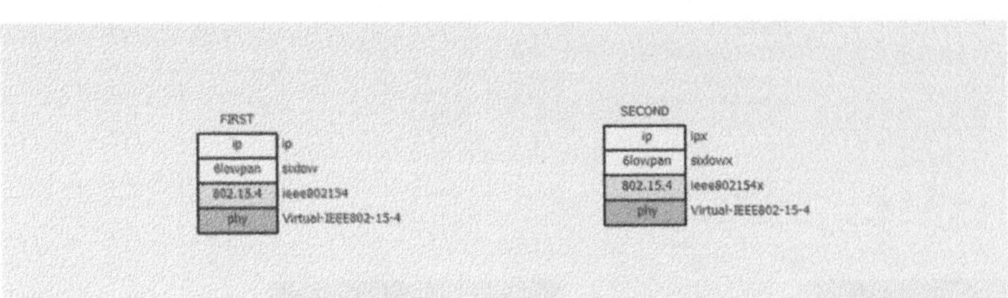

Figure 7.19 Suite Execution.

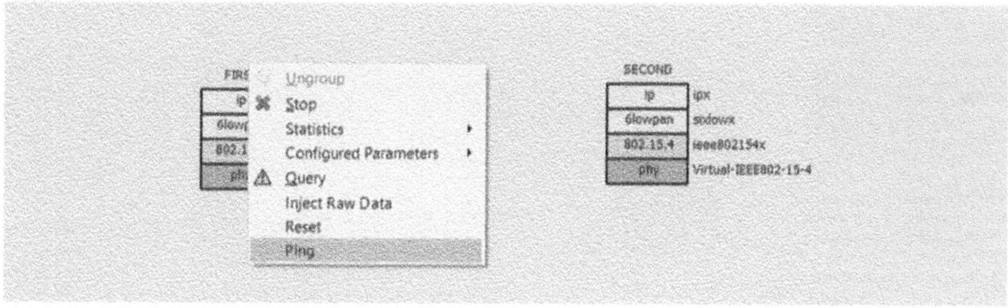

Figure 7.20 Ping Command.

Figure 7.21 shows Netualizer's output during the *Ping* execution. A dialog window dynamically displays the RTT associated with the transmission of ICMPv6 echo traffic. This RTT, also known as latency, is measured in milliseconds. Allow the traffic to run for approximately 10 seconds to gather sufficient data. Halt the Configuration next. Although the suite has completed its execution, the configuration remains active. To terminate the configuration, locate the red cross button within the project agent. Click the red cross button to stop the configuration.

To access and analyze captured traffic, navigate to the project folder located at *Netualizer/Projects/ieee802154*. Locate the PCAP file named *ieee802154.cap*, which contains the captured network traffic (shown in Figure 7.22). Ensure to have Wireshark installed on the system. Double-click the *ieee802154.cap* file to launch it in Wireshark for in-depth analysis.

Figure 7.23 displays an open PCAP trace in Wireshark. The trace contains multiple datagrams. In frames 1 and 2, ICMPv6 Multicast Listener Reports are sent for multicast group registration. These are transmitted by default but have no specific function in this scenario. In frame 3, ICMPv6 ND Neighbor Solicitation request is sent by the IP layer to resolve the link layer address for *2001::41:11* before sending an ICMPv6 echo request. This message is transmitted to a specific broadcast address beginning with *FF02:*. In frame 4, ICMPv6 echo request is sent from *2001::41:10* to *2001::41:11*. In frame 5, ICMPv6 ND neighbor solicitation request is sent by the *ipx* layer to obtain the link layer address of *2001::41:10* before

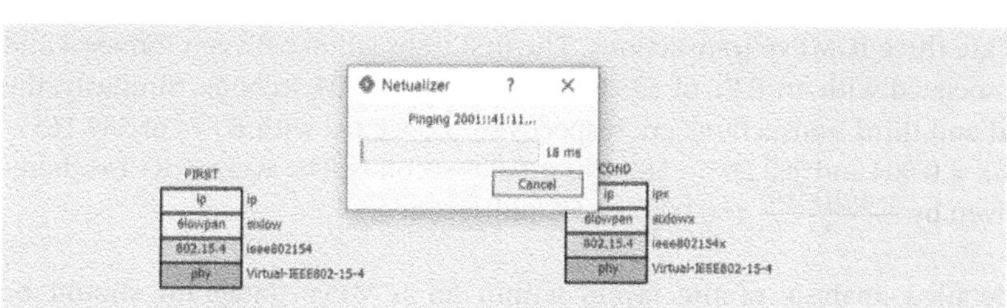

Figure 7.21 Ping Execution on *SECOND* Stack.

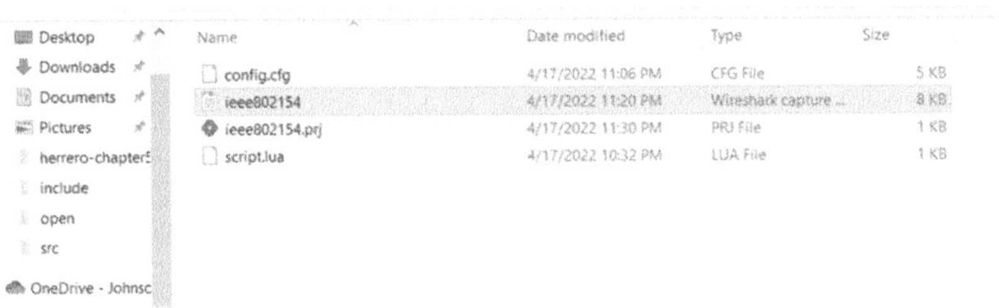

Figure 7.22 Trace Location.

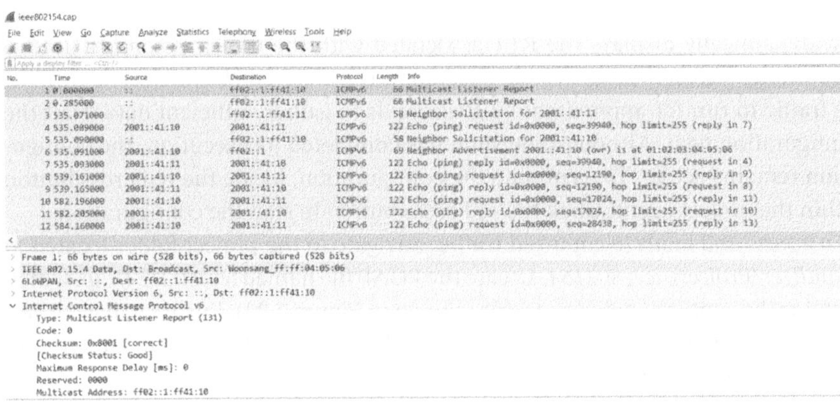

Figure 7.23 Captured ICMPv6 Packets.

sending an ICMPv6 echo reply. In frame 6, ICMPv6 ND neighbor advertisement message is sent by the IP layer to provide the link layer address in response to the previous solicitation. In frame 7, the first ICMPv6 echo reply. Subsequent frames show the periodic transmission of ICMPv6 echo requests and replies.

Example 7.1 Refer to Figure 7.23, what is the average RTT?

Solution:
There are three ICMPv6 transactions. The first transaction between frames 4 and 7 is associated with an RTT of $535.093 - 535.089 = 0.004$ seconds. Similarly, the second and third transactions are respectively associated with RTTs of $539.165 - 539.161 = 0.004$ and $582.205 - 582.196 = 0.009$ seconds. The average RTT is therefore given by $\frac{2\times0.004+0.009}{3} = 0.0057$ or 57 milliseconds.

A detailed analysis of the layers within an ICMPv6 datagram should be conducted. Begin by selecting one of the ICMPv6 *echo* requests. Figure 7.24 illustrates the fields associated with the IEEE 802.15.4 layer. Refer to Figure

4.18 for a comprehensive explanation of the encoding of these fields. They include the 16-bit *Frame Control Field* (FCF), the 8-bit sequence number, and the address field. The address field incorporates the default destination ICMPv6 *echo* requests. Set to *0x23*, aligning with the value specified in Figure 7.15. The source and destination link layer 64-bit addresses, which are *01:02:03:FF:FE:04:05:06* and *01:02:03:FF:FE:04:05:07*, respectively. It is important to note that IEEE 802.15.4 security fields are not transmitted due to their disabled state.

In a similar manner, Figure 7.25 shows the fields associated with the 6LoWPAN layer. Refer to Section 4.4.1.2 for a detailed explanation of the encoding of these fields. Compression is stateless, although the mechanism is grounded in IPHC. The TF field, responsible for encoding both the traffic class and flow label, is set to *11* to denote that they are not transmitted. The 6LoWPAN header that follows is transmitted in-line, as indicated by the next header flag. Due to the stateless nature of the encoding, the context identifier extension is set to false, and both source and destination address compression are disabled. Both the source and destination addresses are sent in-line as *2001::41:10* and *2001::41:11*, respectively.

```
v IEEE 802.15.4 Data, Dst: Woonsang_ff:ff:04:05:06, Src: Woonsang_ff:ff:04:05:06
  v Frame Control Field: 0xcc61, Frame Type: Data, Acknowledge Request
      .... .... .... .001 = Frame Type: Data (0x1)
      .... .... .... 0... = Security Enabled: False
      .... .... ...0 .... = Frame Pending: False
      .... .... ..1. .... = Acknowledge Request: True
      .... .... .1.. .... = PAN ID Compression: True
      .... .... 0... .... = Reserved: False
      .... ...0 .... .... = Sequence Number Suppression: False
      .... ..0. .... .... = Information Elements Present: False
      .... 11.. .... .... = Destination Addressing Mode: Long/64-bit (0x3)
      ..00 .... .... .... = Frame Version: IEEE Std 802.15.4-2003 (0)
      11.. .... .... .... = Source Addressing Mode: Long/64-bit (0x3)
    Sequence Number: 9
    Destination PAN: 0x0023
    Destination: Woonsang_ff:ff:04:05:06 (01:02:03:ff:ff:04:05:06)
    Extended Source: Woonsang_ff:ff:04:05:06 (01:02:03:ff:ff:04:05:06)
    FCS: 0x7f91 (Correct)
```

Figure 7.24 IEEE 802.15.4 Header on Wireshark.

```
v 6LoWPAN, Src: 2001::41:11, Dest: 2001::41:10
  v IPHC Header
      011. .... = Pattern: IP header compression (0x03)
      ...1 1... .... .... = Traffic class and flow label: Version, traffic class, and flow label compressed
      .... .0.. .... .... = Next header: Inline
      .... ..11 .... .... = Hop limit: 255 (0x3)
      .... .... 0... .... = Context identifier extension: False
      .... .... .0.. .... = Source address compression: Stateless
      .... .... ..00 .... = Source address mode: Inline (0x0000)
      .... .... .... 0... = Multicast address compression: False
      .... .... .... .0.. = Destination address compression: Stateless
      .... .... .... ..00 = Destination address mode: Inline (0x0000)
    Next header: ICMPv6 (0x3a)
    Source: 2001::41:11
    Destination: 2001::41:10
```

Figure 7.25 6LoWPAN Header on Wireshark.

```
v Internet Protocol Version 6, Src: 2001::41:11, Dst: 2001::41:10
    0110 .... = Version: 6
  v .... 0000 0000 .... .... .... .... .... = Traffic Class: 0x00
      .... 0000 00.. .... .... .... .... .... = Differentiated Services Codepoint
      .... .... ..00 .... .... .... .... .... = Explicit Congestion Notification
    .... .... .... 0000 0000 0000 0000 0000 = Flow Label: 0x00000
    Payload Length: 64
    Next Header: ICMPv6 (58)
    Hop Limit: 255
    Source Address: 2001::41:11
    Destination Address: 2001::41:10
    [Source Teredo Server IPv4: 0.0.0.0]
    [Source Teredo Port: 65535]
    [Source Teredo Client IPv4: 255.190.255.238]
    [Destination Teredo Server IPv4: 0.0.0.0]
    [Destination Teredo Port: 65535]
    [Destination Teredo Client IPv4: 255.190.255.239]
v Internet Control Message Protocol v6
    Type: Echo (ping) reply (129)
    Code: 0
    Checksum: 0x92a4 [correct]
    [Checksum Status: Good]
    Identifier: 0x0000
    Sequence: 39940
    [Response To: 4]
    [Response Time: 4.000 ms]
  v Data (56 bytes)
```

Figure 7.26 IPv6 and ICMPv6 Headers on Wireshark.

Wireshark software reveals the decompressed IPv6 layer, arising from the decoding of the IPv6 layer, along with the ICMPv6 packet (Gupta and Conta 2006). Figure 7.26 presents a representation of these two layers in unison. The IPv6 layer follows the encoding outlined in Figure 2.11. Parameters remain identical for both layers, as this process involves a mapping from 6LoWPAN to IPv6. The ICMPv6 packet undergoes in-line transmission, consequently, it does not undergo compression.

7.2.3 Transport Layer

The preferred transport layer commonly used for 6LoWPAN adaptation is UDP (RFC 1980). In this section, UDP layers are placed on top of the adapted *ip* and *ipx* IPv6 layers. Note that UDP transport is compressed using the NHC 6LoWPAN header. To continue building stacks with UDP support, make sure to save the current *ieee802154* project as *ieee802154udp*.

To create two UDP layers, click the blue UDP layer in Figure 7.27. Name the layers *udp* and *udpx*, respectively. Place the layers on top of the previously created *ip* and *ipx* layers. Remember to choose any unique name for a layer. Figure 7.28 shows the UDP-compliant *FIRST* and *SECOND* stacks.

To simplify the configuration, change the port number of the *udpx* layer in the *SECOND* stack. This is done because, even though both *udp* and *udpx* layers share the same default port numbers, the stacks *FIRST* and *SECOND* have different IP addresses (*2001::41:10* and *2001::41:11*, respectively). To change the port number, right-click the *udpx* layer. Select the *Port* attribute in the *Parameters* menu, as shown in Figure 7.29.

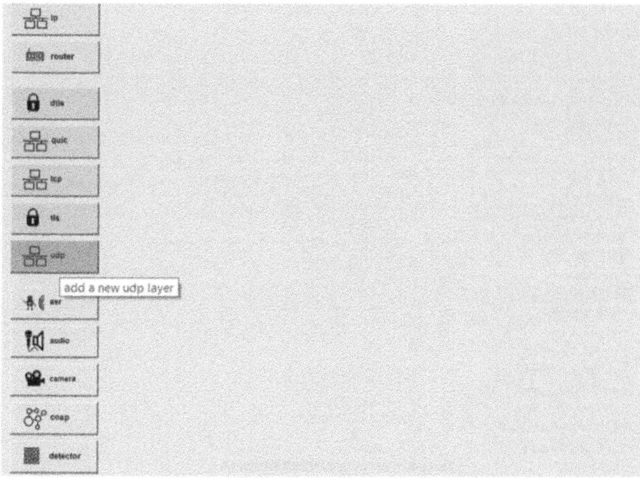

Figure 7.27 UDP Layer Selection.

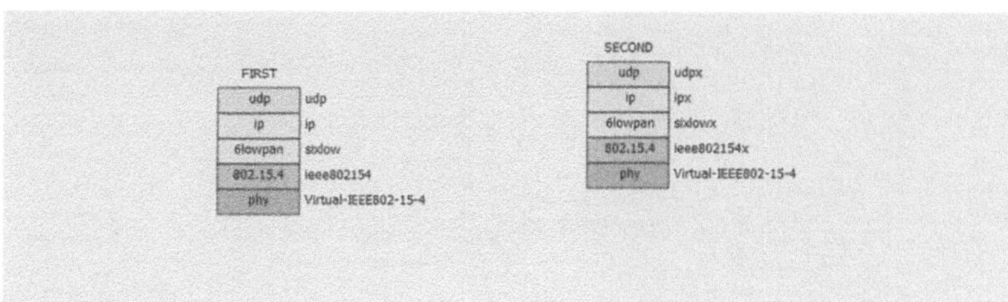

Figure 7.28 UDP Protocol Stacks.

To align the port numbers between the stacks, verify the default port number of the *udp* layer in the *FIRST* stack. Do this by right-clicking the *udp* layer and confirming the number. Set the same port number on the *udpx* layer in the *SECOND* stack. Figure 7.30 shows the user interface when 4000 is entered as the port number for the *udpx* layer. UDP is a connectionless protocol. This means that to send segments from the *FIRST* stack to the *SECOND* stack, manually configure the destination UDP ports and IPv6 addresses. To configure the UDP destination port number on the *udp* layer, right-click the *udp* layer, and select the *Dest Port* option in the *Parameters* menu. Set the port number to 4000.

To configure the destination IPv6 address, right-click the IP layer named *ip* in the *FIRST* stack. Select the *Dest Address* option in the *Parameters* menu. Set the destination address to *2001::41:11*. Deploy the configuration to the agent. Click the *Run Suite* option in the *Scripts* menu. Generate messages and analyze PCAP trace. The goal is to observe 6LoWPAN fragmentation behavior in Wireshark.

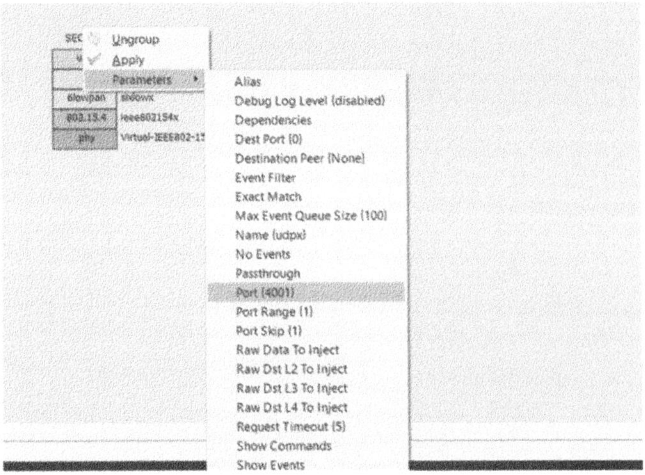

Figure 7.29 UDP Port Number Selection.

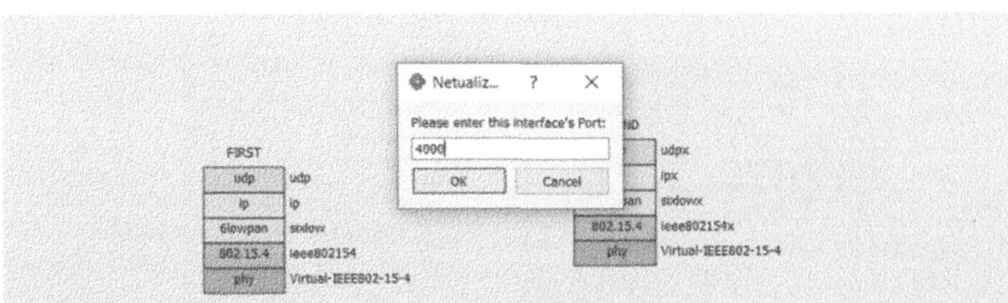

Figure 7.30 UDP Port Number Configuration.

Inject raw data. Right-click the *udp* layer in the *FIRST* stack. Select the *Inject Raw Data* action.

In the field labeled *Raw Data* (shown in Figure 7.31), type *Hello World!* (12 bytes). Click the *OK* button to proceed. Skip destination prompts. Specifically, if prompted for *What's the destination L3 address* or *What's the destination L4 port*, click *OK* again to skip them. This is because the default destination port and address have been already configured.

Access the *Inject Raw Data* option: Right-click the *udp* layer in the *FIRST* stack and select *Inject Raw Data*. Type the message, in the *Raw Data* field, type the sequence *0123456789* fifteen times, creating a 150-byte message in Figure 7.32. Click the *OK* button to proceed. This message size exceeds the MTU of the IEEE 802.15.4 physical layer, intentionally triggering fragmentation by the 6LoWPAN layer.

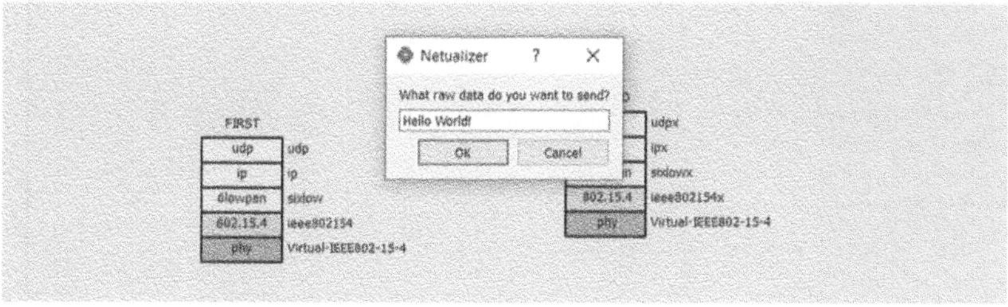

Figure 7.31 *Hello World!* Message Injection.

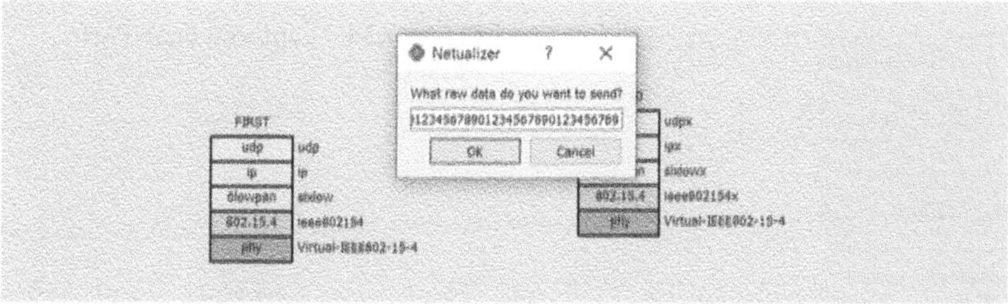

Figure 7.32 150-byte Message Injection.

Stop the configuration as this is necessary to access the PCAP file. Locate the PCAP file. Navigate to the *Netualizer/Projects/ieee802154udp* directory. Find the file named *ieee802154.cap*. Open the PCAP file in Wireshark by double-clicking the ieee802154.cap file to open it using Wireshark. Observe the two UDP packets in frame numbers 4 and 6. Note that Wireshark automatically decompresses the packets for display. The UDP segment in frame number 4 contains the 12-byte payload. The UDP segment in frame number 6 contains the 150-byte payload. Frame number 6 is assembled from two separate 6LoWPAN fragments. Frame number 5 contains the first fragment of the reassembled packet in frame number 6.

Example 7.2 Consider Figure 7.33, what is the fragmentation overhead of frame 6?

Solution:
The actual data length is 150 bytes long. The sizes of frames 5 and 6 are 124 and 122 bytes long, respectively. The fragmentation overhead is then $124 + 122 - 150 = 96$ bytes.

Figure 7.34 displays the 6LoWPAN packet for the initial 12-byte *Hello World!* message. It incorporates an IPHC header and an NHC header that compresses UDP segments (explained in Section 4.4.1.2). The fields align with those described

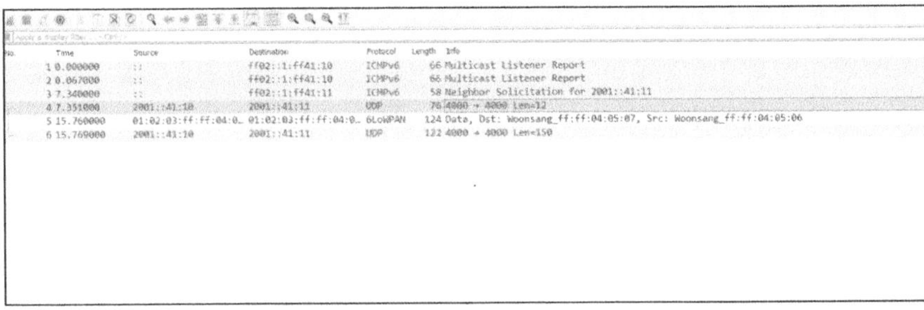

Figure 7.33 Captured IEEE 802.15.4 Packets.

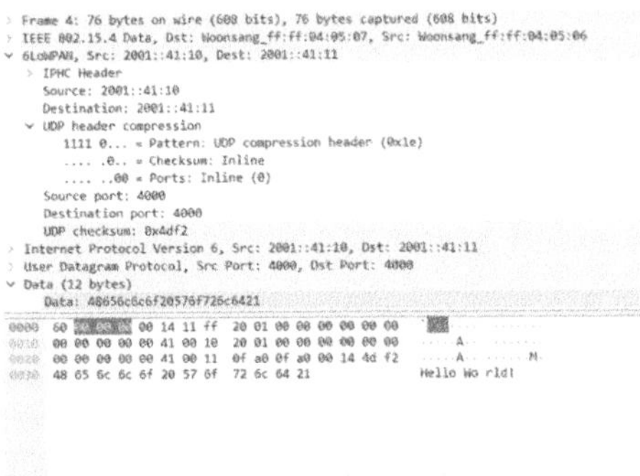

Figure 7.34 Captured 6LoWPAN Packet.

in Section 4.4.1.2. The UDP header indicator begins with the sequence *11110*, signifying the presence of a UDP header. UDP checksum and port information are carried as in-line parameters. Source and destination ports are both 4000, as expected.

The transmission of the second 150-byte message (the sequence of fifteen *0123456789*) undergoes fragmentation (Montenegro et al. 2007). Figure 7.35 presents the initial fragment. The fragmentation header precedes the IPHC header, as indicated in Section 4.4.1.2. The fragmentation header begins with

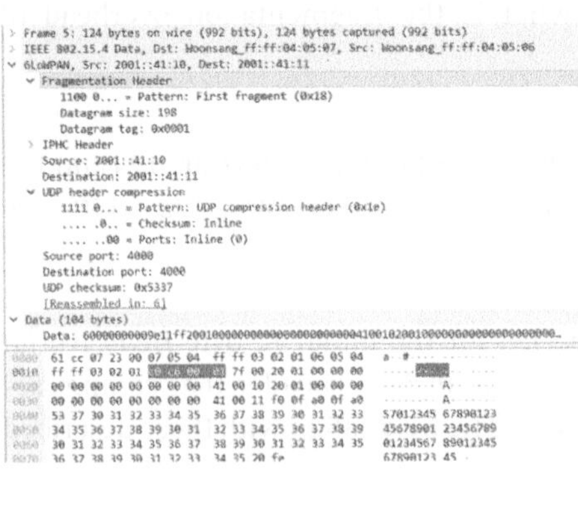

Figure 7.35 Initial 6LoWPAN Fragment.

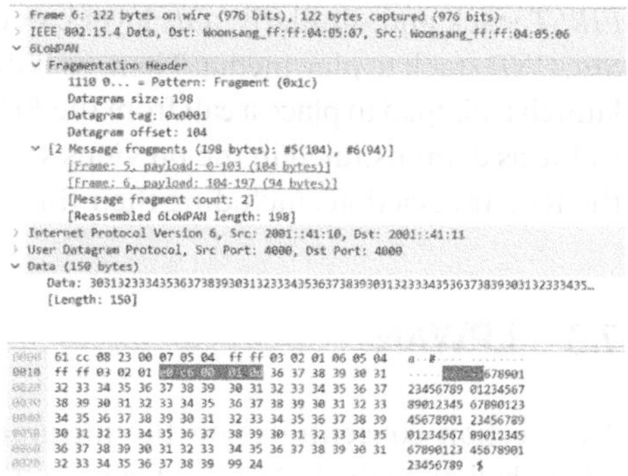

Figure 7.36 Non-Initial 6LoWPAN Fragment.

a *11000* dispatch value. The header indicates a total length of 198 bytes. The fragment has a tag of 1. As the initial fragment, it lacks an offset field, which is assumed to be zero.

Figure 7.36 displays the subsequent non-initial fragment. It shares the following characteristics with the first fragment format. The dispatch value is *11100*. The overall datagram length is 198 bytes. The tag is 1. A distinctive feature of this fragment is the offset of 104, that indicates its position within the reassembled datagram. Note that Wireshark automatically reassembles the fragments and presents them as a unified UDP layer.

7.2.4 WPAN RTC

RTC can be enabled on the protocol stacks by adding SIP and RTP protocols in accordance with Section 6.3. On top of the UDP stacks, add an RTP layer and enable the ITU-T Recommendation G.711 codec on it. Then create an additional UDP layer (i.e., named *udp2*) and place it over the corresponding IP layer. On top of this latter layer, add an SIP layer. Link the SIP layer to the RTP layer. Add SIP and RTP layers on both stacks *FIRST* and *SECOND*. This is shown in Figure 7.37. Note that in addition, an audio layer can be added on top of the RTP layer in the

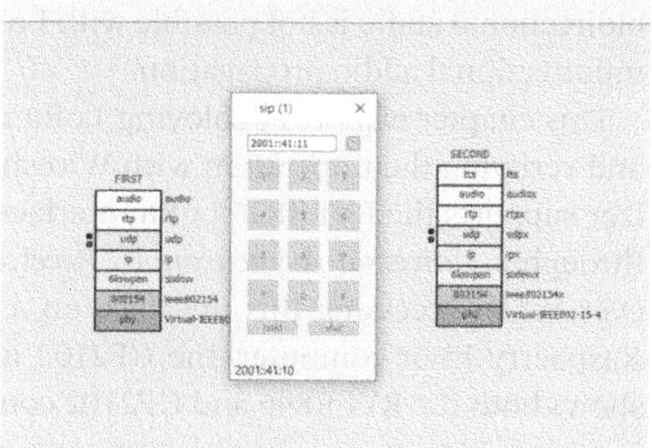

Figure 7.37 SIP and RTP over IEEE 802.15.4.

FIRST stack and, similarly, a *tts* layer can be added on top of the RTP layer in the *SECOND* stack to play media. Then, as indicated in Section 6.3, start the suite and launch a dialpad to place a call from the *FIRST* to the *SECOND* stack. By capturing traffic as demonstrated in the previous sections of this chapter, it is possible to see the RTC transactions including their timing.

7.3 LPWAN

This section concentrates on the integration of LoRa physical and link layers, which belong to the LPWAN category, with a 6LoBTLE adaptation layer to enable IPv6 and UDP connectivity.

7.3.1 Physical and Link Layers

LoRa, a technology within the LPWAN category, offers a stack that facilitates access communication between devices and gateways. These gateways, in turn, enable connectivity with core-side applications (Lavric and Popa 2017; Shanmuga Sundaram et al. 2020; Saari et al. 2018; Herrero 2021). While LoRa was originally not designed to support end-to-end IPv6 connectivity, which is essential for full IoT support, this section explores its integration with 6LoBTLE to bridge this gap. Sensor generated media is encapsulated over LoRaWAN following three different classes. Devices transmit multiple copies of sensor readouts to multiple gateways to ensure delivery. The network core infrastructure removes redundant copies of each message.

To enable fully compliant IoT scenarios, this chapter explores the utilization of LoRa physical and link layers for encapsulating IPv6 traffic. This approach aligns with the IPv6 adaptation employed for WPAN technologies, aiming to achieve end-to-end IP support over LoRa. LoRa maintains low transmission rates to extend signal coverage, but this can introduce significant latency for IoT applications, as even adapted IPv6 datagrams require substantial transmission time. Although bidirectional audio is not possible with LoRa, VLBR codecs can be used to support unidirectional audio propagation.

This chapter explores deploying LoRa networks using the Netualizer software and verifying their operation with Wireshark (see Wireshark for details). Netualizer supports the RYLR896 LoRa interface, which connects to a Netualizer agent through a *Universal Asynchronous Receiver Transmitter* (UART) interface. However, when accessing this interface on a platform with only a USB port, like a Raspberry Pi or computer, the CP2102 module can bridge the gap. Figure 7.38 shows both the RYLR896 and CP2102 components.

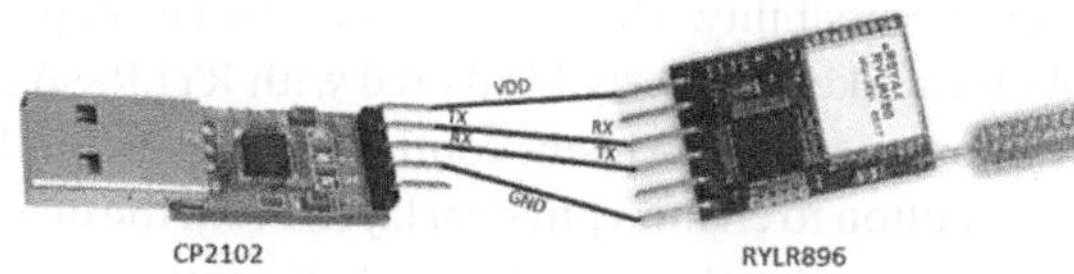

Figure 7.38 RYLR896-based Radio.

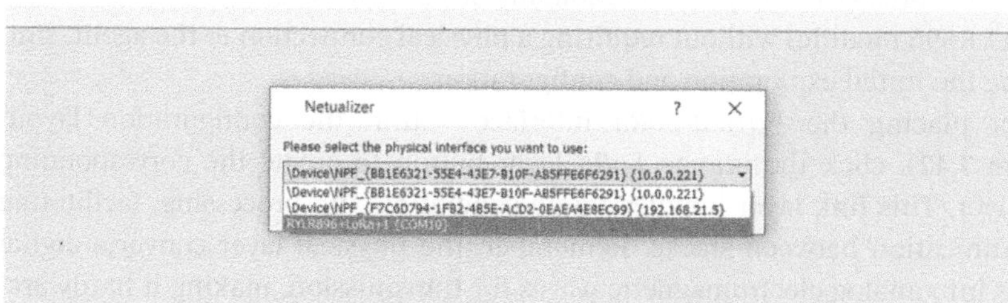

Figure 7.39 Network Interface Selection.

Upon running a Netualizer agent on a hardware platform that has an RYLR896 module connected via a UART interface Ltd, select this interface when creating a regular physical interface. In this scenario, the LoRa interface will be displayed as *RYLR896+LoRa+1 (COM10)* within the available options, as shown in Figure 7.39. The *1* index signifies that this is the first RYLR896 LoRa interface. Should additional RYLR896 interfaces be present, they would be listed with increasing indexes on their respective COM ports.

Building on its virtual device support for WPAN technologies, Netualizer now offers this capability for LoRa devices, streamlining network deployment processes. This means that before physically deploying a LoRa RYLR896 device, leverage a virtual LoRa device to configure, test, and emulate the planned network configuration. Once fully functional virtual deployment is achieved, it can be seamlessly transitioned to a real RYLR896 device physically attached to the agent. To activate virtual device support, navigate to the *Agents* menu and enable the *Virtual Hardware Support* option, as illustrated in Figure 7.3.

After restarting Netualizer, launch project creation for the LoRa topology by clicking the New Project option in the File menu. Name the project lora and remember to check the Attach Local Agent option in Figure 7.40 to ensure local

Figure 7.40 Project Name Selection.

agent accessibility. This project can later be adapted to support remote agents on platform like Raspberry Pis paired with RYLR896 devices.

Upon opening the agent configuration panel in Netualizer, click the green *phy* layer button to create a physical layer. From the displayed list of available interfaces in Figure 7.41, select the *Virtual-LoRa* option. This versatile interface, as already mentioned, offers the complete functionality of a real physical LoRa interface (like the RYLR896 module) without requiring a physical connection to the agent, simplifying the initial exploration and configuration.

After placing the *Virtual-LoRa* interface within the configuration layout (Figure 7.42), click the orange *LoRa* layer button to create the corresponding link layer. This link layer handles frame generation and processing, facilitating communication between stacks. Remember, the physical layer converts digital frames into analog electromagnetic waves for transmission, making it hardware-dependent but fortunately emulatable through the agent's LoRa virtualization, eliminating the need for immediate physical hardware connection.

Following Netualizer's agent configuration conventions, assign a unique identifier (name) to the link layer to enable access from the Lua project script. This practice remains essential even in scenarios that do not involve scripting. For this specific configuration, adopt the name *lora* for the link layer, as illustrated in Figure 7.43.

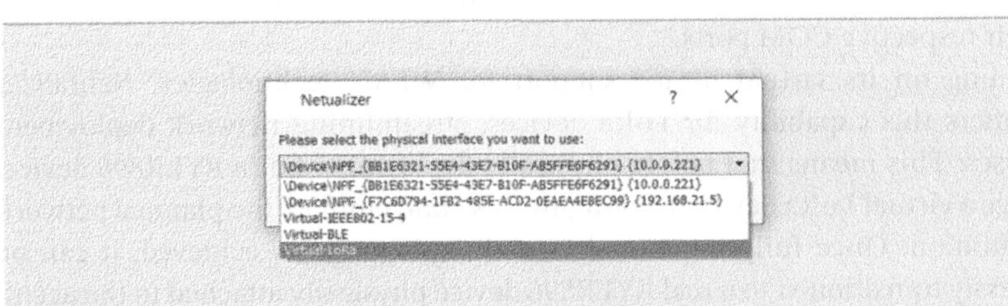

Figure 7.41 The LoRa Interface.

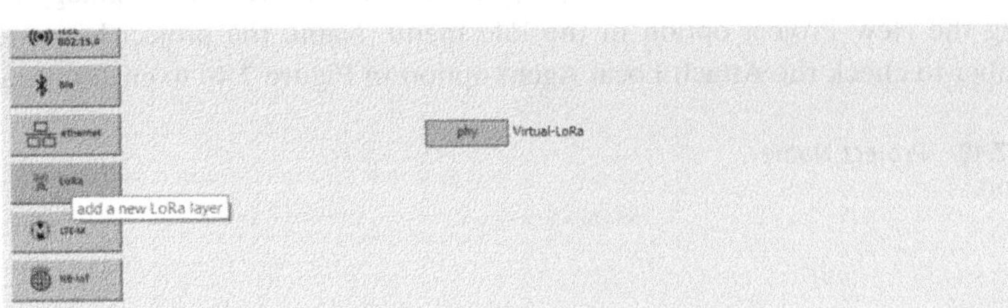

Figure 7.42 Lora Layer Creation.

Figure 7.43 LoRa Layer Name Selection.

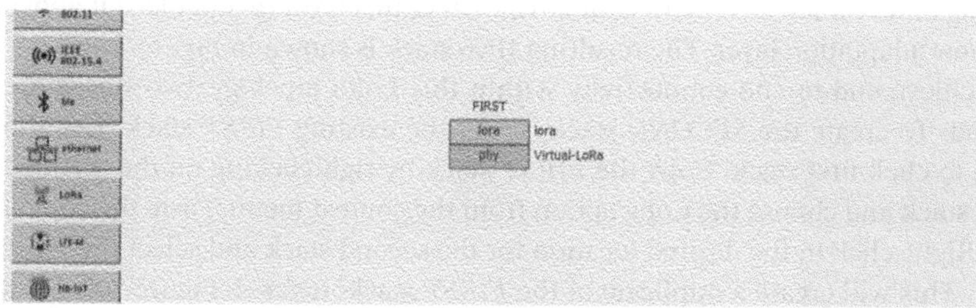

Figure 7.44 LoRa 2-Layer Stack.

To build a 2-layer stack, position the lora link layer directly above the lora physical layer. Given that this configuration involves two stacks, rename the default *STACK* label to *FIRST* by double-clicking on it. The resulting stack is shown in Figure 7.44.

7.3.2 Network Layer

By bypassing reliance on LoRaWAN and its upper layers for building LoRa stacks, integration with IPv6-compatible layers becomes key. While standardized protocols for IPv6 over LoRa are still under development, adopting one of the established WPAN IPv6 adaptation mechanisms can be explored. 6LoBTLE, introduced in Section 4.4.1.2, presents a viable option with its support for IPv6 address compression and other important 6Lo features like encapsulating upper networking layers. Though primarily intended for BLE topologies, 6LoBTLE's inherent flexibility allows its application in other scenarios, such as with LoRa. Click the beige-colored *6LoBTLE* layer button in Figure 7.45 to create an IPv6 adaptation layer.

Since layer names cannot begin with numbers, name the adaptation layer *sixlow* and carefully position it atop the *lora* link layer (Decuir 2015; Montenegro et al. 2007). Create an IPv6 layer by clicking the beige-colored *IP* layer button

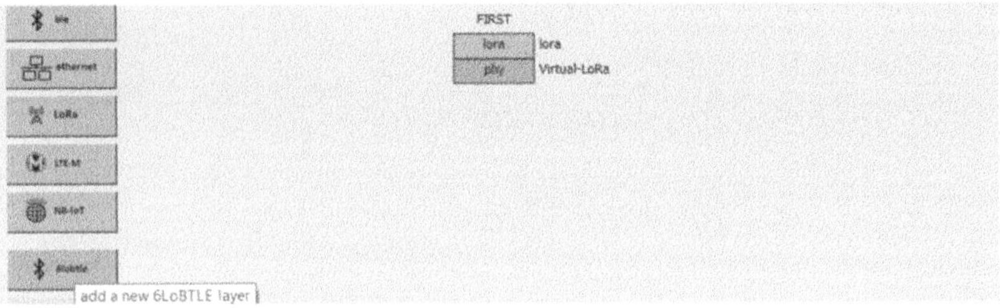

Figure 7.45 6LoBTLE Layer Creation.

(Deering and Hinden 2017; Graziani 2012). Label this layer *ip* and place it above the *sixlow* adaptation layer. The resulting IPv6 stack is shown in Figure 7.46.

To achieve end-to-end connectivity within this LoRa topology, two stacks are required. To create the *SECOND* stack, select the existing *FIRST* stack: Use the mouse to click first stack. Copy the *FIRST* stack by rightclicking on the selected *FIRST* stack and choose the *Copy* option from the context menu. Paste the copied stack: Right-click in the desired location for the second stack and select the *Paste* option. This will create a duplicate of the *FIRST* stack. Refer to Figure 7.47 for a representation of these steps.

To position the second stack, right-click within the configuration panel, near the original *FIRST* stack, and select the *Paste* option from the menu. Rename the

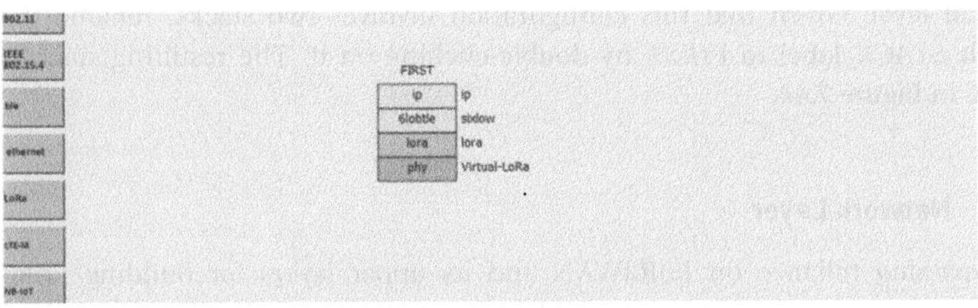

Figure 7.46 LoRa Protocol Stack.

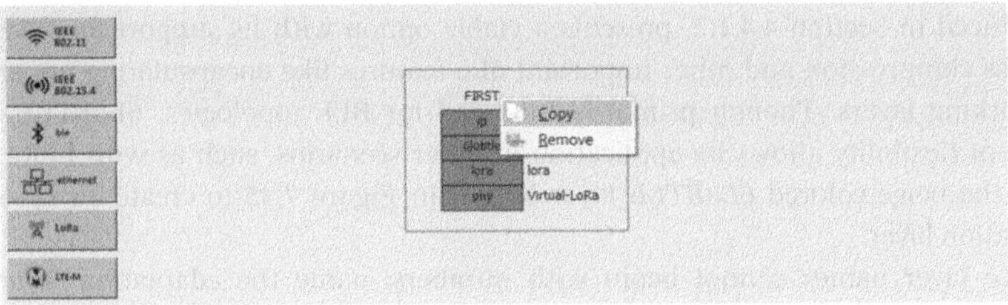

Figure 7.47 Stack Copy.

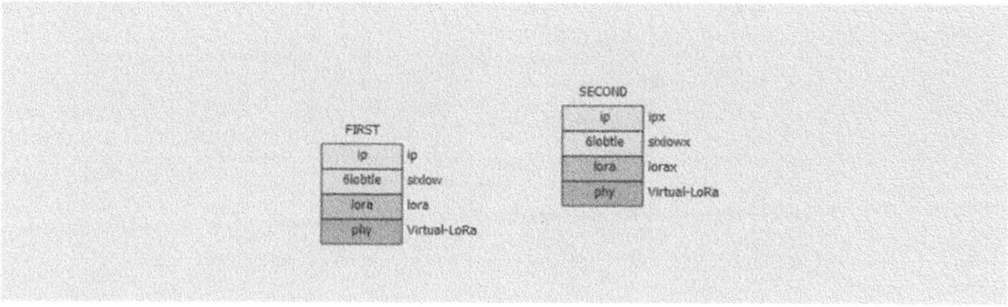

Figure 7.48 Copied Stacks.

copied stack: Double-click on the label *FIRST* of the pasted stack and change it to *SECOND*. Both stacks can be observed in Figure 7.48. The layers within the *SECOND* stack have names identical to those in the *FIRST* stack, but with the addition of the ending *x*. This ensures unique identification, preventing overlapping and guaranteeing accessibility to all layers from the project Lua script.

Due to the *ipx* network layer being a duplicate of the *ip* network layer, its IPv6 address must be modified to guarantee end-to-end connectivity. Follow these steps to alter the IPv6 address of the *ipx* layer. Right-click on the *ipx* layer. Choose the *Address* option within the *Parameters* menu (Refer to Figure 7.49 for a representation). This modification is essential because the 6LoBTLE adaptation layer mandates an IPv6 upper layer.

By default, the *ipx* layer is assigned the IPv6 address *2001::41:10*. To ensure no address conflicts occur, change this address to *2001::41:11*, as shown in Figure 7.50.

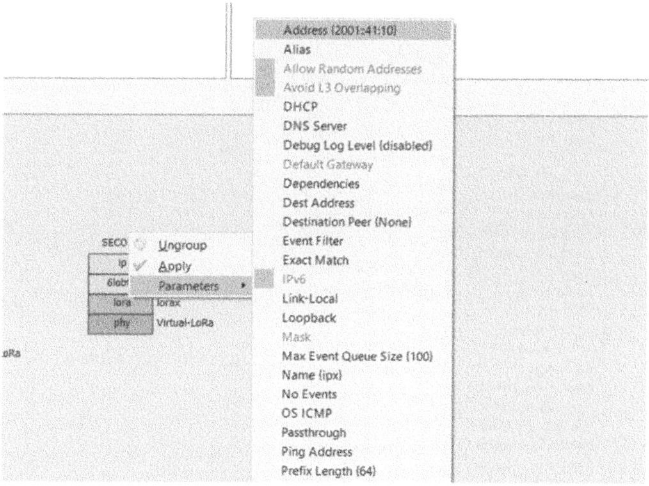

Figure 7.49 IP Address Parameter.

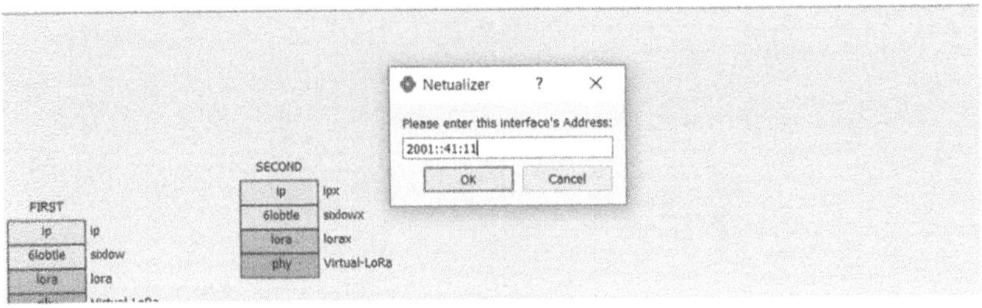

Figure 7.50 IPv6 Address Configuration.

Please note that in some instances, copying a network interface may inadvertently enable DHCP on the copied interface. To avert this, verify that the DHCP parameter within the *Parameters* menu of the *ipx* layer is disabled.

Because the link layer address of the LoRa interface is replicated when stacks are copied, it is important to modify this address to maintain uniqueness. Right-click on the *lorax* layer. Select the *Address* option within the *Parameters* menu (guidance for this process is provided in Figure 7.51).

Figure 7.52 displays the Netualizer prompting for the new address of the *lorax* layer. As shown in the figure, replace the default address value of *10* with *11*. Remember to configure other hardware-specific LoRa parameters identically on both layers, as illustrated in Figure 7.51. These parameters include *Bandwidth*, *Coding Rate*, *ISM Band*, and *Network Id*. This step is important, especially when setting up a real RYLR896 device.

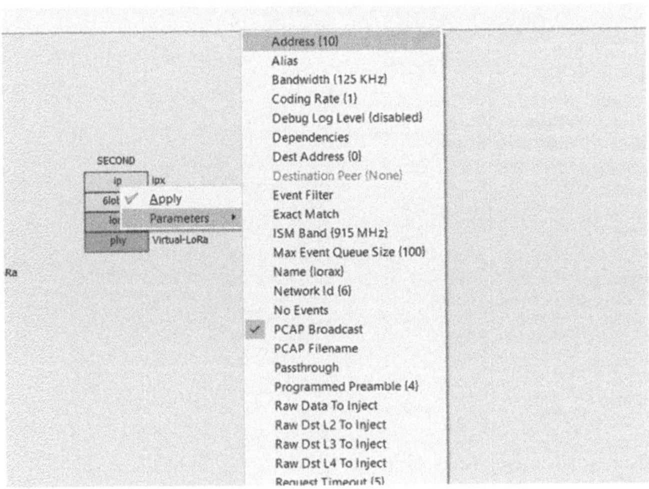

Figure 7.51 LoRa Address Parameter.

Figure 7.52 LoRa Address Configuration.

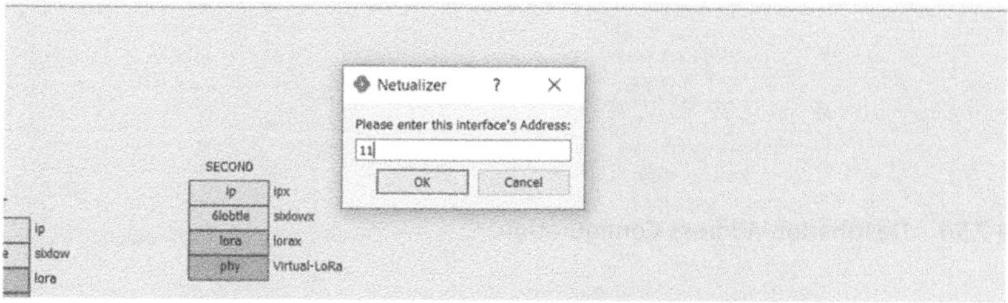

Figure 7.53 LoRa Address Configuration on *SECOND* Stack.

To ensure consistency, set the LoRa address to *10* within the *lora* layer of the *FIRST* stack. Access the *Parameters* menu of the *lora* layer. Choose the *Address* option. Enter the value *10* as the new address (shown in Figure 7.53).

To establish direct communication between the LoRa layers within the *FIRST* and *SECOND* stacks, right-click on the *lorax* layer within the *FIRST* stack. Select the *Dest Address* field from the *Parameters* menu (Refer to Figure 7.54 for guidance). Enter the value *11* as the destination address, aligning it with the address of the *lorax* layer in the *SECOND* stack. Continuing with the configuration process, enter *11* as the destination address within the *lora* layer, as shown in Figure 7.54. Repeat this step for the *lora* layer in the *SECOND* stack, setting its destination LoRa address to *10* to enable reciprocal communication.

Since Wireshark cannot directly capture traffic generated at virtual LoRa interfaces, the *lora* layer requires configuration to sniff LoRa frames and store them within a PCAP trace. These PCAP files can then be opened and analyzed in Wireshark like any other trace. Right-click on the *lora* layer. Access the *Parameters* menu and set the *Start PCAP* flag to *True*. Within the same menu, select the *PCAP Filename* option (Refer to Figure 7.54 for guidance). Enter the filename *lora* to save the trace as *lora.cap* within the current project directory.

With the configuration complete, the suite can be executed. As the configuration does not require executing any specific commands, the default script remains suitable without modifications. To initiate the configuration and run the default script,

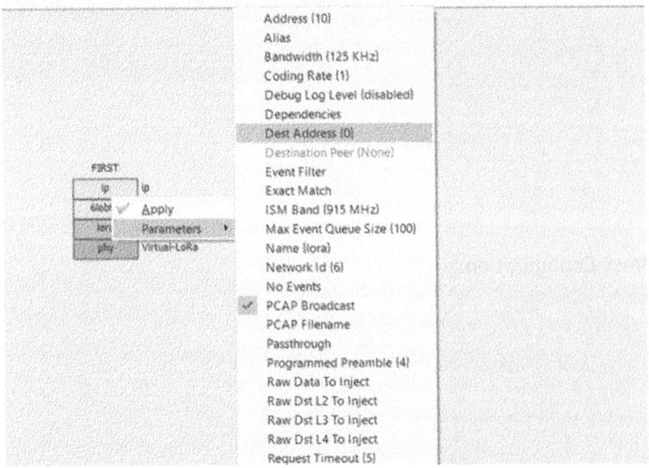

Figure 7.54 Destination Address Configuration.

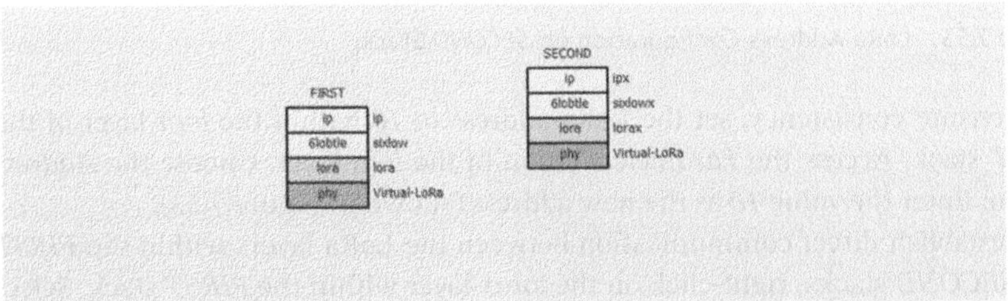

Figure 7.55 Suite Execution.

access the *Scripts* menu. Select the *Run Suite* option. Figure 7.55 shows the configuration panel after suite execution, displaying both the *FIRST* and *SECOND* stacks in an active state. To initiate the most basic form of traffic between the *FIRST* and *SECOND* stacks, ICMP packets will be used. The *ip* layer within the *FIRST* stack possesses the capability to generate ICMP ping echo requests, which are subsequently received and responded to by the *ipx* layer within the *SECOND* stack, as outlined in IETF RFC 4443 (Gupta and Conta 2006). To execute this process, right-click on the *ip* layer within the *FIRST* stack. Select the *Ping* command from the context menu.

To initiate the pinging process, a specific string that serves two purposes needs to be provided. It designates the destination IPv6 address of the *ipx* layer, ensuring the echo requests reach the intended target and it determines the interval, measured in seconds, between the transmissions of the echo requests. These two

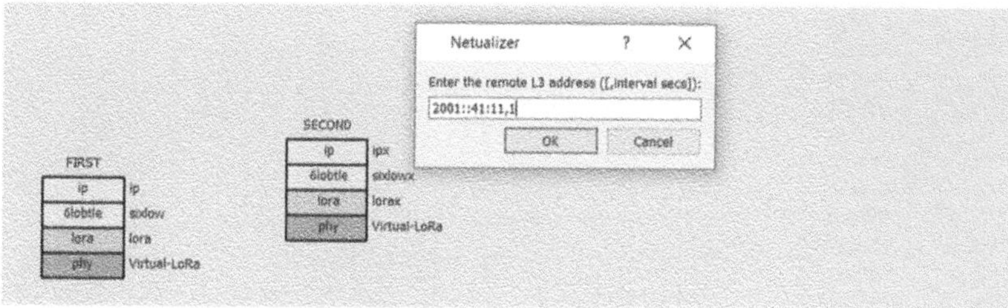

Figure 7.56 Ping Address Configuration.

parameters are entered in a comma-separated format. As shown in Figure 7.56, the string *2001::41:11,1* instructs the system to direct the echo requests to the IPv6 address *2001::41:11* and transmit them once every second.

Within this context, the primary purpose of the *Ping* command is to measure the RTT between the *FIRST* and *SECOND* stacks. This measurement, expressed in milliseconds, indicates the duration required for a signal to travel from the *FIRST* stack to the *SECOND* stack and return back. Figure 7.57 shows the output generated by executing the *Ping* command. It is essential to note that when working with a real RYLR896 device, the observed RTT values are typically higher and exhibit a significant dependence on the physical distance between the transmitting and receiving devices.

Allow the configuration to proceed for approximately 10 seconds. Afterward, initiate its termination by clicking the red cross button, as shown in Figure 7.58. This manual step is important because, while the default Lua script ceases execution automatically, the configuration itself persists until explicitly terminated. Navigate to the current project directory, which is typically located at *Netualizer/Projects/lora*. Locate the generated PCAP trace, named *lora.cap*. Double-click on the *lora.cap* file. If Wireshark is installed on the system, the operating system will automatically launch it and open the PCAP trace for inspection.

Figure 7.59 showcases the opened PCAP trace in Wireshark. However, the software does not have the ability to decode 6LoBTLE traffic encapsulated over LoRa.

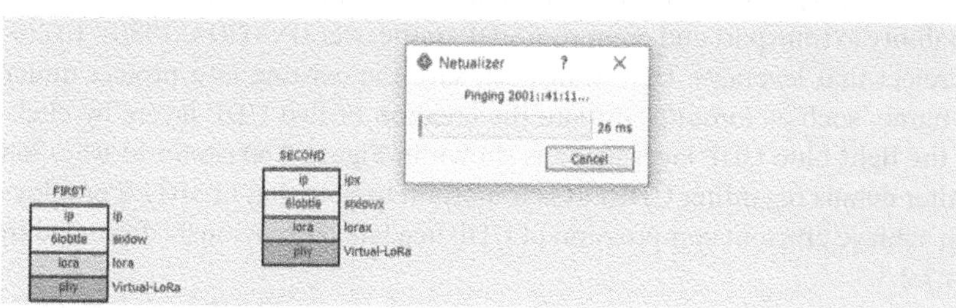

Figure 7.57 Ping Execution.

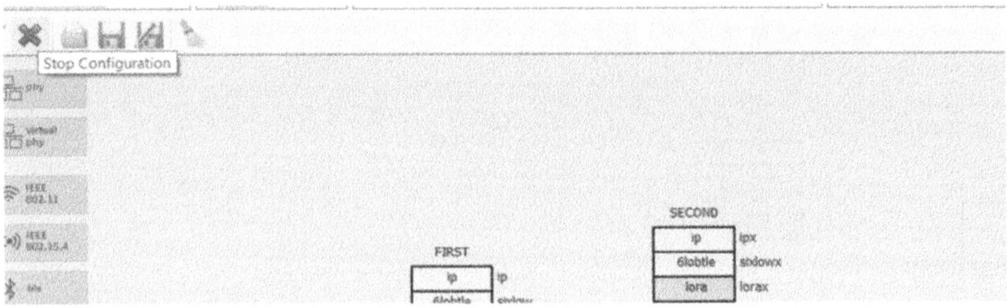

Figure 7.58 Configuration Termination.

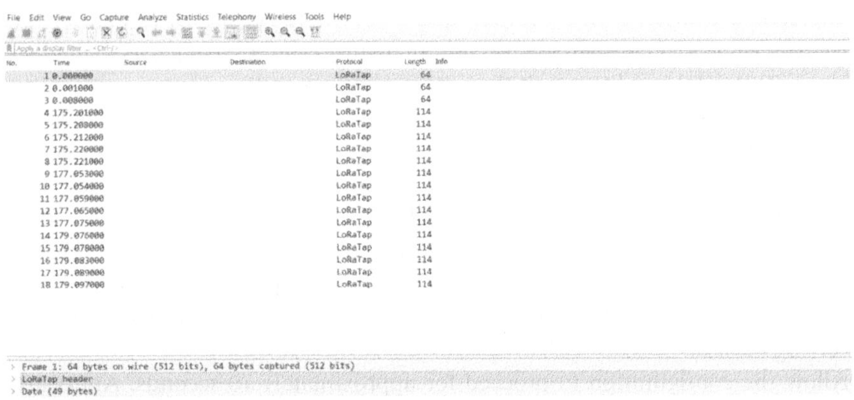

Figure 7.59 Captured LoRa Packets.

This is why the content of the upper layers remains hidden within the captured frames. Put simply, while the LoRa frames contain IPv6 datagrams, their actual contents remain inaccessible due to Wireshark's decoding limitations.

7.3.3 Transport Layer

While 6LoBTLE's primary purpose is to facilitate IPv6 adaptation, it also possesses the capability to transport and compress UDP traffic over IPv6 (RFC 1980). To create a project that leverages UDP transport, save the existing *lora* project under a new name, such as *loraudp*. Initiate the creation of two UDP layers by clicking on the light-blue UDP layer icon, as shown in Figure 7.60 (Refer to RFC 768 for further details regarding UDP). It is important to note that 6LoBTLE employs NHC to achieve efficient compression of UDP headers, as previously discussed in Section 7.2.3.

Assign the names *udp* and *udpx* to the two newly created UDP layers. Position them directly above the *ip* and *ipx* IPv6 layers, respectively, as demonstrated in

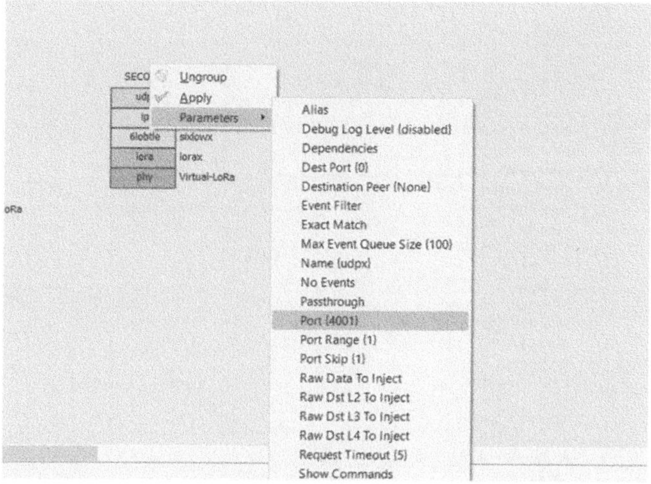

Figure 7.60 UDP Layer Parameters.

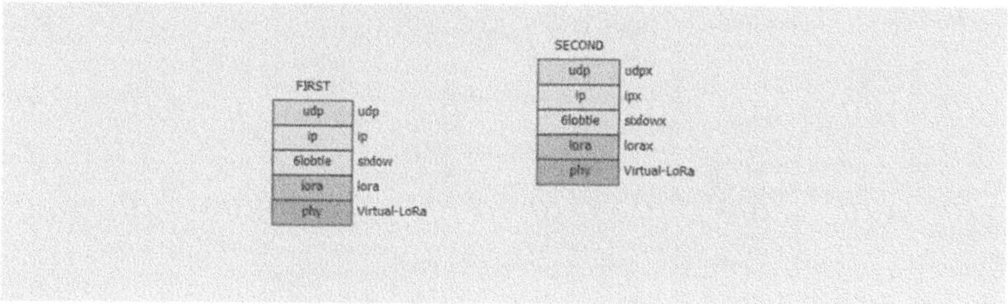

Figure 7.61 LoRa-based UDP Stacks.

Figure 7.61. This configuration results in the *FIRST* stack that comprises the *ip* and *udp* layers and the *SECOND* stack that contains the *ipx* and *udpx* layers. The assignment of unique names to each layer serves an important purpose: It ensures that the layers can be unambiguously accessed and controlled from within the project's Lua script by utilizing dedicated APIs.

To streamline both testing and traffic generation, it is recommended to modify the default port number of the *udpx* layer within the *SECOND* stack from 4001 to 4000. This configuration aligns the transport port numbers of the *FIRST* and *SECOND* stacks, enabling them to communicate effectively while maintaining distinct network IPv6 addresses (*2001::41:10* and *2001::41:11*, respectively). To execute this change, right-click on the *udpx* layer situated within the *SECOND* stack. Access the *Parameters* menu and select the *Port* attribute (Refer to Figure 7.62 for guidance).

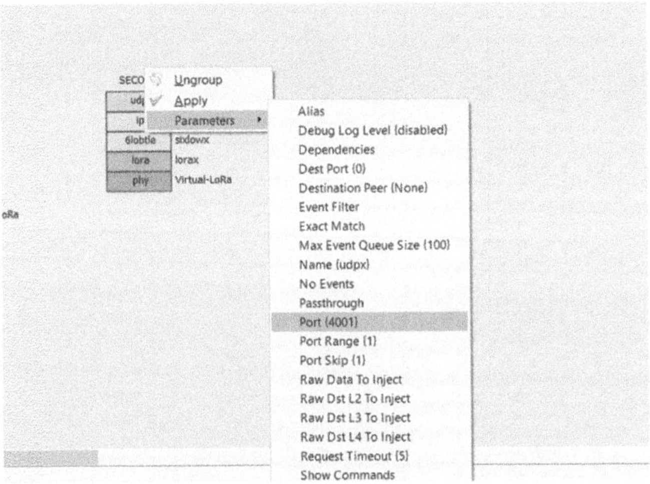

Figure 7.62 Selecting the *udpx* Port Number.

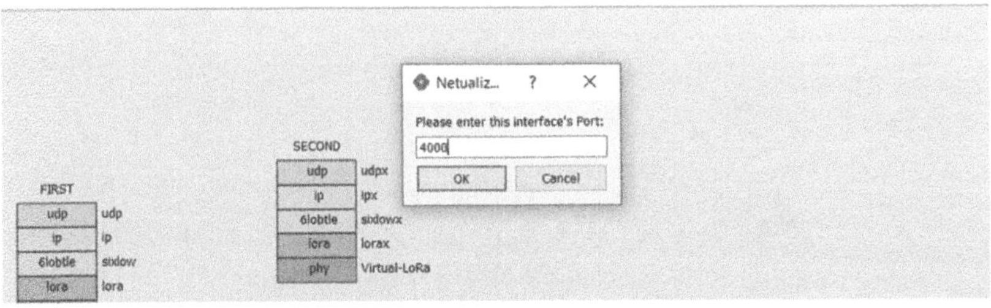

Figure 7.63 Port Number Configuration.

As illustrated in Figure 7.63, input the value 4000 into the UDP port field of the *udpx* layer within the *SECOND* stack. It is important to note that this choice of port number is not a strict requirement; any valid 16-bit port number can be used instead, depending on specific needs or preferences. To ensure that all application traffic generated within the *FIRST* stack is correctly routed to the *SECOND* stack, an additional configuration step is necessary. Due to UDP's connectionless nature, the destination transport layer port needs to be explicitly specified. Right-click on the *udp* layer within the *FIRST* stack. Access the *Parameters* menu and select the *Dest Port* option. Enter the value 4000 as the destination UDP port. This configuration directs all UDP traffic originating from the *udp* layer toward the *udpx* layer, which is listening on port 4000.

To complete the configuration process, proceed to modify the destination IPv6 address of the *ip* layer within the *FIRST* stack. Right-click on the *ip* layer. Access the *Parameters* menu and select the *Dest Address* option, as shown in Figure 7.64.

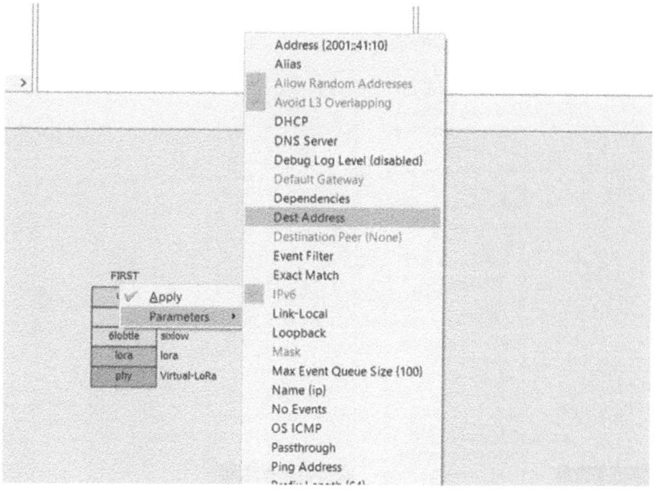

Figure 7.64 Destination Address Parameters.

Enter the value *2001::41:11* as the new destination IPv6 address. This ensures that all IP traffic originating from the *ip* layer will be directed toward the *ipx* layer, which possesses the specified address. To initiate the automatic deployment of the configuration onto the designated agent, access the *Scripts* menu. Select the *Run Suite* option. Upon execution, the default Lua script will promptly conclude its task, which solely involves clearing the output window. To generate a message that will traverse from the *FIRST* to the *SECOND* stack, right-click on the *udp* layer within the FIRST stack. Choose the *Inject Raw Data* option.

Proceed by entering the 12-character string *Hello World!* into the input field, as shown in Figure 7.65. Click the *OK* button twice in response to the subsequent prompts. This action preserves the previously configured destination L3 IPv6 address and L4 UDP port, ensuring correct transmission.

To confirm the successful arrival of the UDP segment (carrying the *Hello World!* string) at the *SECOND* stack, right-click on the *udpx* layer. Navigate to the *Statistics* menu. Choose the *Upstream Last Packet* option, as illustrated in Figure 7.66.

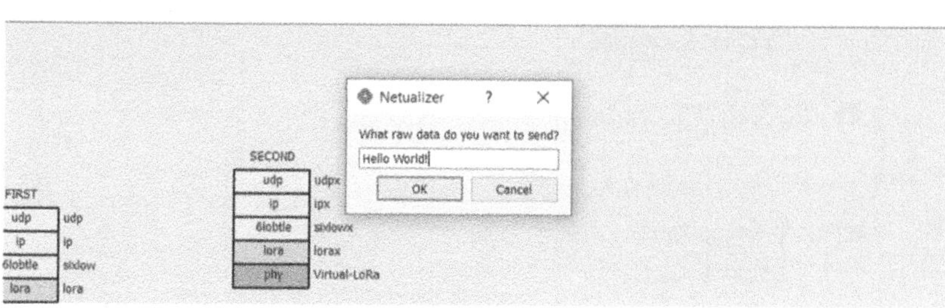

Figure 7.65 *Hello World!* Message Injection.

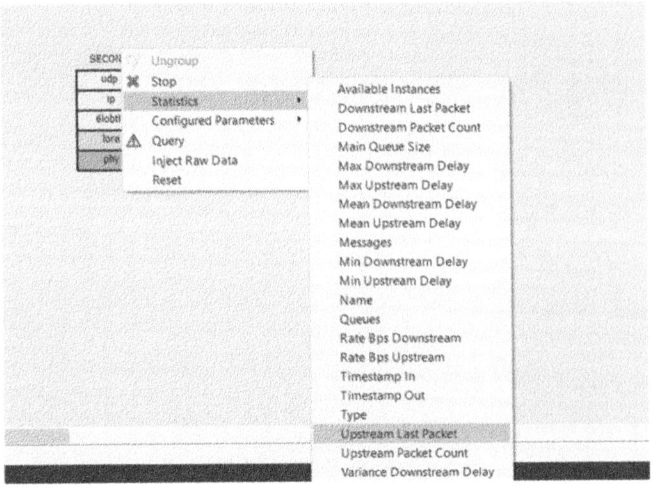

Figure 7.66 Upstream Last Packet Option.

This action will reveal the most recent packet received by the *udpx* layer, allowing to verify its content and ensure that the UDP segment has indeed reached its destination.

Figure 7.67 presents a dual representation of the segment's payload, displaying it as both a string and a hexadecimal data structure. This visualization includes all twelve bytes of the payload, starting at position *0* and concluding at position hexadecimal *C* (equivalent to decimal 12).

Example 7.3 Based on Figure 7.68, what is the average interframe duration?

Solution:

This can be obtained by adding the delay between all frames leading to $\frac{0.001+0+0.006+0.002+9.045}{5} = 1.81$ seconds.

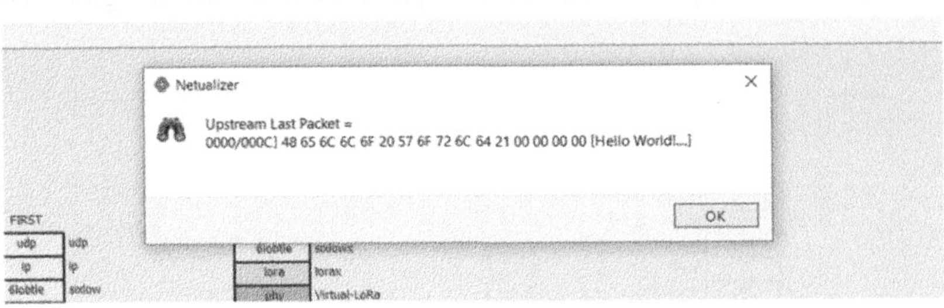

Figure 7.67 Received Message.

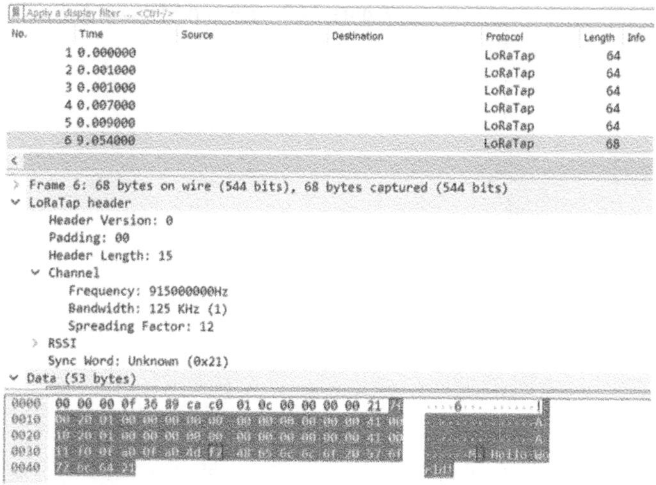

Figure 7.68 Captured LoRa Packet.

To examine the captured LoRa frames within Wireshark, navigate to the current project directory, typically found at *Netualizer/Projects/loraudp*. Locate the generated PCAP trace, named *lora.cap*. Double-click on the *lora.cap* file to launch it within Wireshark. While Wireshark provides a representation of the captured LoRa frames, as shown in Figure 7.68, it is important to note that it lacks the capability to decode 6LoBTLE traffic that is encapsulated over LoRa. This limitation prevents the display of the actual IPv6 datagrams embedded within the payload. However, the hexadecimal dump situated in the lower portion of the segment does reveal the presence of the *Hello World!* string encapsulated within it, confirming its successful transmission.

7.3.4 LPWAN RTC

Support of RTC over LoRa (illustrated in Figure 7.69) is similar to the support of RTC over IEEE 802.15.4 presented in Section 7.2.4. The big difference is the transmission rate, which requires the selection of a codec with high compression like AMR configured at 4.75 Kbps. The 64 Kbps of ITU-T Recommendation G.711 is not an option. In this context, follow the steps detailed in Section 7.2.4 and configure the codec from ITU-T Recommendation G.711 to AMR and set the transmission of the codec to the lowest rate. Then, after starting the suite, open up the dialpad and place a call. The captured traffic is unfortunately not fully decoded by Wireshark but it is good enough to analyze the packet transmission cadence and the interframe latency.

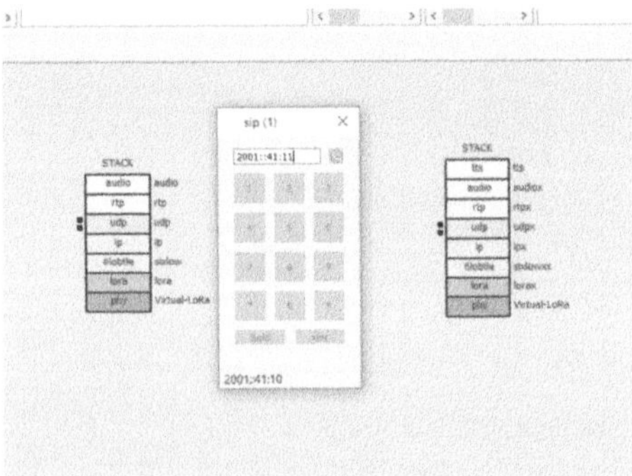

Figure 7.69 SIP and RTP over LoRa.

7.4 Summary

IoT networks are crucial to enabling Cyber-Physical Systems. In this context, and with Netualizer, it is possible to build physical and link layers that support the two main IoT topologies: WPAN and LPWAN. Under WPAN, a basic IEEE 802.15.4 stack that relies on IPv6 adaptation includes a 6LoWPAN layer in between the link and network layers. With 6LoWPAN, the network IPv6 layer is compressed by means of IPHC-compression while the transport UDP layer is compressed by means of NHC. 6LoWPAN is key to support fragmentation when the actual messages are larger than the IEEE 802.15.4 MTU size. In LPWANs, however, Netualizer can be used to build a LoRa physical and link layer stack that enables IPv6 adaptation by means of 6LoBTLE. Like 6LoWPAN, 6LoBTLE relies on IPHC and NHC to respectively encapsulate IPv6 and UDP traffic. With the UDP layer in place, sip and rtp can be used to establish, maintain, and tear down media sessions.

7.5 Homework Problems and Questions

Problem 7.1 Consider Figure 7.23 and look at the matching ICMPv6 echo requests and replies. What is the RTT? What is the transmission rate in each direction?

Problem 7.2 What is the purpose of the *Neighbor Solicitation* (NS) and *Neighbor Advertisement* (NA) messages in Figure 7.23?

Problem 7.3 In the IEEE 802.15.4 header in Figure 7.24, what are the source and destination MAC addresses? Is security-enabled?

Problem 7.4 In the 6LoWPAN header in Figure 7.25, what are the source and destination IPv6 addresses? If stateful IPHC compression is used, why is the compression stateless?

Problem 7.5 Look at Figure 7.36, what is the data length? What is the fragmentation overhead?

Problem 7.6 What is the transmission rate in Figure 7.68?

7.6 Lab Exercises

Problem 7.7 Build a scenario in Netualizer with the following characteristics:

- Two Stacks: *EP1* and *EP2*
- Physical Layer/Link Layer: IEEE 802.15.4 (Emulated)
- Client IP address: 2001::21:10
- Server IP address: 2001::21:12
- Session Layer: SIP/RTP (over UDP)
- Codec: ITU-T Recommendation G.711
- Text-To-Speech on EP1: *Hello Everyone*
- Audio Player on EP2

Capture traffic using Wireshark and place a call from EP2 to EP1. Let the media run for 1 minute. Answer the following:

1. How many RTP packets can be seen?
2. How many SIP packets can be seen?
3. What is the SIP throughput and goodput in each direction?
4. What is the RTP throughput and goodput in each direction?
5. How these results compare to those of the exercises in Problem 10?

Problem 7.8 Introduce an impairment layer between the UDP and the RTP layer in the EP1 stack. Set the transmission loss to 20%. Let the media run for 1 minute.

Answer the following:

1. How many RTP packets can be seen?
2. How many SIP packets can be seen?
3. What is the SIP throughput and goodput in each direction?
4. What is the RTP throughput and goodput in each direction?
5. How these results compare to those of the exercises in Problem 11?

Problem 7.9 Repeat Problem 7.7 but on a LoRa architecture. How does IEEE 802.15.4 compare to LoRa?

Problem 7.10 Now repeat Problem 7.8 but on a LoRa architecture. How does IEEE 802.15.4 compare to LoRa?

Bibliography

Carrara, E., Norrman, K., McGrew, D., Naslund, M., and Baugher, M. (2004). The secure real-time transport protocol (SRTP). RFC 3711, March. Available from: https://www.rfc-editor.org/info/rfc3711. (Accessed March 15th, 2024)

Decuir, J. (2015). Bluetooth smart support for 6LoBTLE: applications and connection questions. *IEEE Consumer Electronics Magazine*, 4(2), pp. 67–70. DOI: 10.1109/MCE.2015.2392955.

Dr., Deering, S.E. and Hinden, B. (2017). Internet protocol, version 6 (IPv6) specification. RFC 8200, July. Available from: https://rfc-editor.org/rfc/rfc8200.txt. (Accessed March 15th, 2024)

Graziani, R. (2012). *IPv6 fundamentals: a straightforward approach to understanding IPv6*. Pearson Education.

Gupta, M. and Conta, A. (2006). Internet control message protocol (ICMPv6) for the Internet protocol version 6 (IPv6) specification. RFC 4443, March. Available from: https://rfc-editor.org/rfc/rfc4443.txt. (Accessed March 15th, 2024)

Herrero, R. (2021). Fundamentals of IoT communication technologies. *Textbooks in telecommunication engineering*. Springer International Publishing. ISBN 978-3-030-70079-9. Available from: https://books.google.com/books?id=k70rzgEACAAJ. (Accessed March 15th, 2024)

Herrero, R. (2023). Practical Internet of things networking. *Textbooks in telecommunication engineering*. Springer International Publishing. ISBN 978-3-031-28443-4.

IEEE. (2020). IEEE standard for low-rate wireless networks. IEEE Std 802.15.4-2020 (Revision of IEEE Std 802.15.4-2015), pp. 1–800.

Ierusalimschy, R. (2006). Programming in Lua.

L7TR. Netualizer: network virtualizer. Available from: https://www.l7tr.com. (Accessed March 15th, 2024)

Lavric, A. and Popa, V. (2017). Internet of things and LoRa low-power wide-area networks: a survey. In 2017 International Symposium on Signals, Circuits and Systems (ISSCS), pp. 1–5.

Montenegro, G., Hui, J., Culler, D., and Kushalnagar, N. (2007). Transmission of IPv6 packets over IEEE 802.15.4 networks. RFC 4944, September. Available from: https://rfc-editor.org/rfc/rfc4944.txt. (Accessed March 15th, 2024)

Reyax Technology Corporation Ltd. Rylr896: 868/915mhz LoRa antenna transceiver module. Available from: https://reyax.com/products/rylr896. (Accessed March 15th, 2024)

RFC. (1980). User datagram protocol. RFC 768, August. Available from: https://www.rfc-editor.org/info/rfc768. (Accessed March 15th, 2024)

Saari, M., bin Baharudin, A.M., Sillberg, P., Hyrynsalmi, S., and Yan, W. (2018). LoRa – a survey of recent research trends. In 2018 41st International Convention on Information and Communication Technology, Electronics and Microelectronics (MIPRO), pp. 0872–0877.

Shanmuga Sundaram, J.P., Du, W., and Zhao, Z. (2020). A survey on LoRa networking: research problems, current solutions, and open issues. *IEEE Communications Surveys Tutorials*, 22(1), pp. 371–388.

Texas Instruments. Ti-2531. Available from: https://www.ti.com/product/CC2531. (Accessed March 15th, 2024)

Wireshark. Wireshark: Network analyzer. Available from: https://www.wireshark.org. (Accessed March 15th, 2024)

Zig. (2015). ZigBee specification. Standard, The ZigBee Alliance, USA, 8.

Index

Hands-On Real Time Communications: A Practical Guide to RTC Protocols in Non-3GPP, 3GPP, 4G/5G/6G and IoT Networks, First Edition. Rolando Herrero.

Companion website: www.wiley.com/go/herrero/RTCProtocols

b

c

d

e

f

j

p

Printed and bound by CPI Group (UK) Ltd, Croydon, CR0 4YY
17/09/2026
14964312-0003